The Solution of the
k(GV)
Problem

ICP Advanced Texts in Mathematics ISSN 1753-657X

Series Editor: Dennis Barden *(Univ. of Cambridge, UK)*

Published

ICP Advanced Texts in Mathematics – Vol. 4

The Solution of the
k(*GV*)
Problem

Peter Schmid
Mathematisches Institut der Universität
Tübingen, Germany

Imperial College Press

Published by

Imperial College Press
57 Shelton Street
Covent Garden
London WC2H 9HE

Distributed by

World Scientific Publishing Co. Pte. Ltd.
5 Toh Tuck Link, Singapore 596224
USA office: 27 Warren Street, Suite 401-402, Hackensack, NJ 07601
UK office: 57 Shelton Street, Covent Garden, London WC2H 9HE

Library of Congress Cataloging-in-Publication Data
Schmid, Peter, 1941–
    The solution of the k(GV) problem / by Peter Schmid.
        p. cm. -- (ICP advanced texts in mathematics ; v. 4)
    Includes bibliographical references and index.
    ISBN-13: 978-1-86094-970-8 (hardcover : alk. paper)
    ISBN-10: 1-86094-970-3 (hardcover : alk. paper)
    1. Kernel functions. I. Imperial College of Science, Technology, and
    Medicine (Great Britain) II. Title.
    QA353.K47S36 2007
    515'.7223--dc22
                                                                    2007039136

British Library Cataloguing-in-Publication Data
A catalogue record for this book is available from the British Library.

Printed in Singapore.

To the memory of Walter Feit

Preface

The $k(GV)$ problem has recently been solved, completing the work of a series of authors over a period of more than forty years. The objective of this book is to describe the developments, the ideas and methods, leading to this remarkable result. All details of the proof will be presented.

Let G be a finite group. The number $k(G)$ of conjugacy classes of G, which is just the number of irreducible complex characters of G, is certainly an invariant of G of special interest. For some families of finite groups, like the symmetric groups and some of the finite classical groups, this invariant is known to some extent. From the point of view of abstract group theory, however, little can be said about $k(G)$ in relation to the group G. Of course $k(G) \leq |G|$, with equality if and only if G is abelian. If N is a normal subgroup of G, then it is easy to see that $k(G) \leq k(N) \cdot k(G/N)$ but there are examples where $k(N) > k(G)$ (e.g. in the dihedral group of order 10). This makes it difficult to bound $k(G)$ by inductive methods.

A unifying notion in group theory is the concept of representation. So finite groups often appear as subgroups of permutation groups or linear groups. The "geometry" of a group (as permutation group or linear group, or as a group of Lie type $etc.$) should be used in order to describe basic invariants. Here we just assume that G is embedded into the linear group $\mathrm{GL}(V)$ of some finite vector space V over some prime field $\mathbb{F}_p = \mathbb{Z}/p\mathbb{Z}$, and we wish to bound $k(G)$ by a function of $|V| = p^m$. A weak form of the $k(GV)$ problem is the question whether $k(G) \leq |V| - 1$ in the case where G is a p'-group ($p \nmid |G|$). Here equality can happen since $\mathrm{GL}(V)$ contains cyclic subgroups of order $|V| - 1$, the so-called Singer cycles. The restriction to p'-groups is essential, because $\mathrm{GL}(V) = \mathrm{GL}_m(p)$ has abelian subgroups of order $p^{\lfloor \frac{m^2}{4} \rfloor}$.

Now consider the semidirect product GV (also written $G \ltimes V$ or $V : G$). Since $k(GV) > k(G)$, one can strengthen the question above by asking whether $k(GV) \leq |V|$ when G is a p'-subgroup of $\mathrm{GL}(V)$. This is known as the $k(GV)$ conjecture.

The $k(GV)$ conjecture is a special case of Brauer's celebrated $k(B)$ problem [Brauer, 1956]. The $k(B)$ conjecture predicts that the number $k(B)$ of ordinary irreducible characters in a p-block B is bounded above by the order of a defect group D of B (Problem (X) in [Feit, 1982]). In the situation of the $k(GV)$ conjecture, the group GV has a single p-block with the unique defect group $D = V$. [Brauer and Feit, 1959] established the general upper bound $k(B) \leq 1 + \frac{1}{4}|D|^2$. This bound has since only been slightly improved. The conjectured estimate $k(B) \leq |D|$ has been proved for cyclic and rank 2 abelian defect groups D, and for some families of finite groups, including the symmetric and finite general linear groups. There is reason to believe that the $k(B)$ conjecture is one of the most difficult problems proposed by Brauer.

[Nagao, 1962] noticed that the $k(B)$ conjecture holds for p-solvable groups provided it holds for semidirect products as discussed above. So the $k(GV)$ theorem implies Brauer's conjecture for such groups. This was the original motivation for treating the $k(GV)$ problem which is, on the other hand, of interest in its own right.

In the proof of the $k(GV)$ conjecture Clifford theory plays an important role. This theory may be loosely described as using normal subgroups to pass from the module of interest to lower-dimensional, more tractable modules. This can be fruitfully applied to many problems in group representation theory. In a typical sequence of reductions one tries to show that the module in question may be assumed to be irreducible, then absolutely irreducible, then primitive, then tensor indecomposable, and finally tensor primitive. If one tries this approach on the $k(GV)$ problem, where V is a faithful coprime $\mathbb{F}_p G$-module, the first two steps are easy, but the third is not. Indeed, the final stage of the proof of the $k(GV)$ conjecture was the rather difficult verification of the problem when V is induced from a certain module with cardinality 5^2.

Nevertheless, Clifford theory is crucial, in a manner we now wish to explain. We consider the centralizers (stabilizers) $C_G(v)$ as v ranges over the vectors in V. It is easy to see that $k(GV)$ is the sum over the $k(C_G(v))$ when v ranges over a set of representatives for the G-orbits on V. Remarkably, however, the main theoretical results in the solution of the $k(GV)$ problem assert that the existence of *one* vector v in V with $C_G(v)$ satisfying suitable conditions is enough to imply that $k(GV) \leq |V|$.

The most elementary of these centralizer criteria asserts that it suffices to find $v \in V$ such that $C_G(v) = 1$ (Theorem 1.5d of this book). Knörr established two more general centralizer criteria which had great influence on later work [Knörr, 1984]. He showed first that $k(GV) \leq |V|$ if $C_G(v)$

is abelian for some v in V (Theorem 3.4d). Even more important was his second criterion, because it allowed one to assume, in many cases, that V is primitive, thus partially overcoming the major obstacle to the direct Clifford-theoretic approach. Knörr's ideas are related to some general techniques developed in [Brauer, 1968].

Gow advantageously reworked Knörr's ideas and proved the result in the case that V is a self-dual G-module [Gow, 1993]. In [Robinson, 1995] it was noticed that it suffices to find $v \in V$ such that the restriction of V to $C_G(v)$ is self-dual. The most powerful criterion then was established in [Robinson and Thompson, 1996]. The Robinson–Thompson criterion (Theorem 5.2b) asserts that $k(GV) \leq |V|$ if there exists v in V such that the restriction of V to $C_G(v)$ contains a faithful self-dual summand (with real-valued Brauer character). Such a vector v will be called a *real vector*.

This criterion is sufficiently easy to verify, and it is compatible with the Clifford-theoretic reduction, and so led to much progress towards a solution of the $k(GV)$ problem. At the end of the reduction steps one is left with a pair (G, V) admitting no real vectors, such that the generalized Fitting subgroup of G has the form $E \cdot Z(G)$, where E is normal in G and absolutely irreducible on V, and where either E is a group of extraspecial type or is quasisimple. Such a pair (G, V) will be called *nonreal reduced*. Robinson and Thompson already showed that this can happen only when the characteristic $p \leq 5^{30}$.

In Chapters 6 and 7, we use counting arguments to show that nonreal reduced pairs do exist only when p is $3, 5, 7, 11, 13, 19$ or 31 and V has dimension $2, 3$ or 4. It follows that the $k(GV)$ conjecture is true when the characteristic p is not one of these seven primes. We give a new treatment of the extraspecial case, using ideas from [Robinson, 1997], [Riese–Schmid, 2000], [Gluck–Magaard, 2002a] and [Riese, 2002]. In the quasisimple case our approach is based on work of [Liebeck, 1996], [Goodwin, 2000], [Riese, 2001] and [Köhler–Pahlings, 2001]. Much is simplified by systematically using properties of characters related to extraspecial groups (Heisenberg groups). The counting techniques usually lead to "small" groups G and modules V of "small" degrees, often affording minimal "Weil characters".

After the analysis of the extraspecial and quasisimple cases was completed around 2000, one had to deal with arbitrary pairs (G, V) admitting no real vectors. In some Clifford-theoretic sense then (G, V) "involves" a nonreal reduced pair, but *a priori* this seemed to be hard to control. So it was striking, and a surprise, that V is just obtained by module induction from a nonreal reduced pair, and that G is close to being a full wreath product with respect to the corresponding imprimitivity decomposition of V (Theorems 8.4 and 8.5c). This crucial result was accomplished in [Riese–Schmid, 2003] using ideas developed in [Gluck–Magaard, 2002b].

In the final stages of the proof of the $k(GV)$ theorem, one therefore had to treat imprimitive modules induced from nonreal reduced modules. Here one requires bounds on the number of conjugacy classes in a permutation group. Liebeck and Pyber have established the general upper bound 2^{n-1} for the number of conjugacy classes in a permutation group of degree n (Theorem 9.3). In this manner the proof was completed by Riese and Schmid for $p \neq 5$, and by Gluck, Magaard, Riese and Schmid for $p = 5$.

Chapter 11 addresses the question of when one can have equality $k(GV) = |V|$, without giving a conclusive answer, however. In the final two chapters we briefly describe some consequences of the $k(GV)$ theorem for general block theory and consider the problem of bounding $k(GV)$ when G is completely reducible on V but $|G|$ and $|V|$ are not coprime.

At present the proof of the $k(GV)$ conjecture relies on the classification of the finite simple groups; see Chapters 7 and 9. But often I argue on the basis of general counting arguments, which do not refer to a certain simple group and which reduce the discussion to groups of low order. Most challenging are indeed the classical groups. For solvable groups the proof is independent of the classification theorem.

I hope this monograph will be comprehensible to a graduate student with some background in group theory and representation theory. Much of this general background is provided by Isaacs' book [Isaacs, 1976]. Some knowledge of the finite simple groups is also needed; the "Atlas of Finite Groups" [Conway et al., 1985] contains much of the necessary information. For the convenience of the reader appendices are included on the cohomology of finite groups, on parabolic subgroups of some finite classical groups, and on the Weil characters of such groups.

I am indebted to Walter Feit for drawing my attention to the $k(GV)$ problem, and for his long-standing support. Thanks also to David Gluck, Kay Magaard and Udo Riese for critical comments and valuable suggestions during the preparation of this book.

March 2007   Peter Schmid, Tübingen

Contents

Chapter 1

Conjugacy Classes, Characters, and Clifford Theory

For the proof of the $k(GV)$ theorem many of the standard methods and techniques from ordinary and modular representation theory will be applied. In this section we describe the necessary concepts and tools from ordinary character theory. The reader is referred to [Isaacs, 1976] for the relevant background and some basic results used but not proved here. This book is cited as [I] in the text.

1.1. Class Functions and Characters

Fix a finite group X of order $|X|$ and exponent $\exp(X) = e$. Let $K = \mathbb{Q}(\varepsilon)$ for $\varepsilon = e^{2\pi i/e}$.

Each complex character χ of X has its values in K, even in its ring of integers $\mathbb{Z}[\varepsilon]$, because for each $x \in X$ there is a basis such that an underlying representation carries x to a diagonal matrix consisting of eth roots of unity, and $\chi(x)$ is the trace of this matrix. This χ is constant on conjugacy classes of X, a *class function*, and if $c = c_x$ is the conjugacy class of X containing x ($c_x = x^X = \{x^t = t^{-1}xt|\ t \in X\}$), we may write $\chi(c) = \chi(x)$. The distinct class sums $\widehat{c} = \sum_{y \in c} y$ form a basis of the centre $Z(\mathbb{C}X)$ of the semisimple group algebra $\mathbb{C}X$ (and of $Z(KX)$). It follows that the set $\mathrm{C}\ell(X)$ of conjugacy classes of X is in bijection with the ordinary (complex) irreducible characters of X, i.e.,

$$(1.1a) \qquad k(X) = |\mathrm{C}\ell(X)| = |\mathrm{Irr}(X)|.$$

If χ, ψ are K-valued class functions on X, their inner product is denoted by $\langle \chi, \psi \rangle = \langle \chi, \psi \rangle_X = \frac{1}{|X|} \sum_{x \in X} \chi(x)\overline{\psi(x)} = \frac{1}{|X|} \sum_{c \in \mathrm{C}\ell(X)} |c| \cdot \chi(c)\overline{\psi(c)}$. One knows that $\mathrm{Irr}(X)$ is an orthonormal basis for the K-vector space of class functions on X [I, 2.17]. The (nonsingular) square matrix $\mathcal{X} = (\chi(c))_{\substack{\chi \in \mathrm{Irr}(X) \\ c \in \mathrm{C}\ell(X)}}$ (rows and columns somehow arranged) is the *character table* of X. For some simple groups (of small order) these tables can be found in [Conway et al., 1985], which we usually refer to as the [Atlas].

1

Orthonormality of the irreducible characters may be expressed by the matrix equation $\mathcal{X} \cdot T \cdot \overline{\mathcal{X}}^t = I$, where T is the diagonal matrix with entries $\frac{1}{|C_X(x)|}$ for $x \in c$ and $C_X(x)$ is the centralizer ($|c| = |X : C_X(x)|$). This gives $\overline{\mathcal{X}}^t \cdot \mathcal{X} = T^{-1}$ and hence the *second orthogonality relations*:

$$(1.1b) \qquad \textstyle\sum_{\chi \in \mathrm{Irr}(X)} \overline{\chi(x)}\chi(y) = \begin{cases} |C_X(x)| & \text{if } x^X = y^X \\ 0 & \text{otherwise} \end{cases}.$$

In particular $|X| = \sum_{\chi \in \mathrm{Irr}(X)} \chi(1)^2$ is the sum over the squares of the degrees of the irreducible characters.

A generalized character χ of X is a rational integer combination of irreducible characters of X ($\chi \in \mathbb{Z}[\mathrm{Irr}(X)]$). A subgroup E of X is called *p-elementary* if $E = P \times Z$ where P is a p-group for some prime p and Z is a cyclic p'-group, and it is elementary if it is p-elementary for some prime p. We have Brauer's characterization of characters:

Theorem 1.1c (Brauer). *Let χ be a complex valued class function on X. Then $\chi \in \mathbb{Z}[\mathrm{Irr}(X)]$ if and only if the restriction $\mathrm{Res}_E^X(\chi) \in \mathbb{Z}[\mathrm{Irr}(E)]$ for any elementary subgroup E of X. In fact, each generalized character of X is a $\mathbb{Z}$-linear combination of characters of X induced from linear characters of elementary subgroups.*

For a proof see [I, 8.10 and 8.12]. Induced characters will be discussed in Sec. 1.2 below. Since linear characters (of degree 1) can be realized over their value fields, using elementary properties of the Schur index we get:

Theorem 1.1d (Brauer). *K is a splitting field for X, that is, for all characters χ of X there is a matrix representation over K or, equivalently, a KX-module affording χ.*

It is immediate, then, that K is a splitting field for all subgroups of X. The *Schur index* will be briefly discussed in Sec. 1.3.

If V and W are (right) KX-modules affording the characters χ, ψ, then $V \oplus W$ affords $\chi + \psi$ (sum), and the KX-module $V \otimes_K W$ (diagonal X-action) affords $\chi\psi$ (product). $V \otimes_K V = \mathrm{Sym}^2(V) \oplus \mathrm{Alt}^2(V)$ decomposes into the *symmetric squares* and *alternating squares*.

1.2. Induced and Tensor-induced Modules

Suppose Y is a subgroup of X and W is a (right) KY-module affording the character θ. Let $n = |X : Y|$ be the index of Y in X, and let $Y\backslash X = \{Yt \,|\, t \in X\}$ denote the (transitive) X-set with respect to right multiplication $(Yt, x) \mapsto Ytx$. Then the *normal core* $N = \mathrm{Core}_X(Y) = \bigcap_{t \in X} Y^t$ is the kernel of this permutation representation. Hence $G = X/N$ is a transitive subgroup of the symmetric group S_n. If $Y \neq X$ $(n > 1)$, then $\bigcup_{t \in X} Y^t \neq X$ and there are at least $n - 1$ permutations in G without fixed points. See also [Serre, 2003] for a recent discussion of this classical result by Jordan.

Theorem 1.2a (Frobenius). *There is an embedding of X into the wreath product $Y \,\mathrm{wr}\, S_n = Y^{(n)} : S_n$, which is uniquely determined up to conjugacy.*

The wreath product is defined by letting S_n act on the nth direct power $Y^{(n)}$ of Y permuting the direct factors (so $(y_i)_i^\pi = (y_i)_{i\pi} = (y_{i\pi^{-1}})_i$, sending an entry in the ith position to the $i\pi$th position). The wreath product may be identified with the group of all monomial $n \times n$-matrices with entries in Y. "The" Frobenius embedding is obtained by choosing a right transversal $\{t_i\}_{i=1}^n$ to Y in X. Associate then to $x \in X$ the element $(x_i) \cdot \pi_x$ in the wreath product, where $\pi_x : i \mapsto ix$ in $G \subseteq S_n$ and $x_i \in Y$ are defined by $t_i x = x_i t_{ix}$. Replacing $\{t_i\}$ by $\{y_i t_i\}$ for certain $y_i \in Y$ leads to an embedding conjugate under $(y_i)^{-1}$.

Now the *base group* $B = Y^{(n)}$ of the wreath product acts *diagonally* onto $W^{(n)} = W \oplus \cdots \oplus W$ (n direct summands) via $(w_i) \cdot (y_i) = (w_i y_i)$, and S_n through $(w_i)_i \cdot \pi = (w_{i\pi^{-1}})_i$. This makes $W^{(n)}$ into a $K[Y \,\mathrm{wr}\, S_n]$-module. The induced KX-module $V = \mathrm{Ind}_Y^X(W)$ is obtained through "the" Frobenius embedding of X into $Y \,\mathrm{wr}\, S_n$ (where conjugate embeddings yield isomorphic module structures). Fixing a transversal $\{t_i\}$ the character of X afforded by V is given by

$$(1.2b) \qquad \mathrm{Ind}_Y^X(\theta)(x) = \textstyle\sum_{i=1}^n \theta(t_i x t_i^{-1}),$$

where we set $\theta(\cdot) = 0$ for elements outside Y.

Induction and restriction are maps on class functions *adjoint* to each other, namely related by the *Frobenius reciprocity*

$$(1.2c) \qquad \langle \mathrm{Ind}_Y^X(\theta), \chi \rangle_X = \langle \theta, \mathrm{Res}_Y^X(\chi) \rangle_Y.$$

In terms of modules this says that if W is a KY-module, V a KX-module, every KY-homomorphism $f : W \to \operatorname{Res}_Y^X(V)$ extends uniquely to a KX-homomorphism $\operatorname{Ind}_Y^X(f) : \operatorname{Ind}_Y^X(W) \to V$.

We often will use *Mackey decomposition* for characters (and modules). Suppose Y and H are subgroups of X. Then H acts on $X\backslash Y$, and if $\{r_j\}$ is a set of representatives for the distinct H-orbits (or double cosets of X mod (Y, H)), for a character θ of Y we have

$$(1.2\text{d}) \qquad \operatorname{Res}_H^X(\operatorname{Ind}_Y^X(\theta)) = \textstyle\sum_j \operatorname{Ind}_{Y^{r_j} \cap H}^H(\operatorname{Res}_{Y^{r_j} \cap H}^{Y^{r_j}}(\theta^{r_j})).$$

Here for $x \in X$ we define the (conjugate) character θ^x of $Y^x = x^{-1}Yx$ by $\theta^x(y^x) = \theta(y)$ for $y \in Y$.

One can define $\operatorname{Ind}_Y^X(W) = W \otimes_{KY} KX$ (viewing KX as a left KY-module). Let now $W^{\otimes n} = W \otimes \cdots \otimes W$ (n factors, the tensors over K). Again $B = Y^{(n)}$ acts diagonally on $W^{(n)}$ via $(\otimes_i w_i) \cdot (y_i) = \otimes_i w_i y_i$, and S_n acts as $(\otimes_i w_i) \cdot \pi = \otimes_i w_{i\pi^{-1}}$. This makes $W^{\otimes n}$ into a $K[Y \operatorname{wr} S_n]$-module. Then the *tensor-induced* KX-module $\widehat{V} = \operatorname{Ten}_Y^X(W) = W^{\otimes n}$ is obtained through "the" Frobenius embedding of X into the wreath product. Mackey decomposition (for modules) carries over to tensor induction in the obvious (multiplicative) way. So if $\{r_j\}$ is a set of representatives for the cycles (orbits) of an element $x \in X$ in its action on $Y\backslash X$, and if the jth cycle has size n_j (so that $r_j x^{n_j} r_j^{-1} \in Y$), then

$$(1.2\text{e}) \qquad\qquad \operatorname{Ten}_Y^X(\theta)(x) = \textstyle\prod_j \theta(t_j x^{n_j} t_j^{-1})$$

describes the character $\widehat{\chi}$ of X afforded by $\widehat{V} = \operatorname{Ten}_Y^X(W)$. In order to prove this it suffices to consider the case where $\widehat{V} = W \otimes Wx \otimes \cdots \otimes Wx^{n-1}$ (with $x^n \in Y$). If $\mathfrak{b}$ is a basis of W consisting of eigenvectors for x^n, then the $w_{i_0} \otimes w_{i_1} x \otimes \cdots \otimes w_{i_{n-1}} x^{n-1}$, with $w_{i_j} \in \mathfrak{b}$, form a basis of $\widehat{V}$ for which the matrix of x is monomial. Only basis vectors of the form $v_i = w_i \otimes w_i x \otimes \cdots \otimes w_i x^{n-1}$, with fixed $w_i \in \mathfrak{b}$, are eigenvectors of x on $\widehat{V}$, and if $w_i x^n = c_i w_i$ with $c_i \in K$, then $v_i x = c_i v_i$. Hence $\widehat{\chi}(x) = \theta(x^n)$, as desired.

1.3. Schur's Lemma

Suppose V and W are KX-modules affording the characters χ, ψ, respectively. Then

$$\langle \chi, \psi \rangle = \dim {}_K\operatorname{Hom}_{KX}(V, W).$$

If χ, ψ are irreducible, $\mathrm{Hom}_{KX}(V, W) = 0$ if $\chi \neq \psi$ and $\mathrm{End}_{KX}(V) \cong K$ if $V = W$ (Schur's lemma). Of course this yields the first orthogonality relations. In particular, if $\rho : X \to \mathrm{GL}_n(K)$ is a (matrix) representation of X affording $\chi \in \mathrm{Irr}(X)$, only the scalar matrices are permutable with all $\rho(x)$, $x \in X$. Extending ρ linearly to KX we get a K-algebra homomorphism into $M_n(K)$ and, by restriction to the centre, a homomorphism $\omega_\chi : Z(KX) \to K$, the *central character* associated to χ. If $c = c_x = x^X$ is the conjugacy class of $x \in X$,

(1.3a) $$\omega_\chi(\widehat{c}) = \frac{\chi(x)|X : C_X(x)|}{\chi(1)} = \frac{\chi(c)|c|}{\chi(1)}.$$

The values of ω_χ are algebraic integers since the product of class sums is a (nonnegative) integer linear combination of class sums. This is important for block theory (Chapter 2). Writing $1 = \langle \chi, \chi \rangle = \frac{\chi(1)}{|G|} \sum_{c \in C\ell(X)} \omega_\chi(\widehat{c}) \overline{\chi(c)}$ we see that $\frac{|G|}{\chi(1)}$ is an integer (being rational and an algebraic integer). Clifford theory will even yield the following (see also [I, 3.12 and 6.15]).

Theorem 1.3b (Itô). *The degree $\chi(1)$ of an irreducible character χ of X divides $|X : V|$ for any abelian normal subgroup V of X.*

We used that irreducible representations over K are *absolutely* irreducible. Replacing K by the field $K_0 = \mathbb{Q}(\chi)$ generated by the values of χ, there is a unique irreducible $K_0 X$-module V, up to isomorphism, whose character contains χ. Then V affords the character $m\chi$ where $m = m(\chi)$ is the *Schur index* of χ (over the rationals). $D = \mathrm{End}_{K_0 X}(V)$ is a K_0-division algebra with centre K_0 and dimension m^2 [I, 9.21].

We give some further examples where Schur's lemma is involved. Let $x \in X$ and $\chi \in \mathrm{Irr}(X)$, and let $\bar{\chi}$ be the complex conjugate character. Then $\bar{\chi}(x) = \overline{\chi(x)} = \chi(x^{-1})$ (see Sec. 1.5). We assert that

(1.3c) $$|\chi(x)|^2 = (\chi\bar{\chi})(x) = \frac{\chi(1)}{|X|} \sum_{y \in X} \chi([x, y]).$$

Here $[x, y] = x^{-1} x^y$ denotes the commutator of x and y. Let ρ be a representation of X affording χ. The sum on the right is the trace of $\rho(x^{-1}) \frac{\chi(1)}{|X|} \sum_{y \in X} \rho(x^y) = \rho(x^{-1}) \frac{\chi(1)}{|X|} \omega_\chi(\widehat{c}_x) |C_X(x)| = \rho(x^{-1}) \chi(x)$, and the assertion follows.

We see that $|\chi(x)| = \chi(1)$ if and only if $\chi([x, y]) = \chi(1)$ for all $y \in X$. Note that $\mathrm{Ker}(\chi) = \{x \in X | \chi(x) = \chi(1)\}$ is the kernel of ρ, and $Z(\chi) = \{x \in X | |\chi(x)| = \chi(1)\}$ consists of those $x \in X$ for which $\rho(x)$ is a scalar matrix.

Suppose next that $X = \langle Y, x \rangle$ for some subgroup Y, and that $\theta = \mathrm{Res}_Y^X(\chi)$ is (still) irreducible. We assert that then

(1.3d) $\sum_{y \in Y} |\chi(xy)|^2 = |Y|.$

The 1-character 1_X is contained in $\chi\bar{\chi}$ with multiplicity 1, because we have $\langle 1_X, \chi\bar{\chi} \rangle = \langle \chi, \chi \rangle = 1$. Similarly $\langle 1_Y, \mathrm{Res}_Y^X(\chi\bar{\chi}) \rangle = \langle 1_Y, \theta\bar{\theta} \rangle = 1$. Clearly $\mathrm{Res}_Y^X(1_X) = 1_Y$. So if $\psi \neq 1_X$ is an irreducible constituent of $\chi\bar{\chi}$, $\mathrm{Res}_Y^X(\psi)$ does not contain 1_Y. Hence if W is a KX-module affording ψ and τ is an underlying representation, then $\sum_{y \in Y} \tau(y) = 0$ as

$$\dim{}_K C_W(Y) = \dim{}_K \mathrm{Hom}_{KY}(K, W) = \langle 1_Y, \psi \rangle_Y = 0.$$

It follows that $\sum_{y \in Y} \psi(xy) = 0$ by considering the trace of $\tau(x) \sum_{y \in Y} \tau(y)$. We conclude that $\sum_{y \in Y} (\chi\bar{\chi})(xy) = \sum_{y \in Y} 1_X(xy) = |Y|.$

1.4. Brauer's Permutation Lemma

Suppose G is a finite group acting on the finite set Ω (from the right, by permutations). Then we write $C\ell(G|\Omega) = \mathrm{orb}(G \text{ on } \Omega)$ for the set of orbits of G on Ω. So $C\ell(X) = C\ell(X|X)$ with X acting by conjugation. By the Cauchy–Frobenius fixed point formula

(1.4a) $|C\ell(G|\Omega)| = \frac{1}{|G|} \sum_{g \in G} |C_\Omega(g)|.$

This is sometimes also called Burnside's lemma; it is easily proved by means of the counting principle or using Frobenius reciprocity [Serre, 2003]. Each orbit is a (transitive) G-set and so isomorphic to $H \backslash G$ for some subgroup H, which is determined up to conjugacy in G (being a point stabilizer). The isomorphism type of the G-set Ω is determined by the "marks" $|C_\Omega(H)|$ for all (nonconjugate) subgroups H of G [I, 13.23]. We associate to the G-set Ω the permutation character π_Ω of G, counting the fixed points of each element.

Theorem 1.4b (Brauer). *Suppose G is a finite group which acts on $\mathrm{Irr}(X)$ and on $C\ell(X)$ such that $\chi^g(c^g) = \chi(c)$ for all $\chi \in \mathrm{Irr}(X)$, $c \in C\ell(X)$ and $g \in G$. Then for each $g \in G$, the number $k_g(X) = |C_{C\ell(X)}(g)|$ of g-invariant conjugacy classes agrees with the number $|C_{\mathrm{Irr}(X)}(g)|$ of g-invariant irreducible characters of X. In particular, the permutation characters $\pi_{C\ell(X)} = \pi_{\mathrm{Irr}(X)}$ of G agree.*

The proof [I, 6.32] is based on the fact that the character matrix of X is nonsingular. It follows, in view of Eq. (1.4a), that G has the same number of orbits on $\mathrm{Irr}(X)$ and on $\mathrm{C}\ell(X)$. Of course, this does not mean that these sets are permutation isomorphic (unless G is cyclic).

1.5. Algebraic Conjugacy

Let $\Gamma = \mathrm{Gal}(K|\mathbb{Q})$ be the Galois group of K over the rationals. We have a natural action of Γ on $\mathrm{Irr}(X)$. We also have a permutation action on X as $\Gamma \cong (\mathbb{Z}/e\mathbb{Z})^{\star}$, where $x^{\sigma} = x^{n}$ if $\sigma \in \Gamma$ corresponds to the coset of n modulo $e = \exp(X)$. This preserves $\mathrm{C}\ell(X)$. Let $\chi \in \mathrm{Irr}(X)$. If $\varepsilon_{i} \in K$ are the eigenvalues of x appearing in a representation to χ, the eigenvalues for x^{n} are the ε_{i}^{n}. It follows that

$$\chi^{\sigma}(x) = \chi(x)^{\sigma} = \chi(x^{n}) = \chi(x^{\sigma}).$$

In order to apply Theorem 1.4b one has to alter the action of Γ on X (say) by assigning $(x, \sigma) \mapsto x^{\sigma^{-1}}$. This works since Γ is *abelian*. Notice that if σ is complex conjugation (restricted to K), $\chi^{\sigma}(x) = \bar{\chi}(x) = \overline{\chi(x)} = \chi(x^{-1})$.

So the number of real-valued irreducible characters of X is equal to the number of *real conjugacy classes* c of X, satisfying $c^{-1} = c$ (Burnside). X is called a *real group* if all its conjugacy classes are real, that is, if all $\chi \in \mathrm{Irr}(X)$ are real-valued.

Suppose X has odd order. Then $\{1\}$ is the unique real class of X. Since $\chi(1) = \bar{\chi}(1)$ is odd, we get $|X| = \sum_{\chi \in \mathrm{Irr}(X)} \chi(1)^{2} = 1 + 2 \sum_{i=1}^{\frac{k(X)-1}{2}} (1 + 2n_{i})^{2}$ for certain integers $n_{i} \geq 1$. Consequently

(1.5a) $$|X| \equiv k(X) \,(\mathrm{mod}\, 16),$$

a well known result due to Burnside.

Recall that a character of X takes only values which are algebraic integers. If $\alpha \neq 0$ is such an algebraic integer with the distinct conjugates α_{i} over the rationals ($1 \leq i \leq n$), then $\sum_{i=1}^{n} |\alpha_{i}| \geq n$, with equality only if α is a root of unity. For by the arithmetic–geometric mean inequality we have

(1.5b) $$\tfrac{1}{n} \sum |\alpha_{i}| \geq (\prod |\alpha_{i}|)^{\frac{1}{n}} = |\mathrm{N}(\alpha)|^{\frac{1}{n}},$$

with equality only if all $|\alpha_i|$ are equal. But the *norm* $N(\alpha) = \prod_i \alpha_i$ is a nonzero rational integer. Hence the assertion holds, with equality only if all $|\alpha_i|$ are equal and $N(\alpha) = \pm 1$. In this case $|\alpha_i| = 1$ for all i, whence α is a root of unity (since only finitely many powers of α are distinct).

Lemma 1.5c (Gallagher). *Suppose $y \in X$ is such that $\chi(y) \neq 0$ for each $\chi \in \mathrm{Irr}(X)$. Let $N = [\langle y \rangle, X]$. Then $k(X) \leq |C_X(y)| - (|X/N| - k(X/N))$.*

Proof. We follow [Gallagher, 1962]. N is the (normal) subgroup of X generated by all commutators $[t, x]$, $t \in \langle y \rangle$, $x \in X$. By the second orthogonality relations (1.1b),

$$|C_X(y)| = \sum_{\chi \in \mathrm{Irr}(X)} |\chi(y)^2| = \Sigma_1 + \Sigma_2,$$

where the first sum is over those χ with $|\chi(y)| = \chi(1)$ and the second sum is over the others. Now $|\chi(y)| = \chi(1)$ if and only if $y \in Z(\chi)$ (as described in Sec. 1.3), and this happens if and only if N is in the kernel of χ and so χ may be viewed as a character of X/N. Thus $\Sigma_1 = |X/N|$, and the number of irreducible characters of X in Σ_1 is equal to $k(X/N) = |\mathrm{Irr}(G/N)|$.

For each σ in $\Gamma = \mathrm{Gal}(K|\mathbb{Q})$ we have $|\chi^2|^\sigma = (\chi \cdot \bar{\chi})^\sigma = \chi^\sigma \overline{\chi^\sigma} = |\chi^\sigma|^2$. Thus Σ_1 and Σ_2 are Galois stable. By hypothesis the average over the Galois class of $|\chi(y)^2|$ is ≥ 1 (and is equal to 1 only if $\chi(y)$ is a root of unity). Consequently $\Sigma_2 \geq k(X) - k(X/N)$, and the result follows. □

Theorem 1.5d. *Suppose X has an abelian normal Sylow p-subgroup, V, for some prime p. Then $X = GV$ for some p-complement G in X, uniquely determined up to conjugacy. For each $v \in V$,*

$$k(GV) \leq |C_G(v)| \cdot |V| - \big(|G| - k(G)\big).$$

In particular, if $C_G(v) = 1$ for some $v \in V$, then $k(GV) \leq |V|$ and equality only holds if G is abelian.

Proof. By a simple cohomological argument $X = GV$ is as claimed (Appendix A1; Schur–Zassenhaus theorem). Let $v \in V$. We assert that $\chi(v) \neq 0$ for each $\chi \in \mathrm{Irr}(X)$. By Theorem 1.3b, $\chi(1)$ is not divisible by p. Letting $\mathfrak{p}$ be a prime ideal above p in the ring of integers of K, we have $\chi(v) \equiv \chi(1) \pmod{\mathfrak{p}}$ (cf. Chapter 2). Hence the assertion. Now $N = [\langle v \rangle, X] = [v, G]$ is a normal subgroup of X contained in V, and

$C_X(v) = C_G(v)V$. By an elementary counting argument, carried out in Theorem 1.7a below, $k(X/N) \leq |V/N| \cdot k(G)$. Thus by the preceding lemma $k(X) \leq |C_G(v)| \cdot |V| - |V/N|(|G| - k(G))$.

From $C_G(v) = 1$ it follows that $k(X) \leq |V|$, and then $k(X) = |V|$ only if $|G| = k(G)$, that is, if G is abelian. □

1.6. Coprime Actions

If G is a finite group and V is a finite G-module of order prime to $|G|$, then all (Tate) cohomology groups $\mathrm{H}^n(G, V)$ vanish (A1). For $n = 1, 2$ this leads to the Schur–Zassenhaus theorem (already mentioned). For $n = 0, -1$ this tells us that the fixed module $C_V(G)$ is the image of the trace map $v \mapsto \sum_{g \in G} vg$ on V and that the commutator module $[V, G] = [V, G, G]$ is its kernel. Then $C_V(G) \cap [V, G] = 0$ and so

$$(1.6a) \qquad V = C_V(G) \oplus [V, G].$$

$\mathrm{Irr}(V) = \mathrm{Hom}(V, \mathbb{C}^\star)$ is the character group of V, and $|C_{\mathrm{Irr}(V)}(G)| = |\mathrm{Irr}(V/[V, G])| = |V/[V, G]| = |C_V(G)|$. The corresponding holds for all subgroups of G. Hence we have the following.

Proposition 1.6b. *If V is a G-module where V and G have coprime order, then V and $\mathrm{Irr}(V)$ are isomorphic G-sets.*

The proposition is true without assuming that V is abelian. In fact, if G acts on X by automorphisms and $|G|$ and $|X|$ are coprime, then $\mathrm{Irr}(X)$ and $\mathrm{C}\ell(X)$ are isomorphic G-sets. This result is due to Isaacs and Dade. Its proof makes use of the Feit–Thompson theorem. So either G is solvable, in which case a proof can be found in [I, 13.24], or X is solvable, where a proof can be found in [Isaacs, 1973]. We do not need this result in this general form.

Theorem 1.6c (Glauberman). *Suppose G is a cyclic group acting on X by automorphisms where $|G|$ and $|X|$ are coprime. Let ξ be a character of the semidirect product $GX = X : G$ for which $\chi = \mathrm{Res}_X(\xi)$ is irreducible. Then there is a unique irreducible constituent θ of the restriction to $Y = C_X(G)$ of χ, a unique linear character μ of G and a unique sign $\pm$ such that*

$$\xi(gy) = \pm\mu(g)\theta(y)$$

for all generators g of G and all $y \in Y$. If G is a p-group for some prime p, the sign is such that $\langle \chi, \theta \rangle_Y \equiv \pm 1 \pmod{p}$.

This is a special case of a more general character correspondence. For a proof we refer to [I, 13.6 and 13.14]. The character $\widehat{\xi} = \xi \cdot \bar{\mu}$ is the so-called *canonical extension* to GX of χ, determined by the fact that its determinantal character $\det(\widehat{\xi})$ has X in its kernel.

1.7. Invariant and Good Conjugacy Classes

Let N be a normal subgroup of X, and let $G = X/N$. Then G acts on $C\ell(N)$ in the natural way, and $k_g(N) = |C_{C\ell(N)}(g)|$ is the number of g-invariant conjugacy classes of N for each $g \in G$. The conjugacy class g^G of g is called "good for N" provided $C_X(x)N/N = C_G(g)$ for any (some) $x \in g$ (with $Nx = g$). This is well-defined. Suppose N is abelian. Then the class of g is good for N if $C_X(x)/N = C_G(g)$ for $x \in g$, that is, whenever a commutator $[x, y] \in N$ for some $y \in X$ then $[x, y] = 1$. In this case each conjugacy class of G is good for N if and only if N is central in X and no nontrivial element of N is a commutator in X.

Theorem 1.7a (Gallagher). *Let Y be a subgroup of X, and let $N = \text{Core}_X(Y)$ be its normal core. Let $G = X/N$.*

(i) *$k(Y) \leq |X : Y| \cdot k(X)$ and $k(X) \leq |X : Y| \cdot k(Y)$, the latter inequality being proper unless $Y = N$. Moreover,*

$$(1.7b) \qquad k(X) \leq k(N) \cdot k(G),$$

where equality holds if and only if each conjugacy class of G is good for N.

(ii) *Let $g = Nx$ for some $x \in X$. The conjugacy class in $C\ell(N)$ of an element $y \in N$ is fixed by g if and only if $C_g(y) \neq \varnothing$, and then $|C_g(y)| = |C_N(y)|$. The number of g-invariant conjugacy classes of N is*

$$(1.7c) \qquad k_g(N) = \tfrac{1}{|N|} \sum_{y \in N} |C_N(xy)|.$$

Proof. We follow [Gallagher, 1970]. By (1.4a) $k(X) = \frac{1}{|X|} \sum_{x \in X} |C_X(x)|$. The first inequality in (i) is immediate from $C_Y(x) \subseteq C_X(x)$. Using the inequality $|C_X(x)| \leq |X : Y| \cdot |C_Y(x)|$ and the counting principle we have

$$\sum_{x \in X} |C_X(x)| \leq |X : Y| \sum_{x \in X} |C_Y(x)| = |X : Y| \sum_{y \in Y} |C_X(y)|,$$

and this is at most equal to $|X : Y|^2 \sum_{y \in Y} |C_Y(y)|$. We have equality if and only if $C_X(x)Y = X$ for all $x \in X$, and in this case any two conjugate elements of X are Y-conjugate. Then Y is normal in X.

Before proving (1.7b) we settle (ii). Let $t \in N$. Then $tx \in C_g(y) \iff y^{-1}txy = tx \iff xyx^{-1} = t^{-1}yt \iff y^{x^{-1}} = y^t$. Hence $C_g(y) \neq \varnothing$ if and only if y^N is fixed by g^{-1} (or g), and then $C_g(y) = C_N(y)tx$ for some $t \in N$. So $|C_g(y)| = |C_N(y)|$ if $y^N \in C_{C\ell(N)}(g)$ and $C_g(y) = \varnothing$ otherwise. We conclude that

$$k_g(N) = \sum_{y \in N : y^N \in C_{C\ell(N)}(g)} 1/|N : C_N(y)| = \frac{1}{|N|} \sum_{y \in N} |C_g(y)|.$$

Counting the pairs $(tx, y) \in g \times N$ satisfying $(tx)y = y(tx)$ we see that $\sum_{t \in N} |C_N(tx)| = \sum_{y \in N} |C_g(y)|$. This proves Eq. (1.7c), and completes the proof of (ii).

For each $x \in X$ we have $C_X(x)/C_N(x) \cong C_X(x)N/N \subseteq C_G(Nx)$. Hence

$$\sum_{x \in X} |C_X(x)| \leq \sum_{x \in X} |C_G(Nx)| \cdot |C_N(x)| = \sum_{g \in G} |C_G(g)| \sum_{t \in g} |C_N(t)|,$$

and $\sum_{t \in g} |C_N(t)| = \sum_{y \in N} |C_g(y)| \leq \sum_{y \in N} |C_N(y)|$ by (ii). Hence

$$\sum_{x \in X} |C_X(x)| \leq \sum_{g \in G} |C_G(g)| \sum_{y \in N} |C_N(y)|,$$

where equality holds if and only if $C_X(Nx) = C_X(x)N$ for each $x \in X$ (and each X-class of N is an N-class). We are done. □

Theorem 1.7d (Keller). *Let Y be a proper subgroup of X, $N = \mathrm{Core}_X(Y)$ and $G = X/N$. Let $\Omega = C\ell(N|X)$ with N acting by conjugation, which is a G-set (with G acting by conjugation). Then we have a partition $\Omega = \biguplus_{g \in G} \Omega_g$ where Ω_g is the set of N-orbits contained in the coset g.*

(i) *For each $g \in G$ the centralizer $C_G(g)$ is the stabilizer in G of Ω_g, and $|\Omega_g| = k_g(N)$.*

(ii) *Let $g_1 = 1, g_2, \cdots, g_r, g_{r+1}, \cdots, g_{k(G)}$ be representatives for the distinct conjugacy classes of G, the first r classes being just those meeting $H = Y/N$. Then*

$$k(X) = \sum_{i=1}^{k(G)} |C\ell(C_G(g_i)|\Omega_{g_i})| \leq k(Y) + (k(G) - r) \cdot M$$

where $M = \max\{k_g(N)| \ g \notin \bigcup_{t \in G} H^t\}$.

Proof. This is a recent result due to [Keller, 2006]. Let $g \in G$. It is obvious that $C_G(g)$ is the stabilizer in G of Ω_g. By the Cauchy–Frobenius formula (1.4a) and part (ii) of the preceding theorem,

$$|\Omega_g| = \frac{1}{|N|} \sum_{y \in N} |C_g(y)| = \sum_{y \in N : y^N \in C_{\Omega_1}(g)} 1/|N : C_N(y)| = k_g(N).$$

This proves (i). Each G-orbit on Ω is of the form $(x^N)^G$ for some unique conjugacy class $x^X \in C\ell(X)$, and determines the conjugacy g^G of G defined by $Nx = g$ or, equivalently, by $x^N \subseteq g$ ($x^N \in \Omega_g$). In particular, $k(X) = |C\ell(G|\Omega)|$. For $h \in G$ we have $|C\ell(C_G(g^h)|\Omega_{g^h})| = |C\ell(C_G(g)|\Omega_g)|$. This yields the identity given in (ii). For $1 \leq i \leq r$ we may pick the representatives $g_i \in H$, belonging then to certain distinct conjugacy classes of H. Of course $r \leq k(H)$, and $k(G) > r$ by Jordan's theorem. By what is already proved (applied to Y),

$$\sum_{i=1}^{r} |C\ell(C_G(g_i)|\Omega_{g_i})| \leq \sum_{i=1}^{r} |C\ell(C_H(g_i)|\Omega_{g_i})| \leq k(Y).$$

For the remaining $k(G) - r$ conjugacy classes g^G of G, for which g is not in $\bigcup_{t \in G} H^t$, we take the trivial estimate $|C\ell(C_G(g)|\Omega_g)| \leq |\Omega_g|$, and use the fact that $|\Omega_g| = k_g(N)$. $\square$

1.8. Nonstable Clifford Theory

Let N be a normal subgroup of X. Then X acts on N via conjugation (as a group of automorphisms), and on $\mathrm{Irr}(N)$. We have induced actions of $G = X/N$ on $C\ell(N)$ and on $\mathrm{Irr}(N)$, and Theorem 1.4b applies. Fix $\theta \in \mathrm{Irr}(N)$. The stabilizer of θ (in X) is called the *inertia group* $T = I_X(\theta)$, and $T/N = I_G(\theta)$. If $\chi \in \mathrm{Irr}(X|\theta)$ is an irreducible character of X *above* θ, that is, θ is a constituent of $\mathrm{Res}_N^X(\chi)$, there are just $s = |X : T|$ distinct X-conjugates $\theta = \theta_1, \cdots, \theta_s$ of θ and

$$(1.8a) \qquad \mathrm{Res}_N^X(\chi) = e_\chi \sum_{i=1}^{s} \theta_i$$

for some integer $e_\chi \geq 1$, the *ramification index* of χ with respect to N.

Theorem 1.8b (Clifford). *Let $T = I_X(\theta)$. The map $\psi \mapsto \chi = \mathrm{Ind}_T^X(\psi)$ is a bijection from $\mathrm{Irr}(T|\theta)$ onto $\mathrm{Irr}(X|\theta)$. The ramification indices $e_\chi = e_\psi$ are divisors of $|T/N|$.*

For a proof we refer to [I, 6.11 and 11.29].

1.9. Stable Clifford Theory

Let again N be a normal subgroup of X, and let $G = X/N$. Suppose that $\theta \in \mathrm{Irr}(N)$ is G-invariant ($I_G(\theta) = G$). This is a necessary condition for the existence of a character χ of X extending θ. If such a χ exists then $\mathrm{Irr}(X|\theta) = \{\chi\lambda = \chi \otimes \lambda|\ \lambda \in \mathrm{Irr}(G)\}$, and this has cardinality $k(G)$ [I, 6.17]. Moreover, then $\{\chi\lambda|\ \lambda \in \mathrm{Irr}(G), \lambda(1) = 1\}$ is the set of all characters of X extending θ, and this has cardinality $|G/G'|$. Here $G' = [G,G]$ is the commutator subgroup of G, the kernel of the characters of G of degree 1.

In general we proceed as follows. Let $K_0 = \mathbb{Q}(\theta)$ be the field generated by the values of θ, and let W be an irreducible K_0N-module affording $m\theta$, where $m = m(\theta)$ is the Schur index. Then $D = \mathrm{End}_{K_0N}(W)$ is a centrally simple K_0-algebra with $\dim_{K_0} D = m^2$ (1.3). Since N is normal in X and θ is stable under G, each conjugate module $Wg = W \otimes g$ (affording θ^g) is a K_0N-module isomorphic to W ($g \in G$). Choose K_0N-isomorphisms $\tau_g : Wg \to W$ (with $\tau_1 = id_W$). We have $\mathrm{Res}_N^X\big(\mathrm{Ind}_N^X(W)\big) = \bigoplus_{g \in G} Wg$, and by Frobenius reciprocity (1.2c) the τ_g extend uniquely to units in the G-graded ring $\mathrm{End}_{K_0X}\big(\mathrm{Ind}_N^X(W)\big) = \bigoplus_{g \in G} D\tau_g$. Then $\tau_g^{-1}D\tau_g = D$ for all $g \in G$. By the Skolem–Noether theorem [Bourbaki, 1958, Chap. 8, §10] we may choose the τ_g such that they centralize D (via conjugation). Then $\tau_{gh}^{-1}\tau_g\tau_h = \tau(g,h) \cdot id_W$ for some nonzero scalar $\tau(g,h) \in K_0$.

We have a *projective representation* of G with 2-cocycle $\tau \in Z^2(G, K_0^\star)$, where the multiplicative group $K_0^\star = (K_0 \setminus \{0\}, \cdot)$ is viewed as a trivial G-module. The cohomology class of τ depends only on θ, K_0 and the group extension $N \rightarrowtail X \twoheadrightarrow G$; it is written $\mu_{K_0G}(\theta)$. This "Clifford obstruction" is functorial in that it maps onto the corresponding cohomology class when replacing K_0 by an extension field.

Proposition 1.9a. *The Clifford obstruction $\mu_{K_0G}(\theta) \in \mathrm{H}^2(G, K_0^\star)$ vanishes if and only if there is a character χ of X extending θ and satisfying $K_0(\chi) = K_0$. The order of $\mu_{K_0G}(\theta)$ is a divisor of the number of $|G|$th roots of unity in K_0. There is a distinguished central group extension $Z \rightarrowtail G(\theta) \twoheadrightarrow G$, where Z is a cyclic group of order $\exp(N)$, whose cohomology class maps onto $\mu_{KG}(\theta)$ through an (appropriate) embedding of Z into $K^\star$. The exponent of $G(\theta)$ is a divisor of $e = \exp(X)$.*

Proof. If $\mu_{K_0G}(\theta)$ vanishes, by definition one can give $W = \widehat{W}$ the structure of a K_0X-module satisfying $\mathrm{End}_{K_0X}(\widehat{W}) = D$. This $\widehat{W}$ affords $m\chi$ where $\chi \in \mathrm{Irr}(X)$ extends θ, with $m(\chi) = m = m(\theta)$. It follows that

$K_0 = \mathbb{Q}(\chi)$. The converse is proved similarly. The order of $\mu_{K_0 G}(\theta)$ divides $|G|$ by an elementary property of cohomology groups (Appendix A1). It also divides the number of roots of unity in K_0 [Dade, 1974] which, however, will not be used here. We briefly discuss the further (basic) statements.

Replacing K_0 by the complex number field we are just concerned with Schur's theory of lifting projective representations. The *Schur multiplier* $\mathrm{M}(G) = \mathrm{H}_2(G, \mathbb{Z})$ of G fits into the natural *universal coefficient* exact sequence (Appendix A5)

$$0 \to \mathrm{Ext}(G/G', K_0^\star) \to \mathrm{H}^2(G, K_0^\star) \to \mathrm{Hom}(\mathrm{M}(G), K_0^\star) \to 0.$$

Passing to the complex number field, and noting that $\mathbb{C}^\star$ is divisible, we see that $\mathrm{H}^2(G, \mathbb{C}^\star)$ is nothing but the dual of $\mathrm{M}(G)$. So there is a (complex) character χ of X extending θ if $\mathrm{M}(G) = 1$. Of course, then $\mathbb{Q}(\chi) \subseteq K$ and so $\mu_{KG}(\theta)$ vanishes. Let $K_1 = \mathbb{Q}(\varepsilon_1)$ where ε_1 is a primitive $\exp(N)$th root of unity, and let $Z = \langle \varepsilon_1 \rangle$. By Theorem 1.1d this K_1 is a splitting field for N. Let W be a $K_1 N$-module affording θ, and let $\tau(g, h) = \tau_{gh}^{-1} \tau_g \tau_h$ be a 2-cocycle with class $\mu_{K_1 G}(\theta)$. We wish to show that there is a unique element in $\mathrm{H}^2(G, Z)$ mapping onto this cohomology class.

Consider the long exact cohomology sequence to $Z \rightarrowtail K_1^\star \twoheadrightarrow K_1^\star / Z$:

$$\mathrm{H}^1(G, K_1^\star / Z) \xrightarrow{\delta} \mathrm{H}^2(G, Z) \to \mathrm{H}^2(G, K_1^\star) \to \mathrm{H}^2(G, K_1^\star / Z).$$

Either $K_1^\star / Z$ is torsion-free or $|Z|$ is odd and $(-1)Z$ is its unique torsion element. At any rate, δ is the zero map, and it suffices to show that $\mu_{K_1 G}(\theta)$ has trivial image in $\mathrm{H}^2(G, K_1^\star / Z)$. In order to prove this, as well as for the proof of the final statement, we may assume that G is a p-group for some prime p (A4). The construction of $G(\theta)$ will show that its exponent divides $\exp(X)$. Arguing by induction on $|X|$ we may also assume that θ is faithful, because $\mathrm{Ker}(\theta)$ is normal in X, and that there is no proper subgroup X_0 of X covering G such that $N_0 = X_0 \cap N$ has a $G \cong X_0/N_0$-invariant irreducible character θ_0 satisfying $\mu_{K_1 G}(\theta_0) = \mu_{K_1 G}(\theta)$ in $\mathrm{H}^2(G, K_1^\star)$. This reduction will lead us, in the case where G is a p-group, to $X = G(\theta)$.

On the basis of Theorem 1.1c, we find a p-elementary subgroup X_0 of X covering G and an X_0-invariant irreducible character θ_0 of $N_0 = X_0 \cap N$ such that $\langle \theta_0, \theta \rangle_{N_0}$ is not divisible by p [I, 8.24]. But this implies that $\mu_{K_1 G}(\theta_0) = \mu_{K_1 G}(\theta)$. In order to see this, let U be a $K_1 N_0$-module affording θ_0, and consider the K_1-space $\mathcal{H} = \mathrm{Hom}_{K_1 N_0}(U, W)$. By Frobenius reciprocity each $f \in \mathcal{H}$ extends uniquely to a $K_1 X_0$-morphism $\mathrm{Ind}_{N_0}^{X_0}(U) \to \mathrm{Ind}_N^X(W)$, preserving the G-gradings. Since the G-graded ring $\mathrm{End}_{K_1 X_0}\left(\mathrm{Ind}_{N_0}^{X_0}(U)\right) = \bigoplus_{g \in G} K_1 \sigma_g$ is a crosssed product, the maps $f \mapsto \sigma_g^{-1} f \tau_g$ may be considered as elements $\psi_g \in \mathrm{GL}(\mathcal{H})$. We obtain that

$$\psi_{gh}^{-1} \psi_g \psi_h = \sigma(g,h)^{-1} \tau(g,h) \cdot id_{\mathcal{H}}$$

where $\sigma(g,h) = \sigma_{gh}^{-1} \sigma_g \sigma_h \in K_1^\star$ is a factor set with class $\mu_{K_1 G}(\theta_0)$. Taking determinants we get that σ^d and τ^d agree modulo the coboundary obtained from $g \mapsto \det(\psi_g)$. Here $d = \dim_{K_1} \mathcal{H} = \langle \theta_0, \theta \rangle_{N_0}$. Since G is a p-group, hence so is $\mathrm{H}^2(G, K_1^\star)$, and since p does not divide d, the cohomology classes of σ and τ agree. Thus by our choice $X = X_0$ is p-elementary.

Let next M be a G-invariant abelian subgroup of N of maximal order. Since X is p-elementary and X/N a p-group, it follows that $C_N(M) = M$. Let $\lambda \in \mathrm{Irr}(M)$ be an irreducible (linear) constituent of $\mathrm{Res}_M^N(\theta)$ and $X_1 = I_X(\lambda)$, $N_1 = I_N(\lambda) = X_0 \cap N$. Let $\theta_1 \in \mathrm{Irr}(N_1|\lambda)$ be the unique character satisfying $\mathrm{Ind}_{N_1}^N(\theta_1) = \theta$ (Theorem 1.8b). By a Frattini argument X_1 covers G and, by the same argument as before, $\mu_{K_1 G}(\theta_1) = \mu_{K_1 G}(\theta)$ since $\langle \theta_1, \theta \rangle_{N_1} = 1$. Thus $X = X_1$ and $N_1 = N$. It follows that $\mathrm{Res}_M^N(\theta) = \theta(1)\lambda$ and that $M \subseteq Z(N)$ as θ is faithful. But $C_N(M) = M$. Hence $N = M$ is central in X and $\theta = \lambda$ is linear.

Now $\mu_{K_1 G}(\theta)$ is the image of the cohomology class of the central extension $N \rightarrowtail X \twoheadrightarrow G$ under the map induced by $\bar{\theta} = \theta^{-1} : N \to K_1^\star$. Indeed, choose $\tau_g : w \otimes t_g \mapsto w$ for some transversal $\{t_g\}_{g \in G}$ to N in X. Letting $t(g,h) = t_{gh}^{-1} t_g t_h$ be the corresponding factor set, $\mu_{K_1 G}(\theta)$ is the class of the factor set $(g,h) \mapsto \theta(t(g,h)^{-1})$ of G, which has its values in Z. $\square$

Definition 1.9b. The group $G(\theta)$ in the preceding proposition is called the *representation group* of θ (with respect to G). The *extended representation group* is defined as the "fibre-product" (pull-back; "diagonal group" in the terminology of the Atlas)

$$X(\theta) = G(\theta) \Delta_G X$$

of $G(\theta)$ and X amalgamating $G = X/N$. Letting $\tau(g,h) = \tau_{gh}^{-1} \tau_g \tau_h$ be a 2-cocycle with values in Z (viewed as a group of scalar multiplications) and class $\mu_{KG}(\theta)$ in $\mathrm{H}^2(G, K^\star)$, we may write $G(\theta)$ as the group consisting of all pairs $(g,z) \in G \times Z$ with multiplication $(g,z)(h,z') = (gh, zz'\tau(g,h))$. Then $X(\theta)$ consists of all elements $((g,z),x)$ for which $Nx = g$. By Proposition

1.9a, $X(\theta)$ has the same exponent as X. Hence K is a splitting field for $X(\theta)$ by Theorem 1.1d.

Suppose W is a KN-module affording θ. By construction $W = \widehat{W}$ gets the structure of a $KX(\theta)$-module through $w((g,z),x) = (zwx)\tau_g = (zw)\tau_g x$ for $w \in W \subseteq \mathrm{Ind}_N^X(W)$, $g \in G$, $z \in Z$ and $x \in X$ (with $Nx = g$). For $x \in N$ we have $v((1,1),x) = vx$. Thus $\widehat{W}$ is an extension of W when viewed as a module for $\mathrm{Ker}(X(\theta) \twoheadrightarrow G(\theta)) \cong N$.

We may replace K by any subfield which is a splitting field for θ. In this manner we find a character $\widehat{\theta}$ of $X(\theta)$ extending θ, when viewed as a character of $\mathrm{Ker}(X(\theta) \twoheadrightarrow G(\theta))$, which can be written in this same field. This applies in particular when the Schur index $m(\theta) = 1$ (which is true in prime characteristic).

Theorem 1.9c (Clifford). *Let N be a normal subgroup of X, let $G = X/N$ and let $\theta \in \mathrm{Irr}(N)$ be stable in X. Let $\widehat{\theta} \in \mathrm{Irr}(X(\theta))$ extend θ in the above sense, and let $\widetilde{\theta}^{-1}$ be the unique irreducible (linear) constituent of $\widehat{\theta}$ on $\mathrm{Ker}(X(\theta) \twoheadrightarrow X) \cong Z$. Then $\zeta \leftrightarrow \chi = \widehat{\theta} \otimes \zeta$ is a 1-1 correspondence between $\mathrm{Irr}(G(\theta)|\widetilde{\theta})$ and $\mathrm{Irr}(X|\theta)$. Moreover $|\mathrm{Irr}(X|\theta)| = |\mathrm{Irr}(G(\theta)|\widetilde{\theta})| \leq k(G)$.*

Proof. Let $\chi \in \mathrm{Irr}(X|\theta)$, and view χ as a character of $X(\theta)$ by inflation. Since $\widehat{\theta}$ extends θ when viewed as a character of $\mathrm{Ker}(X(\theta) \twoheadrightarrow G(\theta)) \cong N$, there is a unique (irreducible) character ζ of $G(\theta)$ such that $\chi = \widehat{\theta} \otimes \zeta$ (see above). But $\mathrm{Ker}(X(\theta) \twoheadrightarrow X) \cong Z$ is in the kernel of χ. It follows that $\zeta \in \mathrm{Irr}(G(\theta)|\widetilde{\theta})$ where $\widetilde{\theta}$ is as described. Conversely, every $\zeta \in \mathrm{Irr}(G(\theta)|\widetilde{\theta})$ gives rise to an irreducible character $\chi = \widehat{\theta} \otimes \zeta$ in $\mathrm{Irr}(X|\theta)$.

For the final statement we may assume that $X = G(\theta)$, $N = Z$ is central in X and $\widetilde{\theta} = \theta$. By Frobenius reciprocity (1.2c) we then have $\mathrm{Ind}_N^X(\theta) = \sum_{\chi \in \mathrm{Irr}(X|\theta)} \chi(1)\chi$, and this vanishes outside N and agrees with $|G|\theta$ on N. For $\chi \in \mathrm{Irr}(X|\theta)$ we have $|\chi(x)|^2 = \frac{\chi(1)}{|X|} \sum_{y \in X} \chi([x,y])$ by (1.3c) and, of course, $|X| = \sum_{x \in X} |\chi(x)|^2$. Hence

$$|\mathrm{Irr}(X|\theta)| = \frac{1}{|X|} \sum_{x \in X} \sum_{\chi \in \mathrm{Irr}(X|\theta)} |\chi(x)|^2 = \frac{|G|}{|X|^2} \sum_{\substack{x,y \in X \\ [x,y] \in N}} \theta([x,y]).$$

This is at most equal to $\frac{|G|}{|X|^2} \sum_{x \in X} \sum_{y \in C_X(Nx)} 1 = \frac{1}{|G|} \sum_{\bar{x} \in G} |C_G(\bar{x})| = k(G)$. $\square$

We now shall discuss Clifford theory of tensor induction. Suppose N is a nonabelian normal subgroup of X which is the central product of the X-conjugates of some (proper) subgroup N_0. Let $X_0 = N_X(N_0)$ be the normalizer, and let $G - X/N$ and $G_0 = X_0/N_0$. Assume $\theta \in \mathrm{Irr}(N)$ is stable in X, and let $\theta_0 \in \mathrm{Irr}(N_0)$ be the unique irreducible constituent of θ on N_0. As θ_0 is absolutely irreducible, θ is the (tensor) product of the $|X : X_0|$ distinct X-conjugates of θ_0 [I, 4.21].

Define $X(\theta)$ and $X_0(\theta_0)$ as before, with the same Z, and let $\widehat{Z} = \mathrm{Ind}_{X_0}^{X}(Z)$ be the (induced) permutation module.

Theorem 1.9d. *Keeping these assumptions, let $\widehat{\theta}_0$ be an irreducible character of $X_0(\theta_0)$ extending θ_0 (as above). Then there is a group extension $\widehat{Z} \rightarrowtail \widehat{X} \twoheadrightarrow X$ mapping onto $X(\theta)$ such that $\widehat{\theta} = \mathrm{Ten}_{\widehat{X_0}}^{\widehat{X}}(\widehat{\theta}_0)$ is a character of $X(\theta)$ extending θ, $\widehat{X}_0$ being the inverse image in $\widehat{X}$ of X_0.*

Proof. Let $\rho_0 : N_0 \to \mathrm{GL}(W_0)$ be a K-representation affording θ_0 (Theorem 1.1d). Let $\{t_i\}_{i=1}^{n}$ be a right transversal to X_0 in X. Let $h = \prod_i h_i^{t_i}$ be an element in N (with all $h_i \in N_0$). Then $\rho(h) = \otimes_{i=1}^{n} \rho_0(h_i^{t_i})$ is a K-representation of N on $W = \bigotimes_{i=1}^{n} W_0 t_i$ affording θ. Since θ_0 is stable in X_0, we may extend ρ_0 to a projective representation $\widehat{\rho}_0 : X_0 \to \mathrm{GL}(W_0)$. We may choose $\widehat{\rho}_0$ such that its factor set $\tau_0 \in Z^2(X_0, K^\star)$, being inflated from G_0, has order dividing $|Z| = \exp(N_0)$. Let $x \in X$, and let $t_i x = x_i t_{ix}$ be as in Sec. 1.2 (with $x_i \in X_0$). Then $\widehat{\rho}(x) = \otimes_{ix=1}^{n} \widehat{\rho}_0(x_i)$ defines a projective representation of X *tensor induced* from $\widehat{\rho}_0$ which extends ρ and has factor set

$$\widehat{\tau}_0(x, y) = \prod_{i=1}^{n} \tau_0(x_i, y_{ix}).$$

This $\widehat{\tau}_0$ is *co-induced* from τ_0 (and inflated from G). We have $h^x = \prod_i (h_i^{x_i})^{t_{ix}}$ and so

$$\rho(h^x) = \otimes_{ix=1}^{n} \rho_0(h_i^{x_i}) = \widehat{\rho}(x)^{-1} \rho(h) \widehat{\rho}(x).$$

Thus the class of $\tau = \widehat{\tau}_0$ in $\mathrm{H}^2(G, K^\star)$ is nothing but $\mu_{KG}(\theta)$, and $\widehat{\rho}$ lifts to an ordinary representation of $X(\theta)$, say affording $\widehat{\theta}$.

The group extension $\widehat{X}$ represents the cohomology class obtained from $X_0(\theta_0)$ under the natural isomorphism $\mathrm{H}^2(X_0, Z) \cong \mathrm{H}^2(X, \widehat{Z})$ underlying Shapiro's lemma (A3). Here the group $X(\theta)$ is the image of $\widehat{X}$ under the map $\widehat{Z} \to Z$ sending $(z_i)_{i=1}^{n}$ to $\prod_{i=1}^{n} z_i$ ($z_i \in Z$). $\qquad \square$

1.10. Good Conjugacy Classes and Extendible Characters

Let N be a normal subgroup of X, and let $G = X/N$. Let $\theta \in \mathrm{Irr}(N)$ be G-invariant. The conjugacy class of an element $g = Nx$ in G is called "good for θ" provided θ can be extended to $\langle N, x, y \rangle$ for all $y \in X$ satisfying $[x, y] \in N$ [Gallagher, 1970]. By virtue of Theorem 1.9c this may be studied by passing to $G(\theta)$. Hence we may assume that $N = Z$ is cyclic and central in X and that $\theta = \tilde{\theta}$ is linear. Assume also that θ is faithful. If there is a (linear) character λ of $Y = \langle N, x, y \rangle$ extending θ, then Y/N is abelian (as $[x, y] \in N$), $Y/\mathrm{Ker}(\lambda)$ is abelian and $N \cap \mathrm{Ker}(\lambda) = \mathrm{Ker}(\theta) = 1$. Hence Y is abelian. Conversely, if Y is abelian, then θ can be extended to Y. We have proved the following.

Theorem 1.10a. *Suppose $\theta \in \mathrm{Irr}(N)$ is stable under $G = X/N$. Then $|\mathrm{Irr}(X|\theta)|$ is the number of conjugacy classes of G which are good for θ.*

Combining this with Theorem 1.8b we obtain the *Clifford–Gallagher formula*

$$(1.10b) \qquad k(X) = \sum_{\theta \in \mathrm{Irr}(N)} k_\theta \big(I_G(\theta) \big) / |G : I_G(\theta)|,$$

where $k_\theta(I_G(\theta))$ is the number of conjugacy classes of $I_G(\theta)$ which are good for θ. Observe that $k_\theta(I_G(\theta)) = k(I_G(\theta))$ whenever θ can be extended to $I_X(\theta)$, that is, whenever $\mu_{K I_G(\theta)}(\theta)$ vanishes. This happens for instance if all Sylow subgroups of $I_G(\theta)$ are cyclic, because then its Schur multiplier is trivial by (A4).

Chapter 2

Blocks of Characters and Brauer's $k(B)$ Problem

Keeping the assumptions and notation of the preceding chapter we proceed now to modular representations and to block theory. The reader is referred to [Feit, 1982] for the relevant background as well as for results not proved here. This book is cited as [F] in the text. We describe some ideas of Brauer and Feit when attacking the $k(B)$ problem. The study of (major) subsections will motivate us (once again) to consider point stabilizers when dealing with the $k(GV)$ problem. On the basis of the Clifford theory of blocks developed by Fong we shall verify the $k(B)$ problem for p-blocks of p-solvable groups, assuming that the $k(GV)$ theorem is already settled.

2.1. Modular Decomposition and Brauer Characters

Let p be a rational prime. Denote by $|X|_p = p^a$ the p-part of the order of X (= order of a Sylow p-subgroup of X), and let $X_{p'}$ be the set of p'-elements (or p-regular elements) of X. Let $\mathbb{Z}_{(p)}$ denote the localization of the integers at the prime $(p) = p\mathbb{Z}$. Then $R = \mathbb{Z}_{(p)}[\varepsilon]$ is a Dedekind ring with only finitely many nonzero prime ideals, all lying above p. Hence R is a principal ideal domain, with quotient field $K = \mathbb{Q}(\varepsilon)$. Letting $\mathfrak{p}$ be any of the (Galois conjugate) nonzero prime ideals of R, the field $F = R/\mathfrak{p}$ is finite of characteristic p, and is a splitting field for all subgroups of X (as follows from Theorem 1.1d, or from the fact that Schur indices are trivial).

If χ, ζ are K-valued class functions on $X_{p'}$, we define

$$\langle \chi, \zeta \rangle_{p'} = \langle \chi, \zeta \rangle_{X_{p'}} = \frac{1}{|X|} \sum_{x \in X_{p'}} \chi(x)\overline{\zeta(x)}.$$

For irreducible characters χ, ζ of X we define $m_{\chi\zeta} = \langle \chi, \zeta \rangle_{p'}$ (which is symmetric), and we say that they belong to the same p-block, B, of X provided their central characters agree mod $\mathfrak{p}$. See Eq. (1.3a); recall that the central characters have their values in $\mathbb{Z}[\varepsilon] \subseteq R$. This defines a partition of $\mathrm{Irr}(X)$ (which turns out to be independent of the choice of the maximal ideal $\mathfrak{p}$). For the time being, the block B is just a certain nonempty set $\mathrm{Irr}(B)$ of irreducible characters of X. Define

$$\omega_B(\widehat{c}) = \omega_\chi(\widehat{c}) + \mathfrak{p}$$

for any $\chi \in \mathrm{Irr}(B)$, and all conjugacy classes c of X.

If χ is a class function on X, we let $\widetilde{\chi} = \widetilde{\chi}(p^a)$ be the class function on X taking the value $p^a\chi(x)$ for $x \in X_{p'}$, and zero otherwise. For χ, ζ in $\mathrm{Irr}(X)$ we have $p^a m_{\chi\zeta} = \langle \widetilde{\chi}, \zeta \rangle = \langle \chi, \widetilde{\zeta} \rangle$.

Lemma 2.1a. *If $\chi \in \mathbb{Z}[\mathrm{Irr}(X)]$ is a generalized character, then so is $\widetilde{\chi}$. Then even $\frac{1}{p^{a-d}}\widetilde{\chi} \in \mathbb{Z}[\mathrm{Irr}(X)]$ where $d \leq a$ is the smallest integer (if any) such that $\frac{p^d}{|C_X(x)|_p}\chi(x) \in R$ for all $x \in X_{p'}$.*

Proof. We apply Theorem 1.1c. Let $E = P \times Q$ be an elementary subgroup of X, where P is a p-group and Q is a p'-group. Then $|P| = p^b$ with $b \leq a$. Let ρ be the character of P afforded by the regular representation. Then $\mathrm{Res}_E^X(\widetilde{\chi}) = (p^{a-b}\rho) \otimes \mathrm{Res}_Q^X(\chi)$ is a generalized character of E. Thus $\widetilde{\chi}$ is a generalized character of X. It follows that $\frac{1}{p^{a-d}}\widetilde{\chi} \in \mathbb{Z}[\mathrm{Irr}(X)]$ if and only if

$$\langle \frac{1}{p^{a-d}}\widetilde{\chi}, \zeta \rangle \in \mathbb{Z}_{(p)} = R \cap \mathbb{Q}$$

for all $\zeta \in \mathrm{Irr}(X)$. But for $x \in X_{p'}$ and $c = x^X$ we have $|c| = |X : C_X(x)|$ and

$$\frac{1}{|X|}\sum_{y \in c}\frac{1}{p^{a-d}}\widetilde{\chi}(y)\zeta(y^{-1}) = \frac{p^d}{|C_X(x)|_p}\chi(x) \cdot \frac{\zeta(x^{-1})}{|C_X(x)|_{p'}} \in R.$$

Use finally the fact that $\widetilde{\chi}$ vanishes on p-singular elements of X. $\qquad\square$

Reduction mod $\mathfrak{p}$ is an isomorphism from the $\exp(X)_{p'}$th roots of unity in K to those in F. Let V be an FX-module. Each p-regular $x \in X_{p'}$ is semisimple (diagonalizable) on V, its eigenvalues being such roots of unity in F. Lifting these eigenvalues, and summing up, we get a class function $\varphi_V : X_{p'} \to K$. This φ_V is called the *Brauer character* of X afforded by V (with respect to $\mathfrak{p}$). If X is a p'-group, φ_V is an ordinary character of X. If V is an irreducible FG-module, φ_V is called an (absolutely) irreducible Brauer character. The set $\mathrm{IBr}(X) = \mathrm{IBr}_p(X)$ of (absolutely) irreducible Brauer characters of X is a basis for the K-vector space of class functions on $X_{p'}$ [I, 15.11], or [F, IV.3.4]. It follows that

(2.1b) $$|\mathrm{IBr}(X)| = |\mathcal{C}\ell(X_{p'})|$$

is the number of p'-classes in X. The reader is referred to [Jansen *et al.*, 1995] for tables of Brauer characters for some simple groups of small orders; this book will be quoted as [B-Atlas].

Theorem 2.1c. *Let $\chi \in \mathrm{Irr}(X)$. Then $\chi_{p'} = \mathrm{Res}^X_{X_{p'}}(\chi)$ is a Brauer character, hence $\chi_{p'} = \sum_{\varphi \in \mathrm{IBr}(X)} d_{\chi\varphi}\varphi$ for some unique nonnegative integers $d_{\chi\varphi}$, called the decomposition numbers of χ.*

Proof. By Theorem 1.1d, there is a KX-module W affording χ. Let $\{w_j\}$ be a K-basis of W, and let U be the R-submodule of W generated by all $w_j x$, $x \in X$. Then U is finitely generated and torsion-free, hence a free R-module. U must have rank $\chi(1) = \dim_K W$, and U is stable under X. Thus U is an RX-lattice affording χ. It follows that $V = U/\mathfrak{p}U$ is an FX-module affording the Brauer character $\chi_{p'} = \mathrm{Res}^X_{X_{p'}}(\chi)$. This $\chi_{p'}$ is independent of the choice of the lattice U affording χ. □

2.2. Cartan Invariants and Blocks

Let B be a block of X. We say that an irreducible Brauer character φ of X belongs to B, $\varphi \in \mathrm{IBr}(B)$, provided $d_{\chi\varphi} \neq 0$ for some $\chi \in \mathrm{Irr}(B)$.

Theorem 2.2a. *Let B be a block of X, and let $\varphi \in \mathrm{IBr}(B)$. Then $d_{\chi\varphi} = 0$ whenever χ does not belong to B, and φ is a $\mathbb{Z}$-linear combination of the $\chi_{p'}$ for $\chi \in \mathrm{Irr}(B)$. Moreover:*

(i) The associated "projective character" $\widehat{\varphi} = \sum_{\chi \in \mathrm{Irr}(B)} d_{\chi\varphi}\chi$ vanishes off p-regular elements, its degree $\widehat{\varphi}(1)$ is divisible by p^a, and it satisfies $\langle \widehat{\varphi}, \varphi \rangle_{p'} = 1$ and $\langle \widehat{\varphi}, \psi \rangle_{p'} = 0$ for $\varphi \neq \psi$ in $\mathrm{IBr}(X)$.

(ii) For φ, ψ in $\mathrm{IBr}(B)$ let $c_{\varphi\psi} = \langle \widehat{\varphi}, \widehat{\psi} \rangle$ be the "Cartan invariant", and let $C_B = (c_{\varphi\psi})_{\varphi,\psi}$. The Cartan matrix C_B of B is symmetric and positive definite, and $C_B^{-1} = (\langle \varphi, \psi \rangle_{p'})_{\varphi,\psi}$.

Proof. Suppose $d_{\chi\varphi} \neq 0$. Let $V = U/\mathfrak{p}U$ be an FX-module affording $\chi_{p'}$. Then some composition factor of V affords φ. Each class sum $\widehat{c}$ acts as a scalar multiplication on V, thus defining a central character ω_φ (with values in F). This ω_φ must be the reduction mod $\mathfrak{p}$ of ω_χ. In other words, $\omega_\varphi = \omega_B$ for some unique p-block B.

For the second statement, observe that modular decomposition may be seen as a map from generalized characters to generalized Brauer characters, which commutes with restriction to subgroups and induction (defined for Brauer characters like for ordinary characters). Hence by Brauer's induction theorem it suffices to consider the case where $X = P \times Q$ is elementary (P a p-group, Q a p'-group). But in this case every $\varphi \in \mathrm{IBr}(X)$ has P in its kernel, because $C_V(P) \neq 0$ for each irreducible FX-module V, and $C_V(P)$ is X-invariant as P is normal in X. It follows that $\varphi = \chi_{p'}$ for some unique $\chi \in \mathrm{Irr}(X)$.

The surjectivity of the decomposition map is equivalent to the statement that the elementary divisors of the *decomposition matrix* $D_B = (d_{\chi\varphi})$ of B, with $\chi \in \mathrm{Irr}(B)$ and $\varphi \in \mathrm{IBr}(B)$, are all 1. For φ, ψ in $\mathrm{IBr}(B)$ the Cartan invariant

$$c_{\varphi\psi} = \langle \widehat{\varphi}, \widehat{\psi} \rangle = \sum_{\chi \in \mathrm{Irr}(B)} d_{\chi\varphi} d_{\chi\psi}$$

and so $C_B = (D_B)^t \cdot D_B$ is a positive definite symmetric matrix with nonnegative integer entries.

By the second orthogonality relations (1.1b), for any regular $x \in X_{p'}$ and any singular $y \in X$ we have

$$\sum_{\varphi \in \mathrm{IBr}(X)} \varphi(x)\overline{\widehat{\varphi}(y)} = \sum_{\chi \in \mathrm{Irr}(X)} \chi(x)\overline{\chi(y)} = 0.$$

Since the irreducible Brauer characters are linearly independent, $\widehat{\varphi}(y) = 0$ for each φ. Thus if P is a Sylow p-subgroup of X, then $\langle \widehat{\varphi}, 1_P \rangle_P = \frac{\widehat{\varphi}(1)}{|P|}$ is an integer. Varying $c = c_x$ over the p-regular conjugacy classes of X the second orthogonality relations show that $(\varphi(c))_{\varphi,c} \cdot (\widehat{\varphi}(c))^t_{\varphi,c}$ is the diagonal matrix with c_xth entry $|C_X(x)|$. It follows that for φ, ψ in $\mathrm{IBr}(B)$,

$$\langle \widehat{\varphi}, \psi \rangle_{p'} = \sum_{c=c_x \in C\ell(X_{p'})} \frac{1}{|C_X(x)|} \widehat{\varphi}(x)\overline{\psi(x)} = \delta_{\varphi\psi}.$$

From $\widehat{\varphi}_{p'} = \sum_{\psi \in \mathrm{IBr}(B)} c_{\varphi\psi}\psi$ we infer that C_B^{-1} is as asserted. □

We define a graph with $\mathrm{Irr}(X)$ as its vertex set by linking χ, ζ in $\mathrm{Irr}(X)$ if there exists $\varphi \in \mathrm{IBr}(X)$ such that $d_{\chi\varphi} \neq 0 \neq d_{\zeta\varphi}$. This is called the *Brauer graph* (mod p). By Theorem 2.2a, $\mathrm{Irr}(B)$ is a union of connected components of the Brauer graph. We also let $e_B = \sum_{\chi \in \mathrm{Irr}(B)} e_\chi$ where $e_\chi = \frac{\chi(1)}{|X|} \sum_{x \in X} \overline{\chi(x)}x$ is the centrally primitive idempotent of KX associated to χ ($\omega_\chi(e_\chi) = 1$ and $\omega_\chi(e_\zeta) = 0$ for $\zeta \neq \chi$).

Theorem 2.2b (Osima). *We have $e_B = \sum_{c \in C\ell(X)} a_B(c)\hat{c}$ for unique elements $a_B(c)$ in R, and we write e_B and $a_B(c)$ also for the reductions mod $\mathfrak{p}$. Then $a_B(c) = 0$ (in F) if c is a p-singular class, and $\omega_B(e_B) = 1$. Also, $\mathrm{Irr}(B)$ is a connected component of the Brauer graph.*

Proof. By Theorem 2.2a we have $a_B(c) = \frac{1}{|X|} \sum_{\varphi \in \mathrm{IBr}(B)} \varphi(1)\overline{\hat{\varphi}(c)} = 0$ if c is not a p'-class. If c is a p'-class, then

$$a_B(c) = \frac{1}{|X|} \sum_{\varphi \in \mathrm{IBr}(B)} \hat{\varphi}(1)\overline{\varphi(c)}$$

is in R, because p^a is divisor of $\hat{\varphi}(1)$ by Theorem 2.2a. Replacing $\mathrm{Irr}(B)$ by a connected component $\mathfrak{B} \subseteq \mathrm{Irr}(B)$ of the Brauer graph, and $\mathrm{IBr}(B)$ by $\{\varphi \in \mathrm{IBr}(X)|\ d_{\chi\varphi} \neq 0$ for some $\chi \in \mathfrak{B}\}$, the corresponding statement holds for $e = \sum_{\chi \in \mathfrak{B}} e_\chi$. Then an irreducible character χ of X belongs to $\mathfrak{B}$ if and only if $\omega_\chi(e) \not\equiv 0 \pmod{\mathfrak{p}}$. We conclude that $\mathfrak{B} = \mathrm{Irr}(B)$. $\qquad\square$

2.3. Defect and Defect Groups

Let $\chi \in \mathrm{Irr}(B)$. By Theorem 1.3b, $\chi(1)_p \leq p^a$. If $\chi(1)_p = p^{a-d}$ is as small as possible in $\mathrm{Irr}(B)$, then $d = d(B)$ is the *defect* of B, and χ is said to be of height zero in B. In general $\chi(1)_p = p^{a-d+h}$ with *height* $h = h_\chi \geq 0$.

Theorem 2.3a. *Let B be a block of X with defect d and let $\chi \in \mathrm{Irr}(B)$ be of height zero. Then for any $\zeta \in \mathrm{Irr}(B)$, $p^d m_{\chi\zeta}$ is a nonzero rational integer whose p-part is equal to the height of ζ. In particular $p^d C_B^{-1}$ is a positive definite symmetric matrix with integer entries.*

Proof. Let $\tilde{\chi} \in \mathbb{Z}[\mathrm{Irr}(X)]$ be as in Lemma 2.1a. From Theorem 2.2a it follows that $m_{\chi\zeta} = 0$ if $\zeta \notin \mathrm{Irr}(B)$. Hence $\tilde{\chi} = \sum_{\zeta \in \mathrm{Irr}(B)} \langle \tilde{\chi}, \zeta \rangle \zeta$ is in $\mathbb{Z}[\mathrm{Irr}(B)]$. Since $\omega_\chi(\hat{c}) = \frac{\chi(x)}{|C_X(x)|} \cdot \frac{|G|}{\chi(1)} \in R$ for $c = x^X$ and $x \in X_{p'}$, $\frac{1}{p^{a-d}}\tilde{\chi}$ is a generalized character (in B) by Lemma 2.1a. But $\frac{1}{p^{a-d+1}}\tilde{\chi}$ is not in $\mathbb{Z}[\mathrm{Irr}(B)]$, because if P is a Sylow p-subgroup of X then

$$\frac{1}{p^{a-d+1}} \langle \tilde{\chi}, 1_P \rangle_P = \chi(1)/p^{a-d+1}$$

is not an integer. It follows that there is $\theta \in \mathrm{Irr}(B)$ such that $\langle \tilde{\chi}, \theta \rangle/p^{a-d}$ is a unit in R. Then $\theta(1)_p = p^{a-d}$. Let $\zeta \in \mathrm{Irr}(B)$ and $\zeta(1)_p = p^{a-d+h}$. There are similar statements for $\tilde{\zeta}$. By definition $p^a m_{\chi\zeta} = \langle \tilde{\chi}, \zeta \rangle = \langle \chi, \tilde{\zeta} \rangle$.

Thus $p^d m_{\chi\zeta}$ is an integer whose p-part equals $\zeta(1)_{p_\zeta}$. Note that $\frac{\langle \bar{\chi}, \zeta \rangle}{\theta(1)} \equiv \frac{\langle \bar{\chi}, \theta \rangle}{\theta(1)}$ (mod $\mathfrak{p}$), because ω_ζ and ω_θ agree mod $\mathfrak{p}$.

For the final statement observe that for $\chi, \zeta \in \mathrm{Irr}(B)$ we have

$$m_{\chi\zeta} = \sum_{\varphi,\psi \in \mathrm{IBr}(B)} d_{\chi\varphi} \cdot \langle \varphi, \psi \rangle_{p'} \cdot d_{\zeta\psi}.$$

Thus $(p^d m_{\chi\zeta})_{\chi,\zeta} = D_B(p^d C_B^{-1})D_B^t$. Now use the fact that by Theorem 2.2a all the elementary divisors of D_B are equal to 1.   □

Let $c = x^X$ be a conjugacy class of X. Then a Sylow p-subgroup of $C_X(x)$ is a *defect group* for c. By Theorem 2.2b there exist c such that $a_B(c) \neq 0$ and $\omega_B(\hat{c}) \neq 0$. Then c is called a *defect class* for the block B.

Lemma 2.3b. *Let B be a block of X and let D be a defect group of some defect class for B. Let c be any conjugacy class of X.*

(a) *If $a_B(c) \neq 0$ (in F), then D contains a defect group for c.*

(b) *If $\omega_B(\hat{c}) \neq 0$, then D is contained in a defect group for c.*

For a proof of this Min-Max lemma see [I, 15.31] .

It follows that the defect groups of the defect classes for B form a single conjugacy class of p-subgroups of X, called the *defect groups* of B. If D is a defect group of B then $|D| = p^d$ where d is the defect of B. Indeed, let $\chi \in \mathrm{Irr}(B)$ be of height zero, whence $\chi(1)_p = p^{a-d}$. Let $c = c_x$ be a defect class for B. Then $\omega_\chi(\hat{c}) \not\equiv 0$ (mod $\mathfrak{p}$) by definition. Since $\chi(x) \in R$, this shows that $|c|/\chi(1) \in R \cap \mathbb{Q} = \mathbb{Z}_{(p)}$ is not divisible by p. Hence $|D| \geq p^d$ as $|c|_p = p^a/|D|$. On the other hand, also $a_B(c) \neq 0$ and so in view of Theorems 2.2b and 2.2a there is some $\zeta \in \mathrm{Irr}(B)$ such that $\overline{\zeta(c)} \neq 0$ (mod $\mathfrak{p}$). Using that $\omega_{\bar{\zeta}}(\hat{c}) \in R$ it follows that $\zeta(1)_p \leq |c|_p$. Hence $|D| \leq p^d$.

The *principal block* of X is the block containing the 1-character 1_X. Its defect groups are the Sylow p-subgroups of X.

Theorem 2.3c. *Let B be a block of X with defect group D, and let P be a normal p-subgroup of X. Then $|D| = p^d$ where d is the defect of B, and $D \supseteq P$. The block idempotent e_B of B is a central idempotent of $FC_X(P)$, and if $C_X(P) \subseteq P$, then B is the unique p-block of X and hence D a Sylow p-subgroup of X.*

Proof. We have already proved the first statement. Let $N = C_X(P)$, which is normal in X. Let $c \in C\ell(X)$. If some $x \in c$ is contained in N, then $c \subseteq N$. Assume $c \not\subseteq N$. Then each P-orbit on c (by conjugation) has size divisible by p. Let V be an irreducible FX-module. Then $P \subseteq C_X(V)$ (as seen above). It follows that $V\widehat{c} = 0$. Consequently $\widehat{c}$ is in the Jacobson radical of $Z(FX)$, hence is nilpotent.

Now let c be a defect class for B with defect group D. Then $\omega_B(\widehat{c}) \neq 0$ and so $\widehat{c}$ is not nilpotent. Thus $c \subseteq N = C_X(P)$ and $P \subseteq D$.

Let $e_B = \sum_c a_B(c)\widehat{c}$ be as in Theorem 2.2b. We have seen that either $\widehat{c} \in J(Z(FX))$ or $c \in C\ell(X)$ is centralized by P. Since e_B is an idempotent, it follows that $e_B \in Z(FX) \cap FN$.

Let b be a block of N *covered* by B, that is, some $\chi \in \mathrm{Irr}(B)$ lies over some $\theta \in \mathrm{Irr}(b)$. It follows from Theorem 1.8b that the blocks of N covered by B form a unique X-conjugacy class of blocks. Thus $e_B \cdot e_b \neq 0$ and e_B is the sum of the distinct X-conjugates of e_b. Define $\omega_b^X(\widehat{c}) = \omega_b(\sum_{x \in N \cap c} x)$ for any conjugacy class c of X. If $c \not\subseteq N$, then $c \cap N = \varnothing$ and $\omega_b^X(c) = 0$ by definition. Then also $\omega_B(\widehat{c}) = 0$, for otherwise $D \subseteq C_X(x)$ for some $x \in c$ by the preceding lemma, whence $x \in C_X(D) \subseteq C_X(P) = N$, a contradiction. If $c \subseteq N$, then $\widehat{c} \in Z(FN)$ and $\omega_B(\widehat{c}) = \omega_b(\widehat{c}) = \omega_b^X(\widehat{c})$, because central characters agree if they agree on central idempotents (knowing that F is a splitting field for $Z(FN)$). We conclude that $\omega_B = \omega_b^X$ is determined by b, i.e., $B = b^X$ is the unique block of X covering b.

In particular, if $C_X(P) \subseteq P$, then the principal block b is the unique p-block of $N = Z(P)$ and so B is the unique p-block of X. $\qquad\square$

2.4. The Brauer–Feit Theorem

In what follows we fix a block B of X with defect group D, $|D| = p^d$. We let $k(B) = |\mathrm{Irr}(B)|$ and $\ell(B) = |\mathrm{IBr}(B)|$. We have $k(B) \geq \ell(B)$ since the Cartan matrix C_B is nonsingular. One even has $k(B) > \ell(B)$ unless $d = 0$, in which case $k(B) = 1 = \ell(B)$ [F, IV.4.19]. Brauer's celebrated $k(B)$ conjecture is the assertion that always $k(B) \leq p^d$.

Theorem 2.4 (Brauer–Feit). *We have $k(B) \leq 1 + \frac{1}{4}p^{2d}$. If B contains an irreducible character of positive height, then even $k(B) \leq \frac{1}{2}p^{2d-2}$.*

Proof. Let $k_h = k_h(B)$ be the number of irreducible characters in B of height h, so that $k(B) = \sum_{h \geq 0} k_h$. Let $\chi \in \mathrm{Irr}(B)$ be of height zero. By

Theorem 2.3a, $\frac{1}{p^{a-d}}\widetilde{\chi} = \sum_{\zeta \in \mathrm{Irr}(B)} n_\zeta \zeta$ where each $n_\zeta = p^d m_{\chi\zeta} = \frac{1}{p^{a-d}}\langle \widetilde{\chi}, \zeta \rangle$ is a nonzero integer with p-part equal to p^{h_ζ}. From

$$p^d n_\chi = p^d(p^d m_{\chi\chi}) = \langle \frac{1}{p^{a-d}}\widetilde{\chi}, \frac{1}{p^{a-d}}\widetilde{\chi} \rangle = \sum_{\zeta \in \mathrm{Irr}(B)} n_\zeta^2$$

we get $p^d n_\chi \geq n_\chi^2 + (k_0 - 1) + \sum_{h \geq 1} k_h p^{2h}$. It follows that

$$k(B) \leq \sum_{h \geq 0} k_h p^{2h} \leq 1 + p^d n_\chi - n_\chi^2 \leq 1 + \frac{1}{4}p^{2d},$$

because $t \mapsto p^d t - t^2$ takes its maximum in $t = \frac{1}{2}p^d$. We also see that $\sum_{h \geq 1} k_h p^{2h} \leq \frac{1}{4}p^{2d}$ and so $\sum_{h \geq 1} k_h \leq \frac{1}{4}p^{2d-2}$.

Suppose there is a character $\zeta \in \mathrm{Irr}(B)$ with height $h = h_\zeta > 0$. Then $\frac{1}{p^{a-d+h}}\widetilde{\zeta} \in \mathbb{Z}[\mathrm{Irr}(B)]$. Arguing as before this yields that $k_0 p^2 \leq p^d u - u^2$ for $u = p^{d-h} m_{\zeta\zeta}$. Hence $k_0 \leq \frac{1}{4}p^{2d-2}$, and the result follows.   □

2.5. Higher Decomposition Numbers, Subsections

In order to improve Theorem 2.4 one is led to pass to certain subgroups of X and blocks related to B. Let Y be a subgroup of X and b be a block of Y. We say that the *induced block* b^X exists provided the map ω_b^X, defined by $\omega_b^X(\widehat{c}) = \omega_b(\sum_{x \in Y \cap c} x)$ for any conjugacy class c of X, is an F-algebra homomorphism on the centre of FX. In this case ω_b^X determines a unique block $B = b^X$ of X.

Example 2.5a. Suppose b is a block of the subgroup Y of X containing an irreducible character θ such that $\chi = \mathrm{Ind}_Y^X(\theta)$ is irreducible. Then, by formula (1.2b) for induced characters, $\omega_\chi(\widehat{c}) = \omega_\theta(\sum_{x \in Y \cap c} x)$ for each conjugacy class c of X. Thus in this case $B = b^X$ is defined, and $\chi \in \mathrm{Irr}(B)$.

If Y contains $DC_X(D)$ for some p-subgroup D of X, then for any block b of Y the induced block $B = b^X$ is defined, and the defect groups of b are contained in certain defect groups of B [F, III.9.4 and 9.6]. If $N_X(D) \subseteq Y$, then by Brauer's *First Main Theorem* on blocks $b \mapsto b^X$ is a bijection from the blocks of Y with defect group D to the blocks of X with defect group D [I, 15.45], or [F, III.9.7].

Let B be a block of X with defect group D, and let $y \in Z(D)$ be a central element of D, say of order p^n. Let $Y = C_X(y)$. Then b^X is defined for each p-block b of Y, and there exists b such that $b^X = B$ and D is a defect group of b [F, V.9.2]. Fix such a block b. Since y is in the centre of Y, for each irreducible character χ of X and any $\varphi \in \mathrm{IBr}(Y)$, there exist unique $d^y_{\chi\varphi} \in \mathbb{Z}[\varepsilon] \subseteq R$ such that

$$\chi(xy) = \sum_{\varphi \in \mathrm{IBr}(Y)} d^y_{\chi\varphi}\varphi(x)$$

for all p'-elements $x \in Y$. The numbers $d^y_{\chi\varphi} = \sum_{\theta \in \mathrm{Irr}(Y)} \langle \chi, \theta \rangle_Y \cdot \lambda_\theta(y) \cdot d_{\theta\varphi}$, where $\mathrm{Res}^Y_{\langle y \rangle}(\theta) = \theta(1)\lambda_\theta$, are called the *higher decomposition numbers* with respect to y. They are algebraic integers in the field of p^nth roots of unity over the rationals. It follows from Brauer's *Second Main Theorem* on blocks [F, IV.6.1] that if $d^y_{\chi\varphi} \neq 0$ for some $\varphi \in \mathrm{IBr}(b)$, then χ belongs to $b^X = B$. Using the second orthogonality relations (1.1b) one checks that $\sum_{\chi \in \mathrm{Irr}(B)} d^y_{\chi\varphi}\overline{d^y_{\chi\psi}}$ is the (φ, ψ)-entry of the Cartan matrix C_b of b.

Definition 2.5b. The set of columns $(d^y_{\chi\varphi})_{\chi \in \mathrm{Irr}(B)}$, with φ varying over $\mathrm{IBr}(b)$, is called the (major) *subsection* (y, b) associated to $b^X = B$ (where the term *major* indicates that b and B have a defect group in common). For χ, ζ in $\mathrm{Irr}(B)$ we define

$$m^{(y,b)}_{\chi\zeta} = \sum_{\varphi, \psi \in \mathrm{IBr}(b)} d^y_{\chi\varphi} \cdot \langle \varphi, \psi \rangle_{Y_{p'}} \cdot \overline{d^y_{\zeta\psi}}.$$

By definition $m^{(y,b)}_{\chi\zeta} = \overline{m^{(y,b)}_{\zeta\chi}}$ and, by Theorems 2.2a and 2.3a, $p^d m^{(y,b)}_{\chi\zeta}$ is an algebraic integer in R and is nonzero if χ is of height zero in B. Let Q_b denote the quadratic form obtained from the Hermitian form defined by the positive definite symmetric matrix $p^d C_b^{-1}$ with integer entries. Then by definition $Q_b(z) = p^d m^{(y,b)}_{\chi\chi}$ if $z = (d^y_{\chi\varphi})_{\varphi \in \mathrm{IBr}(b)}$.

Lemma 2.5c. *Let (y, b) be a major subsection to $B = b^X$, with defect d. Then $\sum_{\chi \in \mathrm{Irr}(B)} m^{(y,b)}_{\chi\chi} = \ell(b)$ and, for each $\chi \in \mathrm{Irr}(B)$, the trace*

$$\mathrm{Tr}_{K|\mathbb{Q}}(p^d m^{(y,b)}_{\chi\chi}) \geq [K : \mathbb{Q}] \cdot \min Q_b(z),$$

where $z = (z_\varphi)$ varies over all nonzero vectors with integral coordinates. In the case that $\sum_{\varphi \in \mathrm{IBr}(b)} d^y_{\chi\varphi}\varphi(1) \not\equiv 0 \pmod{\mathfrak{p}}$ for all $\chi \in \mathrm{Irr}(B)$, it suffices to take the minimum over those vectors for which $\sum_{\varphi \in \mathrm{IBr}(b)} z_\varphi \varphi(1) \not\equiv 0 \pmod{p}$.

Proof. Consider the $k(B) \times k(B)$-matrix $M = (m^{(y,b)}_{\chi\zeta})_{\chi, \zeta}$. By direct computation, using that $C_b^{-1} = (\langle \varphi, \psi \rangle_{Y_{p'}})_{\varphi, \psi}$ by Theorem 2.2a, one obtains that $M^2 = M$ [F, V.9.4]. Since the rank of M equals the rank of $p^d C_b^{-1}$, which is $\ell(b)$, we infer that the trace of M is equal to $\ell(b)$. It follows that $\sum_\chi m_{\chi\chi}^{(y,b)} = \ell(b)$. For the statement concerning the field traces we may replace K by $K_0 = \mathbb{Q}(\varepsilon_0)$ where ε_0 is a primitive p^nth root of unity such that all $m_{\chi\zeta}^{(y,b)} \in K_0$. Then use that $\mathrm{Tr}_{K_0|\mathbb{Q}}(\varepsilon_0^j) = -p^{n-1}$ if j is divisible by p^{n-1} but not by p^n, and zero otherwise [F, V.9.14]. $\qquad\square$

Theorem 2.5d (Brauer). *Let (y, b) be a major subsection to B (defect d).*

(i) *Assume that $Q_b(z) \geq \ell(b)$ for each nonzero vector $z = (z_\varphi)$ with integer coordinates; if $\sum_{\varphi\in\mathrm{IBr}(b)} d_{\chi\varphi}^y \varphi(1) \not\equiv 0 \pmod{\mathfrak{p}}$ for all $\chi \in \mathrm{Irr}(B)$, consider only those vectors for which $\sum_{\varphi\in\mathrm{IBr}(b)} z_\varphi \varphi(1) \not\equiv 0 \pmod{p}$. Then $k(B) \leq p^d$.*

(ii) *If B contains no irreducible character of positive height, then we have $k(B) \leq p^d \sqrt{\ell(b)}$.*

Proof. (i) By hypothesis and Lemma 2.5c, $\mathrm{Tr}_{K|\mathbb{Q}}(p^d m_{\chi\chi}^{(y,b)}) \geq [K : \mathbb{Q}]\ell(b)$ for each $\chi \in \mathrm{Irr}(B)$. Using that $\sum_{\chi\in\mathrm{Irr}(B)} m_{\chi\chi}^{(y,b)} = \ell(b)$ we get the estimate $p^d[K : \mathbb{Q}]\ell(b) \geq k(B)[K : \mathbb{Q}]\ell(b)$. Hence the result.

(ii) Now $m_{\chi\zeta}^{(y,b)} \neq 0$ for all χ, ζ in $\mathrm{Irr}(B)$. Therefore, varying σ over the Galois group $\Gamma = \mathrm{Gal}(K|\mathbb{Q})$ and ζ over $\mathrm{Irr}(B)$, by the arithmetic–geometric mean inequality (1.5b) we have

$$1 \leq \Big(\prod_{\zeta,\sigma}(p^d|m_{\chi\zeta}^{(y,b)}|^\sigma)^2\Big)^{\frac{1}{k(B)|\Gamma|}} \leq \frac{1}{k(B)|\Gamma|}\sum_{\zeta,\sigma} p^{2d}|(m_{\chi\zeta}^{(y,b)})^\sigma|^2.$$

The term on the right equals $\frac{1}{k(B)|\Gamma|}\sum_\sigma(p^{2d}m_{\chi\chi}^{(y,b)})^\sigma$. Use finally once more that $\sum_\chi m_{\chi\chi}^{(y,b)} = \ell(b)$. $\qquad\square$

Remark. One knows that the $k(B)$ conjecture is true for blocks with cyclic defect groups, even for abelian defect groups of rank at most 2. The best general result so far for p-blocks B of defect $d \geq 2$ is the Brauer–Feit bound $k(B) \leq p^{2d-2}$, which follows by combining Theorems 2.4 and 2.5b [F, VII.10.13 and 10.14].

2.6. Blocks of p-Solvable Groups

X is called *p-solvable* provided each composition factor of X either is a p-group or a p'-group. Then X contains a p-complement, G, and each p'-subgroup of X is contained in a conjugate of G (P. Hall; the proof is easily worked out by induction applying the Schur–Zassenhaus theorem). The block theory for p-solvable groups has been developed in [Fong, 1962].

A p-solvable group X is p-*constrained*, that is, if P is a Sylow p-subgroup of $O_{p'p}(X)$, then $C_X(P) \subseteq O_{p'p}(X)$ (Hall–Higman lemma). Here $O_p(X)$ and $O_{p'}(X)$ are the largest normal p-subgroup and p'-subgroup of X, respectively, and $O_{p'p}(X)/O_{p'}(X) = O_p(X/O_{p'}(X))$.

Theorem 2.6a. *Suppose X is p-solvable with $O_{p'}(X) = 1$. Then X has a unique p-block B. Assume the $k(GV)$ theorem has already been proved. Then $k(B) \leq |D|$. We even have $k(B) < |D|$ unless $X = Z(X) \times GV$ where G is a p-complement in X acting faithfully on the elementary abelian p-group V, and if $k(GV) = |V|$.*

Proof. Let $P = O_p(X)$. Since $O_{p'}(X) = 1$, we know that $C_X(P) \subseteq P$. Hence by Theorem 2.3c the principal block is the unique p-block of X. We first show that $k(X) \leq |D|$ and that this inequality is proper if D is nonabelian. We argue by induction on $|X|$, following [Robinson, 2004]. Let G be a p-complement in X. Suppose $GP \neq X$. Then by induction $k(GP) \leq |P|$ and either P is abelian or $k(GP) < |P|$. From part (i) of Theorem 1.7a it follows that

$$k(X) \leq k(GP) \cdot |X : GP| = k(GP) \cdot |D : P| \leq |D|,$$

and that equality only holds if $k(GP) = |P|$, $N = GP$ is normal in X, $k(X/N) = |X : N|$ and $k_X(N) = k(N)$. Hence if equality holds, then P and $X/N \cong D/P$ are abelian, and the p-group D/P fixes each conjugacy class of the p'-group $N/P \cong G$. Hence D/P centralizes N/P. But D/P acts faithfully on N/P as $O_p(X/P) = 1$. Thus from $k(X) = |D|$ it follows that $X = GP$. We therefore may assume that $D = P$ and $X = GP$.

Choose a minimal normal subgroup M of X, which is an elementary abelian p-group. By induction $k(X/M) \leq |D/M|$ and $k(X/M) < |D/M|$ if D/M is nonabelian. Now the estimate (1.7b) yields that $k(X) \leq |M| \cdot k(X/M) \leq |D|$, and equality only holds if D/M is abelian, M is central in X and no nonidentity element of M is a commutator in X (see the beginning of Sec. 1.7). But since D is nonabelian, $M = D'$ has order p and is generated by commutators in D.

So let $D = P$ be abelian (and normal in $X = GD$). We have $Z(X) = C_D(G)$ as $O_{p'}(X) = 1$, and $D = Z(X) \times V$ where $V = [D, G]$ by Eq. (1.6a). Let $\bar{V} = V/U$ be the Frattini quotient of V, which is a faithful $\mathbb{F}_p G$-module. By the $k(GV)$ theorem and (1.7b)

$$k(X) \leq |Z(X)| \cdot |U| \cdot k(G\bar{V}) \leq |D|,$$

and equality only holds if $k(G\bar{V}) = |\bar{V}|$ and U is central in X. But then $U = 1$ and $V = \bar{V}$, completing the proof.   □

Theorem 2.6b (Fong). *Let X be p-solvable, $N = O_{p'}(G)$ and $G = X/N$. Let B be a p-block of X. Suppose B covers some $\theta \in \mathrm{Irr}(N)$.*

(i) *There is a unique block b of $T = I_X(\theta)$ covering θ such that $b^X = B$, and b, B have a defect group in common. Induction defines a bijection from $\mathrm{Irr}(b)$ and $\mathrm{IBr}(b)$ to $\mathrm{Irr}(B)$ and $\mathrm{IBr}(B)$, respectively.*

(ii) *Suppose θ is X-invariant ($T = X$). Then the Clifford correspondence of Theorem 1.9c, restricted to $\mathrm{Irr}(B)$ resp. $\mathrm{IBr}(B)$, describes a bijection onto $\mathrm{Irr}(B_0)$ resp. $\mathrm{IBr}(B_0)$ for some unique p-block B_0 of the representation group $G(\theta)$ preserving decomposition numbers. In fact, $\mathrm{Irr}(B_0) = \mathrm{Irr}(G(\theta)|\tilde{\theta})$. The defect groups of B and B_0 are Sylow p-subgroups.*

Proof. (i) Let $\chi \in \mathrm{Irr}(B)$ and let $\psi \in \mathrm{Irr}(T)$ occur in $\mathrm{Res}^X_T(\chi)$. Then $\psi \in \mathrm{Irr}(T|\theta)$ and so $\mathrm{Ind}^X_T(\psi)$ is irreducible by Theorem 1.8b. Thus $\mathrm{Ind}^X_T(\psi) = \chi$ by Frobenius reciprocity. Note that $\chi(1) = |X : T|\psi(1)$. Let b be the block of T containing ψ. By Example 2.5a, $B = b^X$. Let c be a defect class for B with defect group D. Then $\omega_B(\hat{c}) \neq 0$ by definition and so $\omega_b(\hat{c_0}) \neq 0$ for some conjugacyc class $c_0 \subseteq c$ of T. By Lemma 2.3b, a defect group D_0 of b is contained in a defect group of c_0, so that we may pick $D_0 \subseteq D$. But by character degrees both b and B have the same defect.

(ii) Let $\hat{\theta}$ be a character of $X(\theta)$ extending θ in the sense of Definition 1.9b. Like θ this $\hat{\theta}$ is irreducible as a Brauer character, and by Theorem 1.9c it defines a bijection from $\mathrm{Irr}(X|\theta)$ onto $\mathrm{Irr}(G(\theta)|\tilde{\theta})$ (where $\tilde{\theta}$ is a linear character of the cyclic central subgroup Z of $G(\theta)$ of order $\exp(N)$). The Clifford correspondence $\chi = \hat{\theta} \cdot \zeta \leftrightarrow \zeta$ extends to Brauer characters preserving decomposition numbers. It follows that $\mathrm{Irr}(G(\theta)|\tilde{\theta})$ is a union of irreducible characters belonging to certain blocks of $G(\theta)$.

We may assume that $p \mid |G|$. Then $P = O_p(G(\theta)) \neq 1$ since $Z = O_{p'}(G(\theta))$ is in the centre of $G(\theta)$ (and $G(\theta)$ is p-solvable). We have $C_{G(\theta)}(P) = P \times Z$. There is a unique block b_0 of PZ covering $\tilde{\theta}$. Hence by Theorem 2.3c, $B_0 = b_0^{G(\theta)}$ is the unique block covering b_0 (and $\tilde{\theta}$). Thus $\mathrm{Irr}(B_0) = \mathrm{Irr}(G(\theta)|\tilde{\theta})$. Let D be a Sylow p-subgroup of $G(\theta)$. There is a character ψ of $D \times Z$ extending $\tilde{\theta}$. Then $p \nmid \psi(1)$, and there is an irreducible constituent ζ of $\mathrm{Ind}^{G(\theta)}_{DZ}(\psi)$ whose degree is not divisible by p. Now ζ lies over $\tilde{\theta}$ and so $\zeta \in \mathrm{Irr}(B_0)$. Hence D is a defect group of B_0.

Of course X, $X(\theta)$, G, $G(\theta)$ have isomorphic Sylow p-subgroups.   □

Theorem 2.6c (Nagao). *Assume that the $k(GV)$ theorem has been already proved. If X is a p-solvable group and B is a p-block of X with defect group D, then $k(B) \leq |D|$ where the inequality is proper if D is nonabelian.*

Proof. We proceed by induction on $|X|_p = p^a$ [Nagao, 1962]. By Theorems 2.6a and 2.6b we may assume that $Z = O_{p'}(X)$ is nontrivial and central in X, and that D is a Sylow p-subgroup of X ($|D| = p^a$). Also $\mathrm{Irr}(B) = \mathrm{Irr}(X|\theta)$ for some irreducible (linear) character θ of Z. Thus

$$k(B) = |\mathrm{Irr}(X|\theta)| \leq k(X/Z)$$

by Theorem 1.9c. Now $O_{p'}(X/Z) = 1$, and so Theorem 2.6a applies. Hence $k(X/Z) \leq |D|$, even $k(X/Z) < |D|$ unless $DZ/Z \cong D$ is abelian. $\square$

2.7. Coprime $\mathbb{F}_p X$-Modules

Let V be an irreducible $\mathbb{F}_p X$-module. By Wedderburn's theorem $F_0 = \mathrm{End}_X(V)$ is a (commutative) finite field, which we may embed into F. Viewing $V = V_0$ as an (absolutely) irreducible $F_0 X$-module, $F_0 \otimes_{\mathbb{F}_p} V = \bigoplus_\sigma V_0^\sigma$ is the direct sum of its Galois conjugates over the prime field. Let $\chi \in \mathrm{IBr}(X)$ be the Brauer character of X afforded by V_0. Let K_I be the subfield of K generated by the $\exp(X)_{p'}$-roots of unity, the *inertia subfield* of K for p (or $\mathfrak{p}$), and let K_D be the *decomposition subfield*. Thus $K_D \subseteq K_I$ and $\mathrm{Gal}(K_I|K_D) \cong \mathrm{Gal}(F|\mathbb{F}_p)$ is cyclic generated by the Frobenius automorphism. So F_0 corresponds to the intermediate field $K_0 = K_D(\chi)$, and we associate to V the trace character $\check{\chi} = \mathrm{Tr}_{K_0|K_D}(\chi) = \sum_{\sigma \in \mathrm{Gal}(K_0|K_D)} \chi^\sigma$.

Now assume X is a p'-group ($K = K_I$). Then χ is an ordinary irreducible character of X. For each $\tau \in \mathrm{Gal}(K|\mathbb{Q})$ there is an irreducible $\mathbb{F}_p X$-module V^τ affording $\check{\chi}^\tau = \mathrm{Tr}_{K_0|K_D}(\chi^\tau)$, which is not isomorphic to V unless $\tau \in \mathrm{Gal}(K|K_D)$. But V, V_0 and V^τ are isomorphic as G-sets. Use the fact that the F_0-dimension of $F_0 \otimes C_V(Y) = C_{F_0 \otimes V}(Y)$ equals

$$\langle \check{\chi}, 1_Y \rangle_Y = [F_0 : \mathbb{F}_p]\langle \chi, 1_Y \rangle_Y = \langle \check{\chi}^\tau, 1_Y \rangle_Y$$

for each subgroup Y of X. Thus $|C_V(Y)| = |C_{V_0}(Y)| = |C_{V^\tau}(Y)|$, which gives the result (cf. Secs. 1.3 and 1.4).

Remark. Let X be p-solvable. The Fong–Swan theorem tells us that then every (absolutely) irreducible Brauer character of X can be lifted to an ordinary character. This is proved via Clifford theory [F, X.2.2], which even yields a p-rational lift (having its values in K_I). Arguing as above one sees that every irreducible $\mathbb{F}_p X$-module has a lift affording a p-rational character.

Chapter 3

The $k(GV)$ Problem

Let p be a rational prime. We consider the situation where $X = GV$ is the semidirect product of a finite p'-group G acting faithfully on an elementary abelian p-group V. Then $O_{p'}(X) = 1$ and so V is the unique defect group of the unique p-block of X (Theorem 2.3c). The $k(GV)$ problem is the special case of the $k(B)$ conjecture asking whether $k(GV) \leq |V|$, or not.

3.1. Preliminaries

We often write the $\mathbb{F}_p G$-module V additively. In this *coprime* situation V is completely reducible (Maschke). All irreducible characters of $X = GV$ have degree prime to p by Theorem 1.3b and so are of height zero in the unique p-block, B. For each $v \in V$ the centralizer $C_X(v) = C_G(v)V$ has a unique p-block b_v as well. We have $b_v^X = B$, and the subsection (v, b_v) is major.

It follows from Theorem 1.5d that $k(X) \leq |V|$ provided there is a vector $v \in V$ such that $C_G(v) = 1$, even $k(X) < |V|$ unless G is abelian. This may also be deduced from Theorem 2.5d. It will turn out that such *regular* vectors v (belonging to regular G-orbits) exist fairly often. In Theorem 3.4d below we shall see that it even suffices to find a vector with abelian point stabilizer, a result due to [Knörr, 1984].

Proposition 3.1a. *Let $V = V_1 \oplus V_2$ be the direct sum of $\mathbb{F}_p G$-modules V_i, and let $G_i = C_G(V/V_i)$. Suppose $k(G_i V_i) \leq |V_i|$ and $k((G/G_i)V_j) \leq |V_j|$ for $i \neq j$. Then $k(GV) \leq |V|$, and if $k(GV) = |V|$ then $G = G_1 \times G_2$ and $k(G_i V_i) = |V_i|$ for $i = 1, 2$.*

Proof. Cleary G_i is faithful on V_i $(i = 1, 2)$, and G/G_i is faithful on V_j for $j \neq i$. Consider the normal subgroups $N_i = G_i V_i$ of $X = GV$, and observe that $N = N_1 \times N_2$ is a normal subgroup of X too. Applying (1.7b) to the N_i yields that $k(X) \leq |V_1||V_2| = |V|$. If we have equality, then necessarily $k(G_i V_i) = |V_i|$ for $i = 1, 2$, and $k(X) = k(N) = |V|$. Moreover, then every conjugacy class of X/N is good for N (Theorem 1.7a). By Brauer's

32

permutation lemma (Theorem 1.4b) then every irreducible character θ of N is invariant in X, and each conjugacy class of X/N is good for θ. Hence

$$|V| = k(X) = \sum_{\theta \in \mathrm{Irr}(N)} k(X/N) - |V| \cdot k(X/N)$$

by the Clifford–Gallagher formula (1.10b). Thus $N = X$. □

Each irreducible (linear) character λ of V can be extended to its inertia group $I_X(\lambda)$; indeed there is a unique extension $\widetilde{\lambda}$ having $I_G(\lambda)$ in its kernel. For each irreducible character θ of $I_G(\lambda)$, inflated to $I_X(\lambda)$, the induced character $\chi_{\lambda,\theta} = \mathrm{Ind}_{I_X(\lambda)}^{X}(\widetilde{\lambda}\theta)$ is irreducible (Theorem 1.8b). By Proposition 1.6b, V and $\mathrm{Irr}(V)$ are isomorphic G-sets. We fix a G-isomorphism $v \mapsto \lambda_v$ between these G-sets, with $\lambda_0 = 1_V$, and we write $\chi_{v,\theta}$ in place of $\chi_{\lambda_v,\theta}$. The Clifford–Gallagher formula gives the following.

Proposition 3.1b. $k(GV) = \sum_i k(C_G(v_i))$ where $\{v_i\}$ is a set of representatives of the G-orbits on V. More precisely, for each $v \in V$ and $\theta \in \mathrm{Irr}(C_G(v))$ the character $\chi_{v,\theta}$ of $X = GV$ is irreducible of degree $|G : C_G(v)| \cdot \theta(1)$, and these are just all the $k(C_G(v))$ distinct irreducible characters of X lying above λ_v.

In this manner the partition of V into G-orbits gives rise to a corresponding partition of $\mathrm{Irr}(GV)$. Since the p-section of any $v \in V$ in $X = GV$, that is, the elements in X whose p-part is conjugate to v, is the union of conjugacy classes of X represented by certain p-regular elements in $C_X(v) = C_G(v)V$, the formula on $k(GV)$ likewise follows from this observation (by conjugacy of the complements to V in $C_X(v)$). (In the non-coprime situation, where G is a complement of the abelian group V in $X = GV$, or where just $G = X/V$, the corresponding formula holds replacing V by $\mathrm{Irr}(V)$ and centralizers by inertia groups.)

We may compute $k(GV)$ also as follows.

Lemma 3.1c. Let $\{g_j\}$ be a set of representatives for the conjugacy classes of G ($1 \le j \le k(G)$). Then $k(GV) = \sum_j |\mathrm{C}\ell(C_G(g_j)|C_V(g_j))|$.

Proof. For $g \in G$ let Ω_g denote the set of V-conjugacy classes contained in the coset Vg. By Theorem 1.7d, $|\Omega_g| = |C_V(g)|$. Since g is a p'-element,

$$V = C_V(g) \oplus [V, g]$$

by Proposition 1.6a, and this is a decomposition of $C_G(g)$-modules. Of course $C_G(g)$ acts on Ω_g as well (by conjugation). Now for each $v \in V$ we have $(vg)^V = vg^V = v[V, g]g$. The assignment $(vg)^V \mapsto v[V, g]$ is a bijection from Ω_g onto $V/[V, g] \cong C_V(g)$ which is compatible with the action of $C_G(g)$. Hence the result. $\square$

Replacing $C_V(g)$ by $V/[V, g]$ Lemma 3.1c also holds in the non-coprime situation.

3.2. Transitive Linear Groups

Suppose G is transitive on $V^{\sharp} = V \smallsetminus \{0\}$. Fix any $v \in V^{\sharp}$. Then $k(GV) = k(G) + k(C_G(v))$ by Proposition 3.1b. Such groups G exist. Identifying V with the additive group of a finite field and G with its multiplicative group we get a cyclic subgroup of $\mathrm{GL}(V)$ of order $|V| - 1$ acting regularly on $V^{\sharp}$. These *Singer cycles* in $\mathrm{GL}(V)$ are conjugate since their generators have the same *irreducible* minimum polynomial over the prime field.

Theorem 3.2. *Suppose G is transitive on $V^{\sharp}$. Then $k(GV) \leq |V|$, and if $k(GV) = |V|$ then either G is a Singer cycle in $\mathrm{GL}(V)$, or $p^m = 2^3$ and G is a Frobenius group of order 21, or $p^m = 3^2$ and G is semidihedral of order 16.*

Proof. In the exceptional cases indeed $k(GV) = |V|$. We shall appeal to [Hering, 1985] for the classification of the transitive linear groups (see also [Huppert–Blackburn, 1982, XII.7.5] for a summary of this work). By Theorem 1.5b we may assume that $H = C_G(v)$ is nontrivial, so that G is nonabelian (and not a Singer cycle). Let $|V| = p^m$. Note that the semidirect product $GV = V : G$ is a 2-transitive permutation group (of degree p^m). Embed G into $\mathrm{GL}(V)$ and let F be a maximal G-invariant subfield of $\mathrm{End}(V)$ containing the identity (with G acting via conjugation). Let $|F| = p^r$ so that V is a vector space of dimension $\frac{m}{r}$ over F. Using that $|G|$ is not divisible by p we obtain that one of the following holds:

(i) $m = 2, r = 1$ and p is equal to $11, 19, 29$ or 59, and G contains a normal subgroup $N \cong \mathrm{SL}_2(5)$.

(ii) $m = 2$ or 4, $r = 1$ (in both cases), and G contains a normal extraspecial 2-subgroup E of order 2^{m+1} such that $C_G(E) = Z(E)$ and G/E acts faithfully on $E/Z(E)$. Moreover, if $m = 2$ then $p = 3, 5, 7, 11$ or 23, and if $m = 4$ then $p = 3$.

(iii) $m = r$ and G is a subgroup of $\Gamma \mathrm{L}_1(p^m)$.

In (i) $N \cong \mathrm{SL}_2(5)$ is a maximal subgroup of the perfect group $\mathrm{SL}_2(p)$ as follows from Dickson's list of subgroups of $\mathrm{PSL}_2(p)$ [Huppert, 1967, II.8.27]. Thus G/N maps injectively into $\mathrm{GL}_2(p)/\mathrm{SL}_2(p) \cong \mathbb{F}_p^*$ and has prime order. Applying the estimate (1.7b) we get $k(G) \le k(N) \cdot k(G/N) \le k(\mathrm{SL}_2(5)) \cdot (p-1) = 9(p-1)$. Since $H = C_G(v)$ is faithful and completely reducible on V, it must be cyclic of order dividing $p-1$. Hence $k(GV) = k(G) + k(H) \le 10 \cdot (p-1) < p^2 = |V|$.

Consider (ii). (Extraspecial groups are briefly discussed in Sec. 4.2 below.) Regard G as a subgroup of $G_0 = N_{\mathrm{GL}(V)}(E)$. If $m = 2$ and $p = 3$ then either $E \cong Q_8$ is a quaternion group, which is regular on $V^\sharp$, or $E \cong D_8$ is dihedral, which is not transitive on $V^\sharp$. In both cases $G_0 \cong \Gamma\mathrm{L}_1(3^2)$ is a (semidihedral) Sylow 2-subgroup of $\mathrm{GL}_2(3)$, and we have $G = G_0$. (This possibility will come up also in (iii).) For $m = 2$ and $p \ge 5$ we must have $E \cong Q_8$, for otherwise G_0 would not be transitive on $V^\sharp$. The cases $p = 5$ and $p = 7$ are of special interest (cf. Sec. 6.1 below):

If $p = 5$, then G_0 is a 5-complement in $\mathrm{GL}_2(5)$ (of order 96) and $C_{G_0}(v)$ is cyclic of order 4. Either $G = G_0$ or $|G_0 : G| = 2$, and $G_0/Z(G_0) \cong S_4$. If $p = 7$, then $G_0 \cong X \circ Z_6 = X \times Z_3$ where X is a Schur cover of S_4, and $|C_{G_0}(v)| = 3$. In both cases $k(GV) \le k(G_0V) < |V|$ by (1.10b).

For $m = 2$ and $p = 11$ we have $G_0 \cong \mathrm{GL}_2(3) \times Z_5$ and $|C_{G_0}(v)| = 2$ for any nonzero $v \in V$. Thus $G = G_0$ by assumption, and $k(GV) < |V|$. The case $m = 2$, $p = 23$ is ruled out since then $G = G_0$ is regular on $V^\sharp$.

Finally let $m = 4, p = 3$. Since $|V^\sharp| = 80$ and $O_4^+(2)$ is a $5'$-group, $E \cong 2_-^{1+4}$ is extraspecial of negative type (and order 2^5) and G/E is a ($3'$-) subgroup of $O_4^-(2) \cong S_5$ of order 5, 10 or 20 (determined up to conjugacy). Hence GV is one of the three exceptional 2-transitive solvable (Bucht) groups [Huppert–Blackburn, 1982, XII.7.4]. Here $H = C_G(v)$ is cyclic of order 2, 4 or 8, respectively. In all cases $k(GV) < |V|$. The largest Bucht group again is of special interest for us (Sec. 6.1).

It remains to consider (iii). So let G be a (nonabelian) subgroup of $\Gamma\mathrm{L}_1(p^m)$, whence $m \ge 2$. Then G is a subgroup of $T = N_{\mathrm{GL}(V)}(S$ for some Singer cycle S in $\mathrm{GL}(V)$. This $T \cong \Gamma\mathrm{L}_1(p^m)$ acts on S like the Galois group of $\mathbb{F}_{p^m}|\mathbb{F}_p$, so that T/S is cyclic of order m. In this case $F = S \cup \{0\}$ is a maximal G-invariant subfield of $\mathrm{End}(V)$. Let $N = G \cap S$. Then G/N is cyclic of order n, say, where $n > 1$ is a divisor of m. Note that n is not divisible by p. Since S is regular on $V^\sharp$ and G is transitive, $H = C_G(v)$ is cyclic of order dividing n. Hence $k(GV) \le k(G) + n$.

Assume $C_T(N) > S$. Then there is a proper divisor d of m such that $|N|$ divides $p^d - 1$. Since G is transitive on $V^\sharp$, we get that $p^m - 1 \leq |G| \leq m(p^d - 1)$ and so $m \geq 1 + p^{m-d}$. This is impossible as $d \leq \frac{m}{2}$ and $m \geq 2$. We conclude that $N = C_G(N)$ is irreducible on V. Just the elements in N of order dividing $p^{m/n} - 1$ are central in G, and if $d \mid n$ then those of order dividing $p^{dm/n} - 1$ are centralized by the subgroup of G/N with index d. Hence using (1.7b) we get the (crude) estimate

$$k(G) \leq \sum_{d|n} \frac{n}{d^2}(p^{dm/n} - 1) \leq \frac{1}{n}(p^m - 1) + n(p^{\lfloor \frac{m}{2} \rfloor} - 1).$$

Here we used that $\frac{m}{n} \leq \lfloor \frac{m}{2} \rfloor$ and $\sum_{n \neq d|n} \frac{n}{d^2}(p^{dm/n} - 1) \leq n \sum_{j=1}^{\lfloor \frac{m}{2} \rfloor} p^j$.

Suppose we have $p^m = |V| \leq k(G) + n$. Then we must have $n = m$. In particular, m is not divisible by p. The resulting inequality

$$p^m \leq \frac{1}{m}(p^m - 1) + m(p^{\lfloor \frac{m}{2} \rfloor} - 1) + m$$

forces that $p \leq 3$, and that $m \leq 2$ for $p = 3$, while $m \leq 4$ for $p = 2$. We conclude that $p^m = 3^2$ or $p^m = 2^3$. If $G \cong \Gamma L_1(3^2)$ then $k(G) = 7$ and $k(G) + m = |V|$, and if $G \cong \Gamma L_1(2^3)$ then $k(G) = 5$ and $k(G) + m = |V|$ likewise. So in these two cases $k(GV) = |V|$, otherwise $k(GV) < |V|$. We are done.   □

3.3. Subsections and Point Stabilizers

In this section let K, R and $\mathfrak{p}|p$ be as in the previous chapter, but we let Γ denote the subgroup of $\mathrm{Gal}(K|\mathbb{Q})$ fixing the p'-roots of unity in K. Since $X = GV$ has the exponent $\exp(G) \cdot p$ and G is a p'-group, $\Gamma \cong \mathrm{Gal}(\mathbb{Q}(\varepsilon_p)|\mathbb{Q})$ where $\varepsilon_p = e^{2\pi i/p}$. The assignment $g^G \mapsto g^X$ is a bijection from $C\ell(G)$ onto $C\ell(X_{p'})$ (Schur–Zassenhaus). Let B be the unique p-block of X. Identifying $\mathrm{Irr}(G)$ with $\mathrm{IBr}(X) = \mathrm{IBr}(B)$ via inflation, for $\varphi \in \mathrm{Irr}(G)$ and $\chi \in \mathrm{Irr}(X)$ the decomposition number $d_{\chi\varphi} = \langle \chi, \varphi \rangle_G$. Thus $\mathrm{Ind}_G^X(\varphi) = \sum_{\chi \in \mathrm{Irr}(X)} d_{\chi\varphi}\chi$ by Frobenius reciprocity (1.2c), and this may be identified with the projective character $\widehat{\varphi}$ in view of Theorem 2.2a.

It follows from (2.1b) and Theorem 2.2a that the projective characters form a basis of the class functions on X vanishing off p-regular elements of X. Hence we have:

Lemma 3.3a. *Induction of class functions yields a bijection $\varphi \leftrightarrow \widehat{\varphi}$ between generalized characters φ of G and those of $X = GV$ vanishing off p-regular elements. For $\varphi \in \mathbb{Z}[\mathrm{Irr}(G)]$ and $g \in G$ we have $\widehat{\varphi}(g) = |C_V(g)|\varphi(g)$.*

Proof. It remains to verify the last statement. Let $\varphi \in \mathbb{Z}[\mathrm{Irr}(G)]$. By formula (1.2b) for induced characters (and class functions), noting that V is a (right) transversal to G in X, we have

$$\widehat{\varphi}(x) = \sum_{v \in V : vxv^{-1} \in G} \varphi(vxv^{-1}) = |C_V(g)|\varphi(g)$$

if $x \in X$ is conjugate to some $g \in G$, and $\widehat{\varphi}(x) = 0$ otherwise. $\square$

Let π_V denote the permutation character of G on the set V (as usual). Let φ, ψ be in $\mathrm{Irr}(G)$. Then Lemma 3.3a and Frobenius reciprocity tell us that the corresponding Cartan invariant of B is given by

$$c_{\varphi\psi} = \langle \widehat{\varphi}, \widehat{\psi} \rangle_X = \langle \varphi \pi_V, \psi \rangle_G.$$

Let $C_B = (c_{\varphi\psi})$ be the Cartan matrix. By Lemma 2.1a the class function on $X = GV$ taking the value $|V|$ on p-regular elements, and the value zero otherwise, is a generalized character of X. Thus by Lemma 3.3a

$$(3.3b) \qquad\qquad \delta_V = |V|/\pi_V$$

is a generalized character of G ($\delta_V(g) = |V : C_V(g)|$ for $g \in G$). This generalized character has been introduced (and studied) by [Knörr, 1984]. It follows from Theorem 2.2a that the (φ, ψ) entry of $|V|C_B^{-1}$, viewing φ, ψ as characters of X/V, is given by

$$\frac{1}{|X|} \sum_{x \in X_{p'}} \varphi(Vx)\overline{\psi(Vx)} = \frac{|V|}{|X|} \sum_{g \in G} \varphi(g)|V : C_V(g)|\overline{\psi(g)} = \langle \varphi \delta_V, \psi \rangle_G,$$

because each element of $X_{p'}$ is conjugate to an element of G and $\delta_V(g)$ is the number of conjugates of $g \in G$ lying in the coset Vg.

We apply this to HV where $H = C_G(v)$ for some $v \in V$.

Theorem 3.3c (Knörr). *Suppose there is $v \in V$ such that for $H = C_G(v)$ we have $\langle \theta \delta_V, \theta \rangle_H \geq k(H)$ for all $\theta \in \mathbb{Z}[\mathrm{Irr}(H)]$ with $\theta(1) \not\equiv 0 \pmod{p}$. Then $k(GV) \leq |V|$, and equality only holds if $\langle \chi \delta_V, \chi \rangle_H = k(H)$ for all $\chi \in \mathrm{Irr}(G)$.*

Proof. Let C_v be the Cartan matrix of the unique p-block b_v of $Y = HV$, and let Q_v be the quadratic form associated to $|V|C_v^{-1}$. As seen above the (φ, ψ) entry of $|V|C_v^{-1}$ is given by $\langle \varphi \delta_V, \psi \rangle_H$. Of course $k(H) = \ell(b_v)$.

Let (v, b_v) be the corresponding (major) subsection to the block B. Define $m_{\chi \zeta}^{(v, b_v)}$ as in Sec. 2.5. For each irreducible character χ of $X = GV$ we have $\chi(v) \equiv \chi(1) \not\equiv 0 \pmod{\mathfrak{p}}$ by Theorem 1.3b. Hence

$$\sum_{\varphi \in \mathrm{Irr}(H)} d_{\chi \varphi}^v \varphi(1) = \chi(v) \not\equiv 0 \pmod{\mathfrak{p}}.$$

Thus $k(X) \leq |V|$ by hypothesis and Theorem 2.5d, and this equality is proper provided $\mathrm{Tr}_{K|\mathbb{Q}}(|V|m_{\chi\chi}^{(v,b_v)}) > [K : \mathbb{Q}] \cdot k(H)$ for some $\chi \in \mathrm{Irr}(X)$.

Now let us investigate $m_{\chi\chi}^{(v,b_v)}$ when $\chi \in \mathrm{Irr}(G)$ (inflated to X). Then the higher decomposition number $d_{\chi\varphi}^v = \langle \chi, \varphi \rangle_H$ is nothing but the multiplicity of $\varphi \in \mathrm{Irr}(H)$ in the restriction to H of χ. Thus

$$|V|m_{\chi\chi}^{(v,b_v)} = \sum_{\varphi, \psi \in \mathrm{Irr}(H)} \langle \chi, \varphi \rangle_H \langle \varphi \delta_V, \psi \rangle_H \langle \chi, \psi \rangle_H = \langle \chi \delta_V, \chi \rangle_H.$$

This completes the proof.   □

The reader is referred to [Knörr, 1984] for a proof of the above result avoiding block theory. Knörr noticed that the hypothesis in Theorem 3.3c is fulfilled if there is a generalized character ψ of $H = C_G(v)$ such that $\psi(h) \in \mathfrak{p}$ for $h \in H^\sharp$ but $\psi(1) \notin \mathfrak{p}$, and such that $\delta_V \geq |\psi|^2$ on H (elementwise). Then $\langle \theta \delta_V, \theta \rangle_H \geq \langle \theta \psi, \theta \psi \rangle_H$ for each $\theta \in \mathbb{Z}[\mathrm{Irr}(H)]$ with $p \nmid \theta(1)$, and

$$|H|\langle \theta\psi, \zeta \rangle_H = \sum_{h \in H} \theta(h)\psi(h)\bar{\zeta}(h) \equiv \theta(1)\psi(1)\zeta(1) \pmod{\mathfrak{p}}$$

for each $\zeta \in \mathrm{Irr}(H)$, where $p \nmid \zeta(1)$ by Theorem 1.3b. So each $\zeta \in \mathrm{Irr}(H)$ is contained in $\theta\psi$, which gives the result. Following [Robinson–Thompson, 1996] we use the following slightly different concept, which will turn out to be fulfilled when v is a so-called real vector for G.

Theorem 3.3d (Robinson–Thompson). *Let $H = C_G(v)$ for some $v \in V$. Assume there is a faithful $\mathbb{F}_p H$-submodule U of V and a rational-valued generalized character ψ of H such that $\psi^2 = \delta_U$ on H, or p is odd and $\psi^2 = \delta_U$ on a subgroup N of H with $|H : N| - 2$, while $\psi(h)^2 = \frac{1}{p}\delta_U(h)$ for $h \in H \smallsetminus N$. Then $k(GV) \leq |V|$, even $k(GV) \leq \frac{p+3}{2p}|V|$ in the latter (odd) case, and in both cases the inequalities are proper unless $[V, H] \subseteq U$.*

Proof. Let $Y = HV$, and observe that $\langle v \rangle$ is in the centre of Y. Suppose $\sigma \in \Gamma$ sends ε_p to ε_p^s. Then for each $h \in H$ and each irreducible character χ of X (or Y) we have $\chi^\sigma(hv) = \chi(hv)^\tau = \chi(hv^s)$ (choosing $s \equiv 1 \pmod{|G|}$ and representing $hv = vh$ by a diagonal matrix with trace $\chi(hv)$).

Suppose first that $\psi^2 = \delta_U$ on H. Then $\psi(h)$ is divisible by p for all $h \in H^\sharp$, as H is faithful on U, and $\psi(1) = \pm 1$. As in Lemma 3.3a let $\widehat{\psi} = \mathrm{Ind}_H^Y(\psi)$, and let $n_\zeta = \langle \widehat{\psi}, \zeta \rangle$ for $\zeta \in \mathrm{Irr}(Y)$. Define

$$\widetilde{\psi} = \sum_{\zeta \in \mathrm{Irr}(Y)} n_\zeta \frac{\zeta(v^{-1})}{\zeta(1)} \zeta.$$

So $\widetilde{\psi}$ and $\Psi = \mathrm{Ind}_Y^X(\widetilde{\psi})$ are pth cyclotomic integer combination of characters of Y and X, respectively. Hence $\langle \Psi, \chi \rangle \in \mathbb{Z}[\varepsilon_p]$ for each $\chi \in \mathrm{Irr}(X)$. We have $\widehat{\psi}(y) = \lfloor C_V(h)|\psi(h)$ if $y \in Y$ is conjugate to hv for some (unique) $h \in H$, and $\widehat{\psi}(y) = 0$ otherwise. Since there are $|V : C_V(h)| \cdot |H : C_H(h)|$ elements in Y conjugate to hv for each $h \in H$, by Frobenius reciprocity

$$|H|\langle \Psi, \chi \rangle_X = |H|\langle \widehat{\psi}, \chi \rangle_Y = \sum_{h \in H} \psi(h)\chi(h^{-1}v^{-1}) \equiv \chi(v^{-1}) \pmod{\mathfrak{p}}$$

for each $\chi \in \mathrm{Irr}(X)$. Thus $\langle \Psi, \chi \rangle \neq 0$ as $p \nmid \chi(1)$ by Theorem 1.3b. Moreover, $\langle \Psi, \chi^\sigma \rangle = \langle \Psi, \chi \rangle^\sigma \neq 0$ for each $\sigma \in \Gamma$. Hence by the arithmetic–geometric mean inequality (1.5b) $\sum_\sigma |\langle \Psi, \chi^\sigma \rangle|^2 \geq p - 1$.

Similarly $\Psi(x) = |C_V(h)|\psi(h)$ if $x \in X$ is conjugate to hv for some $h \in H$, and zero otherwise. It follows that

$$\langle \Psi, \Psi \rangle_X = \langle \widetilde{\psi}, \widetilde{\psi} \rangle_Y = \frac{1}{|H|} \sum_{h \in H} |C_V(h)|\psi(h)^2.$$

Now $\psi^2 = \delta_U$ on H, and $\delta_V = \delta_U \cdot \delta_{U'}$ if $V = U \oplus U'$ as an H-module. We conclude that $\langle \Psi, \Psi \rangle \leq |V|$ and that this inequality is proper unless $\delta_{U'} = 1_H$, that is, $[V, H] \subseteq U$. This gives the result in the first case, as $(p-1)k(X) \leq \sum_{\sigma \in \Gamma} \sum_{\chi \in \mathrm{Irr}(X)} |\langle \Psi, \chi^\sigma \rangle|^2 = (p-1)\langle \Psi, \Psi \rangle \leq (p-1)|V|$.

Let now p be odd and $\psi^2 = \delta_U$ on N but $\psi(h)^2 = \frac{1}{p}\delta_U(h)$ for $h \in H \smallsetminus N$ ($|H : N| = 2$). Let μ be the linear character of H with kernel N, and let $\varphi = \psi + \mu\psi$. Then $\varphi(h) = 2\psi(h)$ for $h \in N$, while φ vanishes on $H \smallsetminus N$. Hence $\varphi(h)$ is an integer multiple of p for each $h \in H^\sharp$, while $\varphi(1) = 2\psi(1)$ is not divisible by p. Like above we define the class function Φ on X by letting $\Phi(x) = |C_V(h)|\varphi(h)$ if x is conjugate to hv for some $h \in H$. Then, as before, $\langle \Phi, \chi^\sigma \rangle = \langle \Phi, \chi \rangle^\sigma \neq 0$ for each $\chi \in \mathrm{Irr}(X)$ and $\sigma \in \Gamma$, so that $\sum_\sigma |\langle \Phi, \chi^\sigma \rangle|^2 \geq p - 1$ by (1.5b). Write $\Phi = \Phi_0 + \Phi_1$, where $\Phi_i(x) = \mu^i(h)|C_V(h)|\psi(h)$ if x is conjugate to hv for some $h \in H$, and 0 otherwise. Then at least one of $\langle \Phi_i, \chi \rangle \neq 0$, $i = 0, 1$, and then $\langle \Phi_i, \chi^\sigma \rangle = \langle \Phi_i, \chi \rangle^\sigma \neq 0$ for each $\sigma \in \Gamma$.

Let S_i be the set of irreducible characters of $X = GV$ which occur with nonzero multiplicity in Φ_i. If $\chi \in S_i$ then $\sum_\sigma |\langle \Phi_i, \chi^\sigma \rangle|^2 \geq p - 1$ by (1.5b). It follows that

$$(p - 1)|S_i| \leq \sum_{\sigma \in \Gamma} \sum_{\chi \in S_i} |\langle \Phi_i, \chi^\sigma \rangle|^2 = (p - 1)\langle \Phi_i, \Phi_i \rangle,$$

where $\langle \Phi_i, \Phi_i \rangle = \frac{1}{|H|} \sum_{h \in H} |C_V(h)|\psi(h)^2$ is the same for $i = 0, 1$ and equals

$$\frac{1}{|H|} \sum_{h \in N} |C_V(h)|\delta_U(h) + \frac{1}{|H|} \sum_{h \in H \smallsetminus N} |C_V(h)|\delta_U(h)/p \leq \frac{|V|}{2} + \frac{|V|}{2p}.$$

Thus $|S_i| \leq \frac{p+1}{2p}|V|$ for $i = 0, 1$ and, as above, these inequalities are proper unless $[V, H] \subseteq U$. We also see that

$$\langle \Phi_0 - \Phi_1, \Phi_0 - \Phi_1 \rangle = \frac{1}{|H|} \sum_{h \in H \smallsetminus N} 4|C_V(h)|\delta_U(h)/p \leq \frac{2}{p}|V|.$$

Hence at most $\frac{2}{p}|V|$ irreducible characters of X occur with nonzero multiplicity in $\Phi_0 - \Phi_1$, and again we only can have equality when $[V, H] \subseteq U$. We see that $|S_0 \smallsetminus S_1| + |S_1 \smallsetminus S_0| \leq \frac{2}{p}|V|$ (arguing as above considering the Γ-classes in $\mathrm{Irr}(X)$). Now $k(GV) = |S_0 \cup S_1| = \frac{1}{2}(|S_0| + |S_1| + |S_0 \smallsetminus S_1| + |S_1 \smallsetminus S_0|) \leq \frac{|V|}{2}(2 \cdot \frac{p+1}{2p} + \frac{2}{p}) = \frac{p+3}{2p}|V|$, completing the proof. $\square$

3.4. Abelian Point Stabilizers

Suppose $H \neq 1$ is an *abelian p'-subgroup* of $\mathrm{GL}(V) = \mathrm{GL}_m(p)$. If H is irreducible, then $F = \mathrm{End}_H(V)$ is a finite field containing H. Hence $H = \langle y \rangle$ is cyclic, $F = \mathbb{F}_p[y]$ and $C_{\mathrm{GL}(V)}(H)$ is a Singer cycle in $\mathrm{GL}(V)$. So $|H|$ divides $p^m - 1$, but $|H|$ does not divide $p^n - 1$ for $1 < n < m$. Also, H acts semiregularly on $V^\sharp$. In this irreducible case define

$$\delta_H = (|H| + 1)1_H - \rho_H$$

where ρ_H is the regular character of H. If H is a Singer cycle, then $\delta_H = \delta_V$ is Knörr's generalized character (on H). Otherwise let $t \geq 2$ denote the number of H-orbits on $V^\sharp$. Then $k(HV) = t + |H| = t + \frac{|V|-1}{t}$ by Proposition 3.1b. From $1 < t < |V| - 1$ we get that $k(HV) < |V|$ and that

$$\delta_V = (t|H| + 1)1_H - t\rho_H \geq (|H| + 1)1_H - \rho_H = \delta_H$$

at each element of H, both functions taking only nonnegative real values. Of course $\langle \delta_V, 1_H \rangle = t(|H| - 1) + 1 > |H| = \langle \delta_H, 1_H \rangle$ by the choice of t.

Lemma 3.4a. *Suppose $H \neq 1$ is an abelian p'-subgroup of $\mathrm{GL}(V)$, and let $[V, H] = \bigoplus_{i=1}^n V_i$ be a decomposition into irreducible $\mathbb{F}_p H$-modules (so that $n \geq 1$ and no V_i is a trivial module). Let $H_i = H/C_H(V_i)$ for each i, and define $\delta_H = \prod_i \delta_{H_i}$, which is a generalized character of H. We have*

$$\delta_V \geq \delta_H$$

at each element of H, and $\langle \delta_V, 1_H \rangle > \langle \delta_H, 1_H \rangle$ unless $\delta_V = \delta_H$ and each H_i is a Singer cycle in $\mathrm{GL}(V_i)$.

Proof. By Proposition 1.6a, $V = C_V(H) \oplus [V, H]$, and $[V, H, H] = [V, H]$. It is obvious that $\delta_V = \prod_i \delta_{V_i}$. The result follows. $\square$

We mention that H has a regular orbit on V, because picking arbitrary $v_i \in V_i^\sharp$ for each i and letting $v = \sum_i v_i$, then

(3.4b) $$C_H(v) = \bigcap_i C_H(v_i) = \bigcap H_i = 1.$$

Proposition 3.4c. *Keep the assumptions of the preceding lemma, embed each H_i uniquely into the corresponding Singer cycle G_i on V_i and let $G = \prod_i G_i$ be the direct product. Then H embeds into G in a natural way. We have $\langle \theta \delta_V, \theta \rangle_H \geq |H|$ for each nonzero generalized character θ of H, and if $\langle \delta_V, 1_H \rangle_H = |H|$ then necessarily $H = G$.*

Proof. Of course G is a p'-group and H is a certain fibre-product of the H_i and so embeds into the direct product $\prod_i H_i$, which is a subgroup of G. For the first statement in the lemma it suffices to show that $\langle \theta \delta_H, \theta \rangle \geq |H|$ for any generalized character $\theta \neq 0$ of H (which will be fixed in what follows).

Let us write $H^* = \mathrm{Hom}(H, \mathbb{C}^*)$ for the character group. Restriction from G to H defines an epimorphism from the character group $G^* = \prod_{i=1}^n G_i^*$ (direct) onto H^*. For each subset $I \subseteq N = \{1, 2, \cdots, n\}$ let $G_I^* = \prod_{i \in I} G_i^*$ (direct), and let H_I^* be its image in H^*. Here each $\lambda \in G_I^*$ is sent to $\lambda_I = \prod_{i \in I} \lambda(i)$ where $\lambda(i)$ is the restriction of the ith component of λ to H_i. Define

$$\gamma_\lambda = \prod_{i \in I} (1_H - \lambda(i)) = \sum_{J \subseteq I} (-1)^{|J|} \lambda_J.$$

We assert that $\delta = \delta_H = \sum_{I \subseteq N} 2^{-|I|} \sum_{\lambda \in G_I^*} |\gamma_\lambda|^2$. By Lemma 3.4a this is true for $n = 1$ (since by definition $\gamma_\varnothing = 1_H$). Proceeding by induction on n, the assertion follows. Using that $\langle \theta |\gamma_\lambda|^2, \theta \rangle = \langle \theta \gamma_\lambda, \theta \gamma_\lambda \rangle = \sum_{\chi \in H^*} \langle \theta \gamma_\lambda, \chi \rangle^2$ we obtain that

$$\langle \theta \delta, \theta \rangle = \sum_{I \subseteq N} 2^{-|I|} \sum_{(\lambda, \chi) \in G_I^* \times H^*} \langle \theta, \gamma_\lambda \chi \rangle^2.$$

We define a map $H^* \to \mathfrak{P}(N)$ assigning to each $\chi \in H^*$ a subset I of N of smallest cardinality, arbitrarily chosen, such that $\langle \theta, \lambda_I \chi \rangle \neq 0$ for some $\lambda \in G_I^*$. Such subsets exist since $\theta \neq 0$ and H is faithful on $[V, H]$. Let $M(I) = M(I, \theta)$ denote the inverse image in H^* of the subset I of N with respect to this map. For $\chi \in M(I)$ we have

$$\langle \theta, \gamma_\lambda \chi \rangle = \sum_{J \subseteq I} (-1)^{|J|} \langle \theta, \lambda_J \chi \rangle = (-1)^{|I|} \langle \theta, \lambda_I \chi \rangle$$

for all $\lambda \in G_I^*$, and there is $\lambda \in G_I^*$ such that $\langle \theta, \lambda_I \chi \rangle \neq 0$. Clearly H^* is the disjoint union of the fibres $M(I)$, and it suffices to show that $2^{-|I|} \sum_{(\lambda, \chi) \in G_I^* \times H^*} \langle \theta, \gamma_\lambda \chi \rangle^2 \geq |M(I)|$ for all I.

Fix $I \subseteq N$, and let $\varphi \in H^*$. The Boolean group $B = \mathfrak{P}(I)$, which is an elementary abelian 2-group with respect to the symmetric difference $J \oplus J' = (J \cup J') \setminus (J \cap J')$, acts on the set $G_I^* \times \varphi H_I^*$ via $(\mu, \zeta)^J = (\mu^J, \mu_J \zeta)$ where $\mu^J \in G_I^*$ is defined by $\mu^J(i) = \mu(i)^{-1}$ for $i \in J$ and $\mu^J(i) = \zeta(i)$ otherwise. Indeed $(\mu^J)^{J'} = \mu^{J \oplus J'}$ and $\mu_J(\mu^J)_{J'} = \mu_{J \oplus J'}$. If $(\lambda, \chi) = (\mu, \zeta)^J$ is in the B-orbit of (μ, ζ) then

$$\gamma_\lambda \chi = \gamma_{\mu^J} \mu_J \zeta = \mu \prod_{j \in J} \mu(j) \prod_{j \in J} (1 - \mu(j)^{-1}) \prod_{i \in I \setminus J} (1 - \mu(i)) = (-1)^{|J|} \gamma_\mu \zeta,$$

so that $\langle \theta, \gamma_\lambda \chi \rangle^2 = \langle \theta, \gamma_\mu \zeta \rangle^2$. Moreover, if C is the stabilizer in B of (μ, ζ) then $|C|^{-1} \gamma_\mu$ is a generalized character of H. In fact, for $J \in C$ and $J' \in B$ we have $\mu_{J \oplus J'} = \mu_J$ and $|J \oplus J'| \equiv |J| + |J'| \pmod 2$, whence $\gamma_\mu = \sum_{J \subseteq I} \mu_J = \sum_{J \in C} (-1)^{|J|} \tau$ for some generalized character τ of H. If not all $J \in C$ have even cardinality, those of even cardinality form a subgroup of C of index 2 and then $\sum_{J \in C} (-1)^{|J|} = 0$. It follows that

$$2^{-|I|} \sum_{(\lambda, \chi) \in (\mu, \zeta)^B} \langle \theta, \gamma_\lambda \chi \rangle^2 = |C|^{-1} \langle \theta, \gamma_\mu \zeta \rangle^2$$

is a nonnegative rational integer. Consider the set $M(I) \cap \varphi H_I^*$, which can be empty. We assert that the nonnegative integer

$$2^{-|I|} \sum_{(\lambda, \chi) \in G_I^* \times \varphi H_I^*} \langle \theta, \gamma_\lambda \chi \rangle^2 \geq |M(I) \cap \varphi H_I^*|.$$

Varying over the subsets I of N and the cosets of H^* mod H_I^* this will show that $\langle \theta \delta, \theta \rangle \geq |H|$.

We may assume that $M(I) \cap \varphi H_I^* \neq \varnothing$. Let $\chi = \chi_1, \cdots, \chi_r$ be its distinct elements. By construction there is $\lambda = \lambda_1 \in G_I^*$ such that $\langle \theta, \gamma_\lambda \chi \rangle = (-1)^{|I|} \langle \theta, \lambda_I \chi \rangle \neq 0$. Since each $\chi \chi_j^{-1} \in H_I^*$ we find $\alpha_j \in G_I^*$ such that $(\alpha_j)_I = \chi \chi_j^{-1}$, and we put $\lambda_j = \lambda \alpha_j$ for $j = 2, \cdots, r$. We have

$$\langle \theta, \gamma_{\lambda_j} \chi_j \rangle = (-1)^{|I|} \langle \theta, (\lambda_j)_I \chi_j \rangle = (-1)^{|I|} \langle \theta, \lambda_I \chi \rangle \neq 0.$$

It now suffices to prove that the (λ_j, χ_j) belong to different orbits under $B = \mathfrak{P}(I)$. Suppose $(\lambda_j, \chi_j) = (\lambda, \chi)^J = (\lambda^J, \lambda_J \chi)$ for some $J \subseteq I$. Then $(\alpha_j)_I = \chi \chi_j^{-1} = \lambda_J^{-1}$ and so the image of $\lambda_j = \lambda \alpha_j$ in H_I^* is the identity on H_J^*. From $\chi_j \in M(I)$ we can conclude that $J = \varnothing$. Hence $\lambda_j = \lambda^J = \lambda$,

$\alpha_j = 1$ and $\chi_j = \chi$. Replacing χ by χ_i we have $\lambda_j = \lambda_i \beta_j$, where $\beta_j = \alpha_j \alpha_i^{-1}$ satisfies $(\beta_j)_I = \chi_i \chi_j^{-1}$. So the same argument applies.

Consider finally the case where $\theta = 1_H$. Then the map $H^* \to \mathfrak{P}(N)$ is uniquely determined, assigning to $\chi \in H^*$ the support of χ, the smallest subset I of N such that $\chi \in H_I^*$. Suppose we have $\langle \delta_V, 1_H \rangle_H = |H|$. By Lemma 3.4a this forces that $\mathrm{Res}_H^G(\delta_V) = \delta$ and that $H_i = G_i$ for each i. Assume that $H \neq G$ (so that there must be a certain amalgamation). Then there is a subset $I \subseteq N$ such that the map $G_I^* \to H_I^*$ is not injective. Consider the set $M(I) = M(I, 1_H)$ of characters in H^* with support I. Then $\varnothing \neq M(I) \subseteq H_I^*$. Keep the notation of the preceding paragraph, picking $\varphi \in H_I^*$. Hence for $\chi = \chi_1$ in $M(I)$ we have $\lambda_I = \chi^{-1}$ in H_I^*, and so on. By assumption there is $\mu \neq \lambda$ in G_I^* with $\mu_I = \lambda_I$. It remains to show that the B-orbit of (μ, χ) is different from the orbits of all (λ_j, χ_j), because this will yield the desired contradiction $\langle \delta, 1_H \rangle > |H|$. If $(\mu, \chi)^J = (\lambda_j, \chi_j)$ for some $J \subseteq I$, then $\chi_j = \mu_J \chi = \lambda_J \chi$ and $\mu^J = \lambda_j$. Pick $\tau \in G_{I \smallsetminus J}^*$ such that $\tau_{I \smallsetminus J} = \lambda_{I \smallsetminus J}$. Then

$$\langle 1_H, \tau_{I \smallsetminus J} \chi_j \rangle = \langle 1_H, \lambda_{I \smallsetminus J} \lambda_J \chi \rangle = \langle 1_H, \lambda_I \chi \rangle \neq 0.$$

We conclude that $I \smallsetminus J = I$, that is, $J = \varnothing$. Hence $\chi_j = \lambda_J \chi = \chi$ and $\mu = \mu^J = \lambda_j = \lambda$, against our choice. This completes the proof. $\square$

Theorem 3.4d (Knörr). *Suppose there is $v \in V$ such that $H = C_G(v)$ is abelian. Then $k(GV) \leq |V|$, and we have equality only if $k(HV) = |V|$ in which case HV is the direct product of certain $H_i V_i$ where H_i either is a Singer cycle on V_i or $H_i = 1, |V_i| = p$.*

Proof. This is immediate from the preceding proposition and Theorem 3.3c. $\square$

Unfortunately the search for *abelian vectors* $v \in V$, for which $C_G(v)$ is abelian, is not compatible with Clifford reduction. So Theorem 3.4d will not be relevant for the proof of the $k(GV)$ theorem. We shall make use of this theorem when discussing the question under which conditions we can have equality $k(GV) = |V|$ (Chapter 11).

Chapter 4

Symplectic and Orthogonal Modules

The ultimate target of this chapter is to show that the assumptions made in Theorem 3.3d are fulfilled if the module carries a nondegenerate G-invariant symplectic or orthogonal form. This makes necessary a discussion of self-dual modules and of automorphism groups (holomorphs) of extraspecial groups, which in turn lead us to a useful new concept of goodness for conjugacy classes, and to Weil characters.

4.1. Self-dual Modules

Let V be a coprime FG-module, where $F = \mathbb{F}_r$ is a finite field of characteristic p. From Proposition 1.6b we know that V is isomorphic to $\mathrm{Irr}(V) = \mathrm{Hom}(V, \mathbb{C}^\star)$ as a G-set. But usually we do not have an isomorphism of G-modules. The FG-module $V^* = \mathrm{Hom}_F(V, F)$, with diagonal action $\lambda^x(v) = \lambda(vx^{-1})$ for $\lambda \in V^*, x \in G, v \in V$, is called the *dual module* to V. The module V is *self-dual* provided $V \cong V^*$ (as FG-modules).

Lemma 4.1a. *V is self-dual if and only if its Brauer character is real-valued.*

Proof. Let ρ be a matrix representation of G on V (to some basis of V). Then the matrix representation of G on V^* with respect to the dual basis is contragredient to ρ, sending $x \in G$ to $\rho(x^{-1})^t$. So if V is self-dual, its Brauer character is real-valued (Sec. 1.5). For the converse use that G is a p'-group and so the Brauer character is an ordinary character (determining the isomorphism type of V). $\square$

So the dual of a matrix representation ρ of G is obtained by applying ρ followed by the inverse transpose automorphism of the linear group. If G is a real group, every representation of G is self-dual. The FG-module V is self-dual if and only if V carries the structure of a nondegenerate G-invariant F-bilinear form, and in the coprime situation these forms are symplectic or symmetric or orthogonal sums of those forms [Gow, 1993]:

Theorem 4.1b (Gow). *Let V be self-dual. Then* $V = U \oplus W$ *where U is a symplectic FG-module (with even F-dimension), affording a G-invariant nondegenerate symplectic form, and where W is an orthogonal FG-module, affording a G-invariant nondegenerate symmetric bilinear form. The action of G on W can be chosen to be trivial when* $p = 2$.

Proof. Without loss of generality $V \neq 0$. By hypothesis there exists an isomorphism $\varphi : V \to V^*$ of FG-modules. Define

$$[v, w] = [v, w]_\varphi = \varphi(v)(w)$$

for $v, w \in V$. This is a nondegenerate F-bilinear form on V. For $x \in G$ we have

$$[vx, wx] = \varphi(vx)(wx) = \varphi(v)^x(wx) = \varphi(v)(wxx^{-1}) = [v, w].$$

Hence the form is G-invariant. The form $(v, w) \mapsto [v, w] + [w, v]$ is symmetric and G-invariant, its radical being an FG-submodule of V.

Suppose first that V is an irreducible FG-module. We assert that V is a symplectic FG-module or that $p \neq 2$ and V is an orthogonal FG-module. The above symmetric form is nondegenerate or is zero. If it is zero and $p \neq 2$, the form $[\cdot, \cdot]$ is symplectic (alternating), and we are done. So let $p = 2$. The symmetric form is now symplectic, too, and we may thus assume that it is zero. Then $[\cdot, \cdot]$ is symmetric and the radical $V_0 = \{v \in V | [v, v] = 0\}$ is an FG-submodule $\neq 0$ of V of codimension at most 1. It follows that $V_0 = V$ and that our form is symplectic.

In the general situation let U be a (proper) irreducible FG-submodule of V. If the form $[\cdot, \cdot]$ is nonzero on U, it is nondegenerate, as U is irreducible. Then U is self-dual, and the preceding paragraph applies. Otherwise U is contained in $U^\perp = \{v \in V | [v, U] = 0\}$, which is an FG-module, and we have $\dim_F V = \dim_F U + \dim_F U^\perp$ as the form is nondegenerate. By Maschke $V = U^\perp \oplus U_0$ for some FG-module U_0. The map $\lambda : U \to U_0^*$ given by

$$\lambda(u)(u_0) = [u, u_0]$$

is an FG-module homomorphism. Its kernel cannot be U for otherwise we had $U_0 \subseteq U^\perp$. Hence λ is injective, and it is surjective since $\dim_F U = \dim_F U_0 = \dim_F U_0^*$. Therefore $U \cong U_0^*$, whence $U^* \cong (U_0^*)^* = U_0$. The form on $U \oplus U_0$, given by

$$(u + u_0, u' + u_0') \mapsto [u_0, u'] - [u_0', u],$$

is symplectic and G-invariant. One verifies that this is nondegenerate using again that $U_0 \not\subseteq U^\perp$. The result follows by induction on $\dim_F V$. $\quad\square$

Lemma 4.1c. *Suppose V is a faithful irreducible FH-module for the abelian group H, and let $n = \dim_F V$. If $n = 1$ and V is self-dual (as an FH-module), then $|H| = 1$ or 2. Let $n \geq 2$. Then V is self-dual if and only if $n = 2m$ is even and $|H|$ a divisor of $r^m + 1$.*

Proof. From Sec. 3.4 we know that $H = \langle y \rangle$ is cyclic and that $C_{\mathrm{GL}(V)}(H)$ is a Singer cycle in $\mathrm{GL}(V)$. We may identify $V = \mathbb{F}_{r^n} = \mathbb{F}_r[y]$; $|H|$ is a divisor of $r^n - 1$ but not of $r^i - 1$ for $1 < i < n$. Clearly $y \in \mathbb{F}_{r^n}$ is an eigenvalue of the F-linear map on the extension field given by multiplication with y. So the eigenvalues of y on V are the different conjugates y^{r^i} of $\mathbb{F}_{r^n} | \mathbb{F}_r$ ($0 \leq i \leq n - 1$). Suppose V is a self-dual FH-module. Then for each eigenvalue the inverse must be an eigenvalue too. If $n = 1$ then necessarily $y = y^{-1}$ and $|H| = 1$ or 2. Let $n > 1$. Then $|H| > 2$, and no eigenvalue to y is equal to its inverse. Hence $n = 2m$ is even, and if $y^{-1} = y^{r^j}$ for some $j \leq n - 1$, then $|H|$ is a divisor of $r^j + 1$. It follows that $|H|$ is a divisor of $r^{2j} - 1$ and so $2j = n = 2m$. We conclude that $|H|$ is a divisor of $r^m + 1$. Conversely, if $n = 2m$ is even and $|H|$ is a divisor of $r^m + 1$, then $y^{-1} = y^{r^m}$. Then for each eigenvalue to y the inverse is an eigenvalue too, whence V is a self-dual FH-module. $\qquad\square$

It follows that V is a self-dual module for a Singer cycle in $\mathrm{GL}(V) = \mathrm{GL}_n(r)$ if and only if $r^n = 2$ or 3. Since $\mathrm{GL}(V)$ contains Singer cycles, the standard module V is self-dual for $\mathrm{GL}(V)$ only when $n = 1$, $r = 2$ or 3, or when $n = 2$, $r = 2$. It is self-dual for $\mathrm{SL}(V)$ only when $n = 1$ or 2, because the intersection of a Singer cycle in $\mathrm{GL}(V)$ with $\mathrm{SL}(V)$ is irreducible and has order $(r^n - 1)/(r - 1)$.

One knows that the symplectic group $\mathrm{Sp}_{2m}(r)$ and the orthogonal group $\mathrm{O}_{2m}^-(r)$ have unique conjugacy classes of *Singer cycles*, which by definition are cyclic and irreducible. They have order $r^m + 1$.

4.2. Extraspecial Groups

Let q be a prime (usually $q \neq p$). A finite nonabelian q-group E is called extraspecial if the centre $Z = Z(E)$ has order q and $U = E/Z$ is elementary abelian, hence a vector space over $\mathbb{F}_q$. Thus E is a central product of nonabelian groups of order q^3. Let $|E| = q^{1+2m}$.

Groups q^{1+2m} : Let q be odd and $E = \Omega_1(E)$, i.e., $\exp(E) = q$. Then $E \cong q^{1+2m} = q_+^{1+2m}$ is determined by its order. Choosing a generator z of Z, E has the generators x_i, x_i^* for $i = 1, \cdots, m$, satisfying the relations $[x_i, x_i^*] = z$, $z^q = x_i^q = (x_i^*)^q = 1$ for all i, the other generators centralizing each other (and z is central). The commutator map $(xZ, yZ) \mapsto [x, y]$ is a (well-defined) nondegenerate symplectic $\mathbb{F}_q$-form on U, identifying Z with $\mathbb{F}_q$. Every symplectic transformation on U can be lifted to an automorphism of E centralizing Z. Let $A = C_{\mathrm{Aut}(E)}(Z)$ and $C(E) = A/\mathrm{Inn}(E) = C_{\mathrm{Out}(E)}(Z)$. We have $\mathrm{Aut}(E) = A : \langle \alpha \rangle$ where α permutes the elements in $Z^\sharp$ transitively, induces on $U \cong \mathrm{Inn}(E)$ a symplectic similitude of order $q - 1$ and on $C(E) \cong \mathrm{Sp}_{2m}(q)$ the unique outer (diagonal) automorphism of order 2. In order to see this, let a be a generator of $\mathbb{F}_q^\star$ and define α by $x_i \mapsto x_i^a$, $x_i^* \mapsto x_i^*$ ($i = 1, \cdots, m$) and $z \mapsto z^a$. This preserves the defining relations and generates $\mathrm{CSp}_{2m}(q)/\mathrm{Sp}_{2m}(q)$.

Groups $2_\pm^{1+2m}$: Let $q = 2$. Then either $E \cong 2_+^{1+2m}$ is the central product of m dihedral groups of order 8, or $E \cong 2_-^{1+2m}$ is the central product of m quaternion groups Q_8. (Observe that $D_8 \circ D_8 \cong Q_8 \circ Q_8$.) The squaring map $xE' \mapsto x^2$ is a (well-defined) nonsingular quadratic form Q on $U = E/Z$ giving, together with the commutator map, a system of defining relations on E. Thus $C(E) = \mathrm{Out}(E) \cong O(U, Q)$, that is, $C(E) \cong O_{2m}^+(2)$ if E is of positive type and $C(E) \cong O_{2m}^-(2)$ otherwise. (This refers to the fact that U has Witt index m (plus) or Witt index $m - 1$ (minus); cf. Appendix B.) Except for $O_4^+(2) \cong S_3 \mathrm{wr}\, S_2$ there is a unique subgroup $\Omega_{2m}^\pm(2)$ in $O_{2m}^\pm(2)$ with index 2, the kernel of the Dickson invariant. The extension $\mathrm{Inn}(E) \rightarrowtail \mathrm{Aut}(E) \twoheadrightarrow C(E)$ is nonsplit when $m > 2$ (Appendix A10).

Groups 2_0^{1+2m} : We have $2_+^{1+2m} \circ Z_4 \cong 2_-^{1+2m} \circ Z_4$, and the isomorphism type of this group is written 2_0^{1+2m}. If $E \cong 2_0^{1+2m}$ is such a group of *extraspecial type*, its centre $Z = Z(E) \cong Z_4$. Let $A = C_{\mathrm{Aut}(E)}(Z)$ and $C(E) = A/\mathrm{Inn}(E)$. An argument similar to before shows that $C(E) \cong \mathrm{Sp}_{2m}(2)$ via the commutator map on $U = E/Z$. We have $\mathrm{Aut}(E) = A : \langle a \rangle$ where α is the noninner central automorphism of E, which inverts the elements of Z and centralizes $U \cong \mathrm{Inn}(E)$ and $C(E)$. Both $U \rightarrowtail A \twoheadrightarrow C(E)$ and $U \rightarrowtail \mathrm{Aut}(E) \twoheadrightarrow C(E) \times \langle \alpha \rangle$ do not split for $m \geq 2$ (Appendix A10).

We write $E \cong q_{\pm,0}^{1+2m}$ to indicate that E is one of the above groups. These groups are, up to cyclic central factors, the q-groups of *symplectic type* all of whose characteristic abelian subgroups are cyclic *and* central (P. Hall); cf. [Aschbacher, 1986, (23.9)].

The character theory of E is easy. E has $\phi(|Z|)$ (Euler function) faithful irreducible characters which are determined on $Z = Z(E)$ and vanish outside the centre. They have value field $\mathbb{Q}(e^{2\pi i/|Z|})$, degree q^m and are conjugate under field *and* group automorphisms. Their Schur index is 1 except when $E \cong 2_-^{1+2m}$, in which case it is 2 and then each field in which -1 is a sum of two squares is a splitting field [I, 6.18 and 10.16].

4.3. Holomorphs

Let $E \cong q_{\pm,0}^{1+2m}$, and let θ be a faithful irreducible character of E. A group X is called a *weak holomorph* of E provided E is a normal subgroup of X such that $C_X(E) = Z(E) = Z(X)$ and $X/E \cong C(E)$ is the resulting (full) symplectic resp. orthogonal group on $U = E/Z(E)$. Then X is determined by E up to *isoclinism*, and E is the unique minimal nonabelian normal subgroup of X unless $q = 2, m = 1$. By definition θ is stable in X, and we call X a *holomorph* of E if, in addition, there is $\chi \in \mathrm{Irr}(X)$ extending θ. Then χ is faithful, of degree $\chi(1) = \theta(1) = q^m$. Since the faithful irreducible characters of E are algebraically conjugate, this does not depend on the choice of θ. By stable Clifford theory each weak holomorph is a holomorph if the Schur multiplier of $C(E)$ is trivial.

Since $C(E)/C(E)'$ is of order 1 or 2 unless $E \cong 3_+^{1+2}$ (where it has order 3) or $E \cong 2_+^{1+4}$ (where it is elementary of order 4), the faithful irreducible characters of degree q^m of a holomorph X of E have the same value field. We say that the holomorph X is *standard* if this character field is as small as possible. It will turn out that standard holomorphs exist and that their isomorphism type is determined by this field (which is the qth cyclotomic field for odd q and contained in the 8th otherwise). Our approach is based on work by [Griess, 1973], [Isaacs, 1973] and [Schmid, 2000].

Proposition 4.3a. *Let X be a standard holomorph of $E \cong q_{\pm,0}^{1+2m}$. Let F be a field of prime characteristic $p \neq q$ such that the character field of X fits into F. Let V be an FE-module affording θ as a Brauer character, embed E into $\mathrm{GL}(V)$ through θ and let $G_0 = N_{\mathrm{GL}(V)}(E)$. Then $G_0 = X \circ Z$ is a central product over $Z(E) = Z(X)$ where $Z = C_{G_0}(E) \cong F^\star$.*

Proof. Let $\chi \in \mathrm{Irr}(X)$ extend θ. This χ is irreducible and faithful as a Brauer character mod p as it is θ ($p \neq q$). Since Schur indices are 1 in prime characteristic, there is an FX-module affording χ as a Brauer character, and this may be identified with V. We may embed X into G_0 through χ. Since E is absolutely irreducible on V, $C_{G_0}(E) = Z \cong F^\star$ consists of scalar multiplications. Of course, G_0 induces on E only automorphisms centralizing $Z(E) = Z \cap E$. Hence the result. □

Examples 4.3b. (i) Let $E = D_8 \cong 2_+^{1+2}$. Then the dihedral and semidihedral groups of order 16 are the unique (weak) holomorphs X of E, up to isomorphism. The two faithful irreducible characters of X have Schur index 1 and the value field $\mathbb{Q}(\sqrt{2})$ if X is dihedral and $\mathbb{Q}(\sqrt{-2})$ otherwise.

(ii) Let $E = Q_8 \cong 2_-^{1+2}$. Then $\mathrm{GL}_2(3) = 2^+ S_4$ and (the binary octahedral group) $2^- S_4$ are the unique (weak) holomorphs of E. These groups are the two Schur covers of S_4. (We use the Atlas notation, writing $2^+ S_n$ for that extension in which the transpositions of S_n lift to involutions.) The two faithful irreducible characters of degree 2 of $2^+ S_4$ have Schur index 1 and value field $\mathbb{Q}(\sqrt{-2})$, those of $2^- S_4$ have index 2 and value field $\mathbb{Q}(\sqrt{2})$.

(iii) Let $E = Q_8 \circ Z_4 \cong 2_0^{1+2}$. Embed E into $\mathrm{GL}_2(5)$ via some faithful irreducible character, and let $X = N_{\mathrm{GL}_2(5)}(E)$. We have seen in Theorem 3.2 that X is transitive on the nonzero vectors of $V = \mathbb{F}_5^{(2)}$. It is a 5-complement in $\mathrm{GL}_2(5)$, unique up to conjugacy ($|X| = 96$). The representation of the $5'$-group X on V can be lifted to a $\mathbb{Q}(i)$-representation. Hence X is a standard holomorph of E, and is the unique one up to isomorphism.

(iv) Let $E = 3_+^{1+2}$. The symplectic group $\mathrm{Sp}_4(3)$ has a subgroup $X = E : \mathrm{Sp}_2(3)$ mapping onto a maximal parabolic subgroup of $\mathrm{PSp}_4(3) \cong \Omega_6^-(2)$. Here the centre $Z = Z(X)$ is generated by elements in the classes $3A_0B_0$, and $|X/X'| = 3$ [Atlas, p. 26]. The image of X is contained in a subgroup of type $E : \mathrm{GL}_2(3)$ of $\mathrm{PSp}_4(3).2 \cong O_6^-(2)$ which is nontrivial on Z. So there is an automorphism α of X inverting the elements of Z. Note that $\mathrm{GL}_2(3)$ and $\mathrm{SL}_2(3)$ have trivial Schur multiplier, and that all their cohomology groups on the standard module vanish. Thus X is a weak holomorph for E. Up to conjugacy under E, there is a unique complement $S \cong \mathrm{Sp}_2(3)$ to E in X which is stable under α.

Let ξ be one of the two faithful irreducible characters of $\mathrm{Sp}_4(3)$ of degree 4, which are fused by α [Atlas]. Evidently $\mathrm{Res}_X(\xi) = \chi + \lambda$ for some faithful irreducible character χ of X (of degree 3) and some linear character λ. We have $\mathbb{Q}(\chi) = \mathbb{Q}(\xi) = \mathbb{Q}(\sqrt{-3})$. Hence X is a standard holomorph of E. There are two further holomorphs of E isoclinic to X. However, these further holomorphs appear in $X \circ Z_9$ and their character value fields require the 9th roots of unity. (Cf. [Atlas, p. xxiii], and observe that χ does not vanish on all elements of $X \setminus X'$.) Thus X is the unique standard holomorph, up to isomorphism. Each faithful irreducible character of E has $|X/X'| = 3$ extensions to X, but these are conjugate under central automorphisms corresponding to $\mathrm{Hom}(X/X', Z(X))$.

Theorem 4.3c. *Let $E = q_+^{1+2m}$, q odd. Up to isomorphism, there is a unique standard holomorph X for E and its value field is $K = \mathbb{Q}(e^{2\pi i/q})$. There is an automorphism α of X permuting the nontrivial elements in $Z - Z(E)$ transitively and leaving invariant a unique, up to conjugacy under E, subgroup $S \cong \mathrm{Sp}_{2m}(q)$ complementing E in X. This α induces on S the outer (diagonal) automorphism of order 2. The faithful irreducible characters of X of degree q^m have Schur index 1 and are conjugate under automorphisms of X.*

Proof. By the above we may assume that $(m, q) \neq (3, 1)$. Then $\mathrm{Sp}_{2m}(q)$ is perfect, and its multiplier is trivial by (A6). Let $A = C_{\mathrm{Aut}(E)}(Z)$. Recall from Sec. 4.2 that $A/\mathrm{Inn}(E) = C(E) \cong \mathrm{Sp}_{2m}(q)$. Now $\mathrm{Sp}_{2m}(q)$ contains nontrivial scalar multiplications, which are fixed point free on $U \cong \mathrm{Inn}(E)$ in coprime action. Hence $\mathrm{H}^n(C(E), U) = 0$ for all n by (A2). So there is a faithful action of $\mathrm{Sp}_{2m}(q)$ on E centralizing Z. Let $X = E : \mathrm{Sp}_{2m}(q)$ be the corresponding semidirect product. Here the complement $S \cong \mathrm{Sp}_{2m}(q)$ to E in X is determined up to conjugacy. X is a weak holomorph of E, and it is the unique one, up to isomorphism, as $X = X'$ is perfect.

X is a standard holomorph of E. For each faithful irreducible character θ of E is stable in X, and $\mathrm{M}(S) = 1$ and $X = X'$ imply that there is a *unique* $\chi \in \mathrm{Irr}(X)$ extending θ. This forces that $K(\chi) = K(\theta) = K$. It remains to show that the automorphism α of E described above can be extended to X. Since it will permute then the nontrivial linear characters of Z transitively, and hence the faithful irreducible characters of E, this will prove that all faithful irreducible characters of X of degree q^m are conjugate under automorphisms of X. Also we may alter it by an inner automorphism, if necessary, such that it fixes S.

Recall that $A \cong X/Z = \bar{X}$. Using that $\bar{X}$ and S are perfect and $\mathrm{M}(S) = 1$, and that $\mathrm{H}^1(S, U^*) = 0$ ($U \cong U^*$ as S-modules) from (A2), (A5) we infer that $\mathrm{H}^2(\bar{X}, Z)$ is the dual group of $\mathrm{M}(\bar{X})$ and that

$$\mathrm{Res} : \mathrm{H}^2(\bar{X}, Z) \to \mathrm{H}^2(U, Z)$$

is injective. Since α fixes the conjugacy class of E in $\mathrm{H}^2(U, Z)$, being an automorphism of E, it therefore fixes the cohomology class of X in $\mathrm{H}^2(\bar{X}, Z)$. This implies that α can be extended to X, as desired. We even see that $X = \hat{A}$ is the (universal) Schur cover of A, because the image of the restriction map is centralized by $S \cong \bar{X}/U$, and $\mathrm{H}^2(U, Z) \cong U^* \oplus \Lambda^2(U)^*$ as an S-module by (A8). We conclude that $|\mathrm{M}(A)| = |\mathrm{H}^2(A, Z)| = |Z|$, which gives the result. $\square$

Theorem 4.3d. *Let* $E = 2_0^{1+2m}$. *Up to isomorphism, there is a unique standard holomorph* X *for* E *and its value field is* $K = \mathbb{Q}(i)$. *The faithful irreducible characters of* X *of degree* 2^m *have Schur index 1 and are conjugate under automorphisms of* X.

Proof. Let $Z = Z(E)$ and $U = E/Z$. Recall from Sec. 4.2 that $\mathrm{Aut}(E) = A : \langle \alpha \rangle$ where $A = C_{\mathrm{Aut}(E)}(Z)$ and where α is an involution inverting the elements of $Z \cong Z_4$ and centralizing $U \cong \mathrm{Inn}(E)$ and $C(E) = A/\mathrm{Inn}(E) \cong \mathrm{Sp}_{2m}(2)$. Note that α is not an inner automorphism on A. Let $\theta \in \mathrm{Irr}(E)$ be one of the two faithful irreducible characters of E. Then $\mathbb{Q}(\theta) = K$.

Let first $m \geq 3$. Then $S = \mathrm{Sp}_{2m}(2)$ is a simple group, and the multiplier $\mathrm{M}(S) = 1$ unless $m = 3$ (where it has order 2). In this situation it suffices to show that a weak holomorph for E exists, and a holomorph when $m = 3$. Embed E into $\mathrm{GL}_{2^m}(K)$ through θ (uniquely up to conjugacy). Let B be the block ideal in the group algebra KE to θ. This B is a centrally simple K-algebra isomorphic to $M_{2^m}(K)$ which is stable under the action A. By the Skolem–Noether theorem there is a function $\tau : A \to \mathrm{GL}_{2^m}(K)$ such that conjugation with $\tau_x = \tau(x)$ is application of $x \in A$. We have $\tau_{xy} = \tau_x \tau_y \cdot \tau(x, y)$ for $x, y \in A$, where $\tau(x, y)$ is a scalar matrix. Thus τ is a projective representation of A with 2-cocycle $\tau \in Z^2(A, K^\star)$. We may arrange matters such that $\langle \tau(\mathrm{Inn}(E)) \rangle \cong E$.

Let X_0 be the subgroup of $\mathrm{GL}_{2^m}(K)$ generated by $\tau(A)$. Since the index $|X_0 : Z(X_0)| = |A|$ is finite, by a transfer argument due to Schur [Huppert, 1967, IV.2.3], $X = X_0'$ is finite. Then $Z(X)$ consists of scalar matrices in $K = \mathbb{Q}(i)$ of finite order, that is, $|Z(X)|$ is a divisor of 4. Since $A = A'$ is perfect and contains $\tau(E) \cong E$, X is the desired holomorph.

The case $m = 1$ is already treated above. For $m = 2$ we have $S \cong S_6$. In this case consider $E_0 = E \circ \widetilde{E}$, where $\widetilde{E} \cong E = 2_0^{1+2}$ and where $\theta_0 = \theta \otimes \widetilde{\theta}$ is a faithful irreducible character of E_0. Let X_0 be the standard holomorph of E_0 and $\chi_0 \in \mathrm{Irr}(X_0)$ be the character extending θ_0. Let $X = C_{X_0}(\widetilde{E})$. Then $X \circ \widetilde{E}$ is a subgroup of X_0 on which χ_0 decomposes as $\chi \otimes \widetilde{\theta}$. Then $\chi \in \mathrm{Irr}(X)$ has its values in K. So X is a standard holomorph of E, and its uniqueness follows by noting that the other extensions $4.A_6.2$ isoclinic to X require the 8th roots of unity. Also, the two extensions of θ to X are interchanged by a central automorphism of X.

It remains to show that α can be extended to an automorphism of X. Let first $m \geq 4$. Then $\mathrm{H}^2(S, Z) = 0$ by (A5) since $S = S'$ and $\mathrm{M}(S) = 1$. Also, $U^* = \mathrm{H}^1(U, Z)$ is the dual module for S, and $\mathrm{H}^1(S, U^*) = \mathrm{H}^1(S, U)$

has order 2 by (A9). From (A2) we infer that the kernel of the restriction map Res : $\mathrm{H}^2(A,Z) \rightarrow \mathrm{H}^2(U,Z)$ has order 1 or 2; its image is in the fixed subspace under S. Now the inclusion $\mathbb{F}_2 \rightarrowtail Z$ induces the zero map $\mathrm{Ext}(U,\mathbb{F}_2) \rightarrow \mathrm{Ext}(U,Z)$. It follows that the universal coefficient exact sequence splits naturally and so by (A8) yields the decomposition

$$\mathrm{H}^2(U,Z) \cong U^* \oplus \Lambda^2(U)^*$$

of S-modules. As S is absolutely irreducible on U, $\mathrm{Hom}_S(U \otimes U^*) \cong \mathrm{End}_S(U)$ has order 2. Hence the commutator form to E in $\Lambda^2(U)^* = \mathrm{Hom}(\Lambda^2(U),Z)$ is the unique nontrivial element in $\mathrm{H}^2(U,Z)$ fixed by S. Consequently $|\mathrm{M}(A)| = |\mathrm{H}^2(A,Z)|$ has order dividing $|Z| = 4$. However, $Z \rightarrowtail X \twoheadrightarrow A$ is a proper extension ($Z \subseteq X'$). Hence $X = X' = \widehat{A}$ is the universal Schur cover of A. By (A5) α can be lifted from A to X, and this lift cannot centralize Z as it is not inner on A.

For $m \leq 3$ consider $E_0 = E \circ \widetilde{E}$ where $\widetilde{E} \cong 2_0^{1+2}$. If X_0 is the standard holomorph of E_0 and α_0 a corresponding automorphism of X_0, then $X = C_{X_0}(\widetilde{E})$ is the standard holomorph of E and α_0 leaves $\widetilde{E}$ and X invariant. Argue first for $m = 3$, then for $m = 2$, $m = 1$.   □

Theorem 4.3e. *Both $E \cong 2_+^{1+2m}$ and $E \cong 2_-^{1+2m}$ have two isomorphism types of standard holomorphs, with value fields $\mathbb{Q}(\sqrt{\pm 2})$ and Schur indices 1 or 2, the latter occurring when E is of negative type and the value field is $\mathbb{Q}(\sqrt{2})$. The two standard holomorphs of either type (positive or negative) agree on the inverse images of $\Omega_{2m}^{\pm}(2)$, and the faithful irreducible characters of degree 2^m are rational-valued on these subgroups. These characters are always conjugate under group automorphisms.*

Proof. Let E be one of 2_+^{1+2m} or 2_-^{1+2m}. We may assume that $m \geq 2$. Let θ be the unique faithful irreducible character of E. Let $K = \mathbb{Q}(\sqrt{-2})$. Observe that K is, in each case, a splitting field for E (as -1 is a sum of 2 squares in K). Arguing as in the proof for Theorem 4.3d we get a subgroup X_0 of $\mathrm{GL}_{2^m}(K)$ for which $Z(X_0)$ consists of scalar matrices (including ± 1) and $X_0/Z(X_0) \cong \mathrm{Aut}(E)$, and we obtain that $Y = X_0'$ represents an extension of $C(E)' \cong \mathrm{O}_{2m}^{\pm}(2)'$ by E.

Now $\mathrm{O}_{2m}^{\pm}(2)' \cong \Omega_{2m}^{\pm}(2)$ is simple except when $E \cong 2_+^{1+4}$. At any rate, let $E_0 = E \circ E_1$ with $E_1 \cong D_8$, and let $Y_0 = Y_0'$ be the corresponding perfect subgroup of $\mathrm{GL}_{2^{m+1}}(K)$. So $Y_0/E \cong C(E_0)'$ has index 2 in $C(E_0) \cong \mathrm{O}_{2(m+1)}^{\pm}(2)$. There is $x \in \mathrm{GL}_{2^{m+1}}(K)$ interchanging two involutions in E_1

and centralizing E such that $\langle E_1, x \rangle$ is the semidihedral holomorph for D_8 (which can be embedded into $\mathrm{GL}_2(K)$). So x induces an orthogonal transvection on $U_0 = E_0/Z(E_0)$, and $x^2 \in E_1$. Since x is determined by this action on E_0 up to multiplication with a scalar matrix, $[Y_0, \langle x \rangle] \subseteq Y_0$ and $X_0 = \langle Y_0, x \rangle$ is a holomorph for E_0. It follows that $X = C_{X_0}(E_1)$ is a weak holomorph for E. The faithful irreducible character of X_0 (given by its embedding into $\mathrm{GL}_{2^{m+1}}(K)$) remains absolutely irreducible on $X \circ E_1$. We conclude that X is a holomorph of E.

By construction there is $\chi \in \mathrm{Irr}(X)$ extending θ which can be written in $K = \mathbb{Q}(\sqrt{-2})$. In particular $\mathbb{Q}(\chi) \subseteq K$. Let T be the inverse image in X of $\Omega_{2m}^{\pm}(2)$. We assert that $\mathrm{Res}_T^X(\chi)$ is rational-valued. This is obvious when $X' = T$, because θ is rational-valued and all extensions to X agree on X'. The case where $E = 2_+^{1+4}$ may be treated using that the elements of $\Omega_4^+(2)$ are just those which are products of an *even* number of transvections or reflections [Aschbacher, 1986, (22.14)], and appealing to Theorem 4.4 below. There are just two extensions of θ to X which agree on T, and these are interchanged by a central automorphism of X.

We claim that X is a standard holomorph of E. Write $E = E_0 \circ \widetilde{E}$ with $\widetilde{E} \cong D_8$. Then $\widetilde{X} = C_X(E_0)$ is a (standard) holomorph of D_8. The restriction of χ to $E_0 \circ \widetilde{X}$ is irreducible and its value field is that of $\widetilde{X}$. Thus $\mathbb{Q}(\chi) = K$ if $\widetilde{X}$ is a semidihedral group and $\mathbb{Q}(\chi) \supseteq \mathbb{Q}(\sqrt{2})$ if it is dihedral. But the latter cannot happen as $\mathbb{Q}(\chi) \subseteq K$. However, replacing X by the isoclinic variant (with respect to T) we obtain the value field $\mathbb{Q}(\sqrt{2})$. $\square$

4.4. Good Conjugacy Classes Once Again

We introduce a concept of "goodness" adapted to extraspecial groups. Suppose X is a finite group containing some $E \cong q_{\pm,0}^{1+2m}$ as a normal subgroup. Let $Z = Z(E)$ and $U = E/Z$. Following [Isaacs, 1973] an element $x \in X$ is called "good for U" provided $C_U(x) = C_E(x)/Z$. In other words, whenever $[x, y] \in Z(E)$ for some $y \in E$ then $[x, y] = 1$. This depends only on the conjugacy class of Zx in X/Z, and also only on $\langle Zx \rangle$.

Theorem 4.4. *Let $x \in X$. There are exactly $|U : C_U(x)|$ cosets of Z in Ex which are good for U, and these are conjugate under E. Assume X has a faithful character χ which is absolutely irreducible on E. Then x is good for U if and only if $\chi(x) \neq 0$, and then $|\chi(x)|^2 = |C_U(x)|$. Furthermore, if x is a q'-element, then x is good for U, and $\chi(x)$ is a rational number provided χ is q-rational.*

Proof. Let $C_U(x) = C/Z$ and $D = C_E(C)$. Then $D/Z = (C/Z)^\perp$ is the orthogonal complement of $C_U(x)$ with respect to the (nondegenerate) symplectic commutator form on U. Hence $|E : D| = |C : Z|$ and, therefore, $|D : Z| = |E : C|$. The group of automorphisms of C centralizing $U = C/Z$ and Z is isomorphic to

$$\operatorname{Hom}(C/Z, Z) \cong C/Z \cong E/D,$$

because C/Z is elementary and Z cyclic. Hence there is $y \in E$ inducing the same central automorphism on C as it does x. It follows that xy^{-1} centralizes C. Clearly $C_U(xy^{-1}) = C_U(x)$. Hence xy^{-1} is good for U (and in the coset Ex). Now suppose x itself is good for U (for simplicity). Then all elements in the coset Dx are good for U. Conversely, to $t \in E \setminus D$ there exists $c \in C$ such that $[t, c] = z$ for some $z \neq 1$ in Z. It follows that

$$[tx, c] = [t, c]^x [x, c] = [t, c] = z,$$

whence t is not good for U. Thus Dx is the set of elements in Ex which are good for U, and its cardinality is equal to $|D| = |Z| \cdot |E : C|$.

Now let $\chi \in \operatorname{Irr}(X)$ be as indicated. Suppose $\chi(x) \neq 0$. Without loss of generality assume that $X = \langle E, x \rangle$. Then $[x, y] = x^{-1}x^y \in E$ for all $y \in X$. If $[x, y] \notin Z$ then $\chi([x, y]) = 0$ (as χ is an extension of a faithful irreducible character of E). If $[x, y] = z$ for some $z \in Z$, then

$$\chi(x) = \chi(x^y) = \chi(xz) = \chi(x)\frac{\chi(z)}{\chi(1)}.$$

We conclude that $\chi(z) = \chi(1)$ and so $[x, y] = z = 1$, whence $y \in C_X(x)$. In particular x is good for U. (The argument even shows that x is good for Z or, equivalently, for the unique linear constituent of χ on Z.) Noting that χ is irreducible on X and applying Eq. (1.3c) we get

$$|\chi(x)|^2 = \frac{\chi(1)}{|X|} \sum_{y \in X} \chi([x, y]) = \frac{\chi(1)}{|X|}\chi(1)|C_X(x)|.$$

Now use that $\chi(1)^2 = |U|$ (by the character theory of groups of extraspecial type) and that $|X : C_X(x)| = |E : C_E(x)|$.

Suppose x is good for U. The set Dx consists of $|E : C|$ distinct cosets mod Z, which are permuted by E via conjugation. But the stabilizer in E

of Zx is C. So these cosets are permuted transitively by E. By virtue of Eq. (1.3d) and using that χ vanishes on elements bad for U we therefore have

$$|E| = \sum_{y \in E} |\chi(yx)|^2 = \sum_{y \in D} |\chi(yx)|^2 = |E : C| \sum_{z \in Z} |\chi(zx)|^2.$$

But for $z \in Z$ we have $|\chi(zx)| = |\chi(x)|$. Thus $|\chi(x)|^2 = |C_U(x)|$.

Suppose x is a q'-element. Then $C_E(x)/Z = C_U(x)$ by coprime action. Moreover, we have an orthogonal decomposition $U = C_U(x) \perp [U, x]$ with respect to the symplectic commutator form, and x acts faithfully and symplectically on $[U, x] = [U, x, x]$. Hence $|C_U(x)| = q^{2n}$ for some integer n, and if E_0 is the inverse image in E of $C_U(x)$, then E_0 is of extraspecial type (and $C_E(E_0)$ maps onto $[U, x]$). It follows that $|\chi(x)| = q^n$, because x is good for U. By Theorem 1.6c

$$\chi(x) = \pm \mu \theta(1)$$

for some sign and some root of unity μ of order dividing $o(x)$. Hence if in addition χ is q-rational, then $\chi(x)$ is a rational number.   □

4.5. Some Weil Characters

The classical groups $\mathrm{GL}_m(q)$, $\mathrm{Sp}_{2m}(q)$ (for odd q) and $\mathrm{GU}_m(q)$ admit Weil characters. Usually these are faithful irreducible characters of smallest possible degree. The Weil characters of the symplectic groups were discovered by [Weil, 1964], [Ward, 1972] and [Isaacs, 1973].

Theorem 4.5a (Weil, Ward, Isaacs). *Let $S = \mathrm{Sp}_{2m}(q^f)$ for some power q^f of the odd prime q and some $m \geq 1$, and let U be its standard module. Then S has a pair of (disjoint) "generic" complex characters $\xi \neq \xi^\circ$, conjugate under an outer diagonal automorphism of S, such that $\xi \bar{\xi} = \pi_U \; (= \xi^\circ \overline{\xi^\circ})$ is the permutation character on U. We have $\xi = \xi_1 + \xi_2$ where the ξ_i are irreducible characters of degree $\xi_1(1) = (q^{fm} - 1)/2$ and $\xi_2(1) = (q^{fm} + 1)/2$ (and similar statement for ξ°). The following hold:*

    (i) *The field of character values for ξ, ξ_1 and ξ_2 is $\mathbb{Q}\left(\sqrt{(-1)^{(q-1)/2}q}\right)$ if f is odd, and $\mathbb{Q}$ otherwise. The characters take only rational values on q'-elements of S.*

    (ii) *Just one of ξ_1, ξ_2 is faithful for S, and ξ_1 is faithful if and only if $q^{fm} \equiv 1 \pmod{4}$. Moreover, ξ_1 is irreducible as a Brauer character in any characteristic different from q, ξ_2 in every characteristic different from q and from 2, and $\xi_2 = 1_S + \xi_1$ on $2'$-elements of S.*

Proof. The trace map $\mathbb{F}_{q^f} \to \mathbb{F}_q$ carries the symplectic $\mathbb{F}_{q^f}$-form on U to one over $\mathbb{F}_q$ which remains nondegenerate and S-invariant. This yields an embedding of $S = \mathrm{Sp}_{2m}(q^f)$ into $\mathrm{Sp}_{2fm}(q)$, hence an embedding of S into the standard holomorph of $E \cong q_+^{1+2fm}$ (Theorem 4.3c). Let $Z = Z(E)$ and identify $U = E/Z$ (as S-modules). The above discussion carries over to the *symplectic holomorph* $X_s = E : S$ (contained in the standard holomorph of E). Recall the definition of the automorphism α of E in Sec. 4.2, via a decomposition $U = W \oplus W^*$ into totally isotropic subspaces. As before we may extend this α to X and pick S to be α-invariant. Then α induces the outer (diagonal) automorphism of order 2 on S and permutes the $q - 1$ nontrivial linear characters of Z transitively. The restrictions to S of (the) $q - 1$ faithful irreducible characters of X of degree q^{fm} lying above these linear characters give rise to two distinct *generic Weil characters* ξ, ξ° of S conjugate under α. For $(m, q^f) = (1, 3)$ we have additional central automorphisms, and here we we pick ξ, ξ° such that they do not contain the 1-character of S (cf. Example (iv) in 4.3b).

Let j be the central involution in S. Let $x \in S$. We claim that x is good for U. Let $t \in E$ such that $Zt \in C_U(x)$. Since j inverts the elements in U, we have $t^j = t^{-1}z$ for some $z \in Z$ and

$$[x, t] = [x, t]^j = [x, t^j] = [x, t^{-1}z] = [x, t]^{-1}.$$

Using that q is odd this implies that $[x, t] = 1$. Hence the claim. It follows that $(\xi\bar{\xi})(x) = |\xi(x)|^2 = |C_U(x)|$ by Theorem 4.4. Thus $\xi\bar{\xi} = \pi_U$ is the permutation character of S on U. In particular $|\xi(j)|^2 = |C_U(j)| = 1$ and so $\xi(j) = \pm 1$ as it is a rational number. Since S is transitive on $U^\sharp$ (by Witt's theorem), it follows that

$$\langle \xi, \xi \rangle = \langle \xi\bar{\xi}, 1_S \rangle = 2.$$

Consequently $\xi = \xi_1 + \xi_2$ for two distinct irreducible characters ξ_i of S. So $\xi_i(j) = \pm\xi_i(1)$ for $i = 1, 2$. It follows that just one of the characters ξ_i has j in its kernel, and that $\xi_i(1) = (q^{fm} \pm 1)/2$. Choose notation such that $\xi_2(1) = \xi_1(1) + 1$.

Let $x \in S$ be a symplectic transvection. Then $|\xi(x)|^2 = |C_U(x)| = q^{f(2m-1)}$. Since S is generated by the symplectic transvections, this gives the result on the character field of ξ, up to the sign. For the precise sign we refer to [Isaacs, 1973] (Theorem 5.7), from which one also infers that $\xi(j) = 1$ if and only if $q^{fm} \equiv 1 \,(\mathrm{mod}\,4)$ (see also [I, 13.32]). Use further that $1_S + \xi_1$ and ξ_2 agree on q-elements (as proved in the next paragraph), and that ξ_1 is faithful if and only if $\xi_1(j) = -\xi_1(1)$, that is, if and only if $\xi(j) = +1$. (We mention that $\xi^\circ = \bar{\xi}$ if and only if f is odd.)

Let us pass to characteristic p for some prime $p \neq q$. Let F be the residue class field of $\mathbb{Z}_{(p)}[e^{2\pi i/q}]$ modulo some maximal ideal, and let V be an FS-module affording ξ as a Brauer character. Then, as before, $V \otimes_F V^*$ is the permutation module of S over F on the set U (with two orbits). Thus

$$\mathrm{End}_{FS}(V) \cong \mathrm{Hom}_{FS}(V \otimes_F V^*, F)$$

has F-dimension 2. If $p \neq 2$ then $V = [V, j] \oplus C_V(j)$ is a proper decomposition into FS-modules, because j is in the kernel of just one ξ_i. So let $p = 2$. Then j acts trivially on each irreducible FS-module, but it is nontrivial on V as it is nontrivial on U. Hence $C_V(j)$ is a proper submodule of V and V is not completely reducible. As before $V/C_V(j) \cong [V, j]$ as FS-modules (via the commutator map $v \mapsto [v, j]$). We cannot have $C_V(j) = [V, j]$ as q is odd. We cannot have $C_V(j) \subset [V, j]$ (properly), because then $C_V(j)$ were irreducible and $[V, j] = C_V(j) \oplus 1_S$, which is impossible. We conclude that $[V, j]$ is irreducible and of codimension 1 in $C_V(j)$, the quotient being the trivial module (affording 1_S). The proof is complete. $\qquad\square$

The characters ξ, ξ_1, ξ_2 of $\mathrm{Sp}_{2m}(q^f)$ obtained in Theorem 4.5a, and their conjugates, are called the *Weil characters* of the symplectic group. For $m \geq 2$ the irreducible Weil characters are the unique faithful irreducible characters of $\mathrm{Sp}_{2m}(q^f)$ of degree less than $q^{fm} - q$ (see Appendix C).

Theorem 4.5b. *Let X be a standard holomorph of $E \cong 2_+^{1+2m}$, $m \geq 3$, and let χ be one (of the two) faithful irreducible characters of X of degree 2^m. Write E as a central product of m dihedral groups $\langle a_i, a_i^* \rangle$ of order 8, with involutions a_i, a_i^*. Let $A = \langle a_1, \cdots, a_m \rangle$ and $A^* = \langle a_1^*, \cdots, a_m^* \rangle$.*

(i) There exists a unique subgroup $L \cong L_m(2)$ of X normalizing both A and A^*, so that A is the standard module for L and A^* its dual. We have $N_X(L) = ZL\langle\tau\rangle$ where $\tau^2 \in Z = Z(E)$ and where τ interchanges A and A^* and induces the inverse transpose automorphism on L.

(ii) $\mathrm{Res}_L^X(\chi) = \pi_A = 2 \cdot 1_L + \xi$ where ξ is an (absolutely) irreducible (Weil) character of L which is rational-valued (with Schur index 1). This ξ is irreducible modulo every odd prime p not dividing $2^m - 1$, and reducible otherwise.

Proof. We use the Atlas notation for the projective special linear groups. So at present $L_m(2) = \mathrm{PSL}_m(2)$. Let $Z = Z(E)$ and $U = E/Z$.

(i) Recall that $\bar{X} = X/E \cong \mathrm{O}_{2m}^+(2)$. Let $W = AZ/Z$ and $W^* = A^*Z/Z$. Then $U = W \oplus W^*$ is a decomposition into maximal totally singular subspaces. Let $P = N_{\bar{X}}(W)$, a maximal parabolic subgroup of $\Omega_{2m}^+(2)$ (see Appendix B). Let $\bar{L} = N_P(W, W^*)$ be a Levi complement, so that $\bar{L} \cong L_m(2)$ and W is the standard $\bar{L}$-module, W^* its dual.

See Appendix (C1) for the existence (and uniqeness) of the subgroup L of X. This L is faithful on A and A^*, and maps isomorphically onto $\bar{L}$. For each i let $e_i = a_i Z$ and $e_i^* = a_i^* Z$, and let $\bar{\tau}_i$ be the orthogonal transvection of U with centre $e_i \cdot e_i^*$. Then $\bar{\tau} = \bar{\tau}_1 \cdots \bar{\tau}_m$ is an involution in $\bar{X}$ interchanging each e_i, e_i^*, hence W and W^*. Also, $\bar{\tau}$ normalizes $\bar{L}$ and induces on it the inverse transpose automorphism.

The $\bar{L}$-modules W and W^* are not isomorphic (Lemma 4.1c). Hence $N_{\bar{X}}(\bar{L})$ permutes $\{W, W^*\}$. Since $\bar{L} = N_{\bar{X}}(W, W^*)$, we see that $N_{\bar{X}}(\bar{L}) = \bar{L}\langle\bar{\tau}\rangle = \mathrm{Aut}(\bar{L})$. We observe that $U = \mathrm{Ind}_{\bar{L}}^{\mathrm{Aut}(\bar{L})}(W)$ as an $\mathrm{Aut}(\bar{L})$-module and so

$$\mathrm{H}^n(\mathrm{Aut}(\bar{L}), U) \cong \mathrm{H}^n(\bar{L}, W) \cong \mathrm{H}^n(\bar{L}, W^*)$$

by Shapiro's lemma (A3). This vanishes for $n = 1, 2$ when $m \geq 6$ by (A9). Then we find $\tau \in X$ mapping onto $\bar{\tau}$ with $\tau^2 \in Z$, and replacing τ by a suitable E-conjugate, if necessary, this τ normalizes L. Note that $N_E(L) = Z$ as L has no fixed points on $U = E/Z$. It follows that $N_X(L) = L\langle\tau\rangle N_E(L) = L\langle\tau\rangle Z$ maps onto $\mathrm{Aut}(\bar{L})$. For $3 \leq m \leq 5$ we argue as follows. Consider $E_0 = E \circ E_1$, with $E_1 \cong 2_+^{1+2(6-m)}$. Let X_0 be the standard holomorph of E_0 having the same value field as X. Then $X = C_{X_0}(E_1)$ (by uniqueness). Let $L_0 \cong L_6(2)$ be the subgroup of X_0 normalizing $A_0 = \langle a_1, \cdots, a_6\rangle$ and $A_0^* = \langle a_1^*, \cdots, a_6^*\rangle$. Then $L = C_{L_0}(E_1)$, because this centralizer is a subgroup of X isomorphic to $L_m(2)$ which normalizes A and A^*. For each i let $\tau_i \in \bigcap_{j \neq i} C_{X_0}(\langle a_j, a_j^*\rangle)$ be an element of the holomorph of $\langle a_i, a_i^*\rangle$ within X_0 mapping onto $\bar{\tau}_i$. Let $\tau_0 = \tau_1 \cdots \tau_6$. Then $L_0 Z$ and $L_0^{\tau_0} Z$ are conjugate in $EL_0 = (EL_0)^{\tau_0}$ (as $\mathrm{H}^1(L_0, U_0) = 0$ for $U_0 = E_0/Z(E_0)$). Hence there are $y_i \in \langle a_i, a_i^*\rangle$ such that for $y_0 = y_1 \cdots y_6$ we have $(L_0 Z)^{\tau_0 y_0} = L_0 Z$, hence $L_0^{\tau_0 y_0} = (L_0^{\tau_0 y_0} Z)' = L_0$. Also $\tau_0 y_0$ normalizes E_1 and hence $L = C_{L_0}(E_1)$. Now $\tau = \tau_1 \cdots \tau_m y_1 \cdots y_m$ maps onto $\bar{\tau}$ and centralizes E_1, hence is in $X = C_{X_0}(E_1)$. Using that τ_i permutes with y_j for $j \neq i$ and that L is centralized by τ_i, y_i when $i > m$ we see that τ normalizes L.

(ii) It follows from Appendix (C1) that $\mathrm{Res}_L^X(\chi) = \pi_W = 2 \cdot 1_L + \xi$ for some irreducible character ξ of L. Since $L = L' \subseteq X'$, the Weil character ξ of L is rational-valued and independent of the choice of χ.

Let p be an odd prime. We investigate the (transitive) permutation character $\pi_{W^\sharp} = 1_L + \xi$, and its reduction mod p. Let M be the permutation $\mathbb{F}_p L$-module affording $\pi_{W^\sharp}$ (as a Brauer character). If $p \nmid 2^m - 1$ then $M = M_0 \oplus 1_L$ where M_0 is (absolutely) irreducible (affording ξ as a Brauer character). So let $p \mid 2^m - 1$. Then 1_L appears at least twice in M, namely in the socle and in the head. (One can show that the "heart" $\widetilde{M}$ of M, the section remaining, is irreducible (and not trivial): Since $L = L'$ is perfect, $\widetilde{M}$ cannot have 1_L in the socle or in the head. By [Seitz–Zalesskii, 1993] each composition factor of $\widetilde{M}$ either is trivial or has dimension $\geq 2^m - m - 1$. We have $2 \cdot (2^m - m - 1) > 2^m - 3 = \dim \widetilde{M}$. Hence $\widetilde{M}$ is irreducible.) $\quad\square$

Remark. Replace $E \cong 2^{1+2m}_+$ by $E_0 = E \circ Z_4 \cong 2^{1+2m}_0$ ($m \geq 3$). There is a corresponding result for the standard holomorph X_0 of E_0. More precisely, let $Y = N_{X_0}(E)$. Then $Y/E_0 \cong O^+_{2m}(2)$ is a maximal subgroup of $\bar{X}_0 = X_0/E_0 \cong \mathrm{Sp}_{2m}(2)$. As before we find $L \cong \mathrm{GL}_m(2)$ in Y normalizing A and A^*. For $\bar{L} = LE_0/E_0$ we have $N_{\bar{X}_0}(\bar{L}) = \bar{L}\langle\bar\tau\rangle$ as before, and we find $\tau \in X_0$ mapping onto $\bar\tau$, normalizing L and interchanging A, A^* such that $\tau^2 \in Z(E_0)$ and $N_{X_0}(L) = L\langle\tau\rangle Z(E_0)$.

4.6. Symplectic and Orthogonal Modules

Now we state and prove two results, due to [Isaacs, 1973] and [Gow, 1993], which will be crucial for our approach to the $k(GV)$ theorem.

Theorem 4.6a (Isaacs). *Let V is a coprime symplectic $\mathbb{F}_p G$-module. There exists a rational-valued character χ of G with $\chi^2 = \pi_V$, and this yields a rational-valued generalized character ψ of G with $\psi(1) = 1$ and $\psi^2 = \delta_V$.*

Proof. We may assume that G is faithful on V. Let $|V| = p^{2m}$. In the odd case let χ be the restriction to G of a generic Weil character of $\mathrm{Sp}_{2m}(p)$ (Theorem 4.5a). Otherwise identify $V = E/Z(E)$ via the commutator form on $E = 2^{1+2m}_0$, and let T be the standard holomorph of E (Theorem 4.3d). By the Schur–Zassenhaus theorem we may embed G into T. Let then χ be the restriction to G of one of the (two) faithful irreducible characters of T of degree 2^m. Apply Theorem 4.4.

Let $X = GV$ be the semidirect product, and view χ as a character of X by inflation. By Lemma 2.1a, $\frac{1}{p^m}\widetilde{\chi}$ is a generalized character of X, because $\chi(1) = p^m$ and $|X|_p = |V| = p^{2m}$. Define $\psi(x) = \frac{1}{p^m}\widetilde{\chi}(x)/|C_V(x)| = p^m\chi(x)/|C_V(x)|$ for $x \in G$. By Lemma 3.3a, ψ is a generalized character of G. We have $\psi(1) = 1$ and $\psi^2 = \delta_V$, and ψ is rational-valued. $\quad\square$

Theorem 4.6b (Gow). *Suppose V is a coprime orthogonal $\mathbb{F}_p G$-module for some odd prime p. Let $N = \mathrm{Ker}(\mu)$ be the kernel of the linear character μ of G afforded by the determinant in its action on V ($|G/N| = 1$ or 2). There is a rational-valued generalized character ψ of G such that $\psi(1) = 1$ and $\psi^2 = \delta_V$ on N, $\psi(x)^2 = \frac{1}{p}\delta_V(x)$ for $x \in G \smallsetminus N$ (if any).*

By virtue of Lemmas 2.1a and 3.3a it suffices to construct a rational-valued generalized character χ of G with the following properties: If $\dim V = 2m$ is even, then $\chi(1) = p^m$, $\chi^2 = \pi_V$ on N and $\chi(x)^2 = \frac{1}{p}|C_V(x)|$ for $x \in G \smallsetminus N$. If $\dim V = 2m + 1$ is odd, then $\chi(1) = p^{m+1}$, $\chi(x)^2 = p|C_V(x)|$ for $x \in N$ and $\chi(x)^2 = |C_V(x)|$ otherwise.

We need two lemmas. Let $\tau = [\cdot, \cdot]$ be the G-invariant, nondegenerate symmetric bilinear form on V. If $\dim V = 2m$ is even, we write $\varepsilon(V) = 1$ if τ has Witt index m and $\varepsilon(V) = -1$ otherwise. Similar notation for nondegenerate subspaces of even dimension ($\varepsilon(0) = 1$ for the zero subspace). We set $d(V) = 1$ if the discriminant of τ is a square in $\mathbb{F}_p^\star$ and $d(V) = -1$ otherwise. Thus $\varepsilon(V)d(V) = (-1)^{(p^m-1)/2}$ if $\dim V = 2m$. In the odd dimensional case $\dim V = 2m+1$, we define $\varepsilon(V) = (-1)^{(p^m-1)/2}d(V)$, and extend this definition to nondegenerate subspaces of odd dimension.

Lemma 4.6c. *Let $I(V)$ be the set of vector $v \in V$ for which $\tau(v, v) = 0$, and let $J(V)$ consist of those with $\tau(v, v) = 1$.*
   (i) If $\dim V = 2m$, then $|I(V)| = p^{2m-1} + \varepsilon(V)(p^m - p^{m-1})$ and $|J(V)| = p^{2m-1} - \varepsilon(V)p^{m-1}$.
   (ii) If $\dim V = 2m+1$, then $|I(V)| = p^{2m}$ and $|J(V)| = p^{2m} + \varepsilon(V)p^m$.

Proof. Recall that $V = U_1 \perp \cdots \perp U_{m-1} \perp V_0$ with hyperbolic planes $U_i = \langle u_i, v_i \rangle$, satisfying $\tau(u_i, u_i) = 0 = \tau(v_i, v_i)$ and $\tau(u_i, v_i) = 1$, where either $V_0 = U_m$ is a hyperbolic plane ($\dim V = 2m$ and $\varepsilon(V) = 1$), or $V_0 = \langle u_m, v_m \rangle$ with $\tau(u_m, v_m) = 0$, $\tau(u_m, u_m) = 1$ and $\tau(v_m, v_m) = -c$ for some nonsquare c in $\mathbb{F}_p^\star$ ($\dim V = 2m$ and $\varepsilon(V) = -1$), or $V_0 = \langle v_m \rangle$ where either $\tau(v_m, v_m) = 1$ or $\tau(v_m, v_m) = c$ is a nonsquare ($\dim V = 2m+1$). It is easy to verify the assertions when $\dim V \leq 2$. Observe that if U is a hyperbolic plane, to every isotropic $u \neq 0$ in U there exists a unique isotropic $v \neq 0$ in U such that $\{u, v\}$ is a hyperbolic pair, and then the set of vectors au, av for $a \in \mathbb{F}_p$ are just all isotropic ones in U (including 0). Hence

$$|I(U)| = 2p - 1 \text{ and } |J(U)| = p - 1.$$

So let $\dim V \geq 3$, and write $V = U \perp W$ where U is a hyperbolic plane. One checks the following recursion formulas ($\dim W = k = \dim V - 2$):

$$|I(V)| = p^{k+1} - p^k + p|I(W)|,$$

$$|J(V)| = (2p - 1)|J(W)| + (p - 1)(p^k - |J(W)|).$$

The lemma follows.　　　　　　　　　　　　　　　　　　　　　　□

Clearly $I = I(V)$ and $J = J(V)$ are stable under the action of G. We consider the permutation characters π_I and π_J of G on these sets, and define

$$\chi = \varepsilon(V)(\pi_I - \pi_J)$$

if $\dim V = 2m$ is even. Then $\chi(1) = p^m$. Let $\dim V = 2m + 1$. Then we let $\chi' = -\varepsilon(V)(\pi_I - \pi_J)$ and define χ by $\chi = \frac{p+1}{2}\chi' + \frac{p-1}{2}\chi' \cdot \mu$, that is, $\chi = p\chi'$ on N and $\chi(x) = \chi'(x)$ for $x \in G \smallsetminus N$. Then $\chi(1) = p^{m+1}$.

The computation of this rational-valued generalized character χ of G will prove that it has the asserted properties, hence will complete the proof of Theorem 4.6b.

Let $x \in G$. Since G is a p'-group, we have an orthogonal decomposition $V = [V, x] \perp C_V(x)$ into nondegenerate subspaces. If $\det(x) = 1$ then $[V, x]$ has even dimension, whence $\dim V$ and $\dim C_V(x)$ the same parity, whereas $\dim [V, x]$ is odd if $\det(x) = -1$. Let $\dim C_V(x) = r_x$ and $\varepsilon_x = \varepsilon([V, x])$.

Lemma 4.6d. (i) *Let* $\dim V = 2m$. *Then* $\chi(x) = \pm \varepsilon_x p^{\lfloor \frac{r_x}{2} \rfloor}$ *where the sign* $+$ *holds if* $\det(x) = 1$ *or* $\det(x) = -1$ *and* $p \equiv 3 \,(\mathrm{mod}\, 4)$.

(ii) *Let* $\dim V = 2m + 1$. *Then* $\chi(x) = \pm \varepsilon_x p^{\lfloor \frac{r_x + 1}{2} \rfloor}$ *where the* $+$ *sign holds if* $\det(x) = 1$.

Proof. Suppose first that $\det(x) = 1$. If $\dim V = 2m$, then r_x is even and by Lemma 4.6c, the very definition of χ, and elementary properties of the Witt index, we get $\chi(x) = \varepsilon(V)\varepsilon(C_V(x))p^{r_x/2} = \varepsilon_x p^{r_x/2}$. If $\dim V = 2m + 1$, then r_x is odd, and we obtain $\chi(x) = \varepsilon(V)\varepsilon(C_V(x))p^{(r_x+1)/2} = \varepsilon_x p^{(r_x+1)/2}$.

Suppose next that $\det(x) = -1$. Then r_x is even precisely when $\dim V$ is odd (and then $\lfloor \frac{r_x + 1}{2} \rfloor = \frac{r_x}{2}$). We obtain that

$$\chi(x) = -\varepsilon(V)\varepsilon(C_V(x))p^{\lfloor \frac{r_x}{2} \rfloor}.$$

It is elementary to show that $\varepsilon(V)\varepsilon(C_V(x)) = \varepsilon_x$ if $\dim V = 2m + 1$ and $\varepsilon(V)\varepsilon(C_V(x)) = (-1)^{(p-1)/2}\varepsilon_x$ if $\dim V = 2m$. We are done.　□

Chapter 5

Real Vectors

The Robinson–Thompson theorem shows that $k(GV) \leq |V|$ provided there is a *real vector* in V for G. This is fundamental for our approach to the $k(GV)$ problem. Here a vector $v \in V$ is called real for G if the restriction to $C_G(V)$ of V contains a faithful self-dual submodule (with a real-valued Brauer character). The search for real vectors is compatible with Clifford reduction and leads to the *nonreal reduced pairs* (which will be classified in the next two chapters).

5.1. Regular, Abelian and Real Vectors

Throughout V is a coprime FG-module, not necessarily faithful, where $F = \mathbb{F}_r$ is a finite field of characteristic p. Let $v \in V$ and let $H = C_G(v)$. The vector v is called regular for G provided $H = C_G(V)$, and it is called abelian (cyclic) if $H/C_G(V)$ is abelian (cyclic). If G is faithful on V, then $k(GV) \leq |V|$ by Theorem 3.4d if there is an abelian vector in V for G. We say that v is *strongly real* for G provided the restriction to H of V is self-dual, and it is called *real* if $\mathrm{Res}_H^G(V)$ contains a self-dual submodule W such that $C_H(W) = C_G(V)$.

Of course regular vectors are abelian and are strongly real, whereas there is no hierarchy between abelian and (strongly) real vectors. There are examples (G, V) where there are neither abelian vectors nor real ones, or vectors of just one kind (see for instance Secs. 6.1, 7.1 below).

Example 5.1a. We are going to describe the *permutation pairs* (G, V). Let $p > d + 1$ and let $E = A_{d+1}$ be the alternating group of degree $d + 1 \geq 5$. Then the deleted (shortened) permutation FE-module V of degree d is coprime, faithful and absolutely irreducible. It is the "heart" of the permutation module $W = \bigoplus_{i=0}^{d} Fw_i$, where $\{w_i\}$ is a permutation basis ($w_i s = w_{is}$ for $s \in E$). It consists of all $\sum_i c_i w_i$ with $\sum_i c_i = 0$ in F (as $p \nmid d + 1$). Embed E into $\mathrm{GL}(V)$, and let $G = N_{\mathrm{GL}(V)}(E)$. Then $G = S \times Z$ where $Z = Z(G) \cong F^\star$, and where we may choose $S \cong S_{d+1}$ such that it acts on V as its natural shortened permutation module. (For $d + 1 = 6$ note that $\mathrm{PGL}_2(9)$ and M_{10} do not have faithful actions on 5-dimensional coprime modules.) There exist vectors of (almost) every kind here.

(i) *There is always a strongly real vector $v \in V$ for G:*

Let $v = dw_0 - \sum_{i=1}^{d} w_i$. If $g = xz$ fixes v ($x \in S$, $z \in Z$), then $zw_{ix} = w_j$ for some positive i, j, hence $z = 1$, and $g = x$ must fix w_0 and permute $\{w_0 - w_i\}_{i=1}^{d}$. Therefore $H = C_G(v) \cong S_d$ and $\mathrm{Res}_H^G(V)$ is the natural permutation over F, which gives the result.

(ii) *There is always an abelian (cyclic) vector in V for G:*

A vector $u = \sum_{i=0}^{d} c_i w_i$ in W is regular for S if and only if all $c_i \in F$ are distinct. So there are $\prod_{i=0}^{d}(r - i)$ regular vectors in W for S. If $\sum_i c_i = c$ ($\neq 0$), there is a (first) index j such that $c_j - c \neq c_i$ for all $i \neq j$, because $p > d + 1$. Hence there are $\prod_{i=1}^{d}(r - i)$ vectors in V regular for S. If u is regular for S, then $C_G(u)$ is cyclic of order dividing $|Z| = r - 1$.

(iii) *There is a regular vector in V for G if and only if $r \geq d + 4$:*

We have seen that there are $\frac{1}{(d+1)!} \prod_{i=1}^{d}(r - i) = \frac{1}{d+1} \binom{r-1}{d}$ regular S-orbits on $V^\sharp$. So we have one such orbit when $r = d + 2$, and $\frac{r-1}{2}$ orbits for $r = d + 3$. In these two cases (where $r = p$) there cannot be a regular G-orbit on $V^\sharp$, because each regular G-orbit on $V^\sharp$ is a disjoint union of $|Z| = r - 1$ regular S-orbits. Let $r \geq d + 4$ in what follows.

Suppose $g = xz$ ($x \in S$, $z \in Z$) is an element in G of prime order s which fixes a vector $u = \sum_i c_i w_i$ in V belonging to a regular S-orbit. Then $c_{ix} = c_i z$ for all i, hence $jx = j$ if some (unique) $c_j = 0$, and the other cycles of x have the same size $s = o(x) = o(z)$, with $c_{ix} = c_i z, \cdots, c_{ix^s} = c_i z^s = c_i$. Hence s is a divisor of $r - 1$ and of either d or $d + 1$, and z is determined by x. Let $\delta = \delta_s \in \{0, 1\}$ be (unique) such that $s \mid d + \delta$. We see that g fixes at most $\prod_{i=0}^{\frac{d+\delta}{s} - 1}(r - 1 - is)$ vectors in V regular for S. There are just $\frac{(d+1)!}{s^{\frac{d+\delta}{s}}(\frac{d+\delta}{s})!}$ elements in $S = S_{d+1}$ of the cycle shape of x [James–Kerber, 1981, 1.2.15]. Using that each group of order s has $s - 1$ generators we get that there are at most $\frac{(d+1)!}{s-1}\binom{(r-1)/s}{(d+\delta)/s}$ vectors in V belonging to regular S-orbits which are fixed by subgroups of order s in G. Summing up over the primes s dividing both $d(d + 1)$, i.e., $d + \delta_s$ for some $\delta_s \in \{0, 1\}$, and $r - 1$ one obtains that

$$(d + 1)! \sum_s \frac{1}{s - 1} \binom{(r - 1)/s}{(d + \delta_s)/s} < \prod_{i=1}^{d}(r - i).$$

This implies that there is a regular vector in V for G, because there is a vector regular for S which is not fixed by any nontrivial element of G.

Confirm the inequality first for $r = d + 4$, where $r = p$ is odd, d is even and only $s = 2,3$ can appear (the latter when $3 \mid r - 1$); use that $r \geq 7$ (even $r \geq 11$). For any s we have $\frac{r-1}{2} - \frac{r-1}{s} \geq \frac{d+\delta_2}{2} - \frac{d+\delta_s}{s}$. Therefore $\binom{(r-1)/s}{(d+\delta_s)/s} \leq \binom{(r-1)/2}{(d+\delta_2)/2}$, blowing up the $(d+\delta_s)/s$-subsets of an $(r-1)/s$-set to $(d+\delta_2)/2$-sets in an underlying $(r-1)/2$-set. Hence it suffices to show that $\binom{(r-1)/2}{(d+\delta)/2} \sum_s \frac{1}{s-1} < \frac{1}{(d+1)!} \prod_{i=1}^d (r-i) = \frac{1}{d+1}\binom{r-1}{d}$. Blowing up the $(d+\delta)/2$-subsets of an $(r-1)/2$-set to d-sets in an underlying $(r-1)$-set we get that $\binom{r-1}{d} > \alpha\binom{(r-1)/2}{(d+\delta)/2}$ where

$$\alpha = \binom{r - 1 - (r-1)/2}{d - (d+\delta)/2} \geq \binom{(r-1)/2}{2} = (r-1)(r-3)/8.$$

Here $(r-1)/2 - [d - (d+\delta)/2] \geq 2$ as we assume that $d+1 < r - 3$. It is thus enough to check that

$$\sum_s \frac{1}{s-1} \leq (r-1)/8,$$

with s ranging over the primes dividing $r - 1$ (say). This is immediate for $r = 11, 13$. If s is an odd prime divisor of $r - 1$, then $2s \mid r - 1$ and $\frac{1}{s-1} \leq \frac{1}{s} + \frac{1}{2s}$. Thus $\sum_s \frac{1}{s-1} \leq \sum_{t|r-1} \frac{1}{t} = \frac{1}{r-1}\sum_{t|r-1} t$, which is known to be $O\big((r-1)^\varepsilon\big)$ for each $\varepsilon > 0$. For $r \geq 17$ we have $\sum_{t|r-1} t \leq (r-1)^2/8$.

Lemma 5.1b. *Let G be faithful on V, and assume that $G = X \circ Z$ is a central product over $Z(X)$ where Z acts on V as a group of scalar multiplications. Let $H = C_X(v)$ for some $v \in V^\sharp$. If $|N_X(H) : H|$ is relatively prime to $|Z : Z(X)|$, or if $N_X(Zv) = H \times Z(X)$, then $C_G(v) = H$. Hence if v is (strongly) real (or abelian, regular) for X, then it is so for G.*

Proof. We write $Zv = vZ$ regarding the elements of Z (acting) as scalars (modules being right modules). Clearly $N = N_X(Zv)$ is contained in $N_X(H)$, and $C_G(v) \subseteq N_G(Zv) = NZ$. To any $y \in N$ there exists a unique $z = z_y$ in Z such that $yz \in C_G(v)$. Conversely, each $x \in C_G(v)$ may be written as $x = yz$ with $y \in X$ and $z \in Z$, and then $y \in N$ and $z = z_y$. The assignment $y \mapsto z_y$ is a homomorphism from N to Z with kernel $H = C_X(v)$. Under either assumption the image of this homomorphism is in $Z(X)$, whence $C_G(v) \subseteq NZ(X) \subseteq X$. $\qquad\square$

Remark. We have $C_G(v) \cap Z = 1$ and $C_G(v)Z/Z = C_{G/Z}(Zv)$ for each nonzero vector $v \in V$. In most situations $Z \cong F^\star$ will be the group of all scalar multiplications. Then $Zv = (Fv)^\sharp$ and $C_{G/Z}(Fv) = C_{G/Z}(Zv)$ is nothing but the stabilizer of the point Fv in the projective 1-space $P_1(V)$.

5.2. The Robinson–Thompson Theorem

We begin by combining Theorems 4.1b, 4.6a and 4.6b.

Lemma 5.2a. *Suppose V is a coprime $\mathbb{F}_pG$-module and $v \in V$ is real for G. Let $H = C_G(v)$ and let U be a self-dual $\mathbb{F}_pH$-submodule of V with $C_H(U) = C_G(V)$. Then there is a rational-valued generalized character ψ of H and a subgroup N of H with $|H : N| \leq 2$ such that $\psi(1) = 1$, $\psi^2 = \delta_U$ on N and $\psi(h)^2 = \frac{1}{p}\delta_U(h)$ whenever $h \in H \smallsetminus N$. We have $N \neq H$ if and only if p is odd and some element of H acts with determinant -1 on U.*

Proof. According to Theorem 4.1b we may write $U = U_0 \oplus W_0$ where U_0 is a symplectic $\mathbb{F}_pH$-module and W_0 is an orthogonal module, the action of H on W_0 being trivial when $p = 2$. By Theorem 4.6a there is a rational-valued generalized character ψ_{U_0} of H satisfying $\psi_{U_0}(1) = 1$ and $\psi_{U_0}^2 = \delta_{U_0}$. If H acts trivially on W_0, then we let $N = H$ and $\psi_{W_0} = 1_H$. Otherwise $p \neq 2$, and we let N be the kernel of the linear character of H afforded by the determinantal action of H on W_0 (and on U). By Theorem 4.6b there is a rational-valued generalized character ψ_{W_0} of H satisfying $\psi_{W_0}(1) = 1$, $\psi_{W_0}^2 = \delta_{W_0}$ on N, and $\psi_{W_0}(h)^2 = \frac{1}{p}\delta_{W_0}(h)$ for $h \in H \smallsetminus N$. Now define $\psi = \psi_{U_0}\psi_{W_0}$. $\square$

Theorem 5.2b (Robinson–Thompson). *Suppose V is a faithful coprime $\mathbb{F}_pG$-module and some $v \in V$ is real for G. Then $k(GV) \leq |V|$ and equality can only hold when v is strongly real for G. Also, $k(GV) < |V|$ if $p \geq 5$ and some element of H acts with determinant -1 on V.*

Proof. Let $H = C_G(v)$, and let U, ψ be as in the preceding lemma. Since V is faithful by hypothesis, U is a faithful (self-dual) $\mathbb{F}_pH$-submodule of V here. Thus U and ψ are as assumed in Theorem 3.3d. It follows that $k(GV) \leq |V|$ and that equality only holds when $[V, H] \subseteq U$. But in this case $V = U \oplus U'$ as an $\mathbb{F}_pH$-module with H acting trivially on U'. Thus V itself then is self-dual for H.

Suppose p is odd and some element of H acts with determinant -1 on V. Assume $k(GV) = |V|$. Then V is self-dual for H, as seen above. So Lemma 5.2a and Theorem 3.3d apply. We conclude that $k(GV) \leq \frac{p+3}{2p}|V|$, and this is less than $|V|$ when $p \geq 5$. $\square$

In [Robinson–Thompson, 1996] the conditions under which the assumptions in Theorem 3.3d are fulfilled have been carefully studied. Let us describe this briefly. Assume (without loss of generality) that V is irreducible (Proposition 3.1a), and faithful, and let $\breve{\chi}$ be the Brauer character of G afforded by V. So, as in Sec. 2.7, $\breve{\chi} = \mathrm{Tr}_{K_D(\chi)|K_D}(\chi)$ for some ordinary absolutely irreducible character χ of G, where K_D is the decomposition field of p in $K = \mathbb{Q}(e^{2\pi i/e})$ for $e = \exp(G)$. This χ is the Brauer character of G afforded by some absolutely irreducible constituent of V, and it is faithful as algebraically conjugate characters have the same kernel. Let $\Gamma = \mathrm{Gal}(K|\mathbb{Q})$ and $\Gamma_D = \mathrm{Gal}(K|K_D)$. Let $H = C_G(v)$ for some $v \in V$.

If θ is an irreducible constituent of $\mathrm{Res}_H^G(\chi)$, then $\breve{\theta} = \mathrm{Tr}_{K_D(\theta)|K_D}(\theta)$ is a Brauer character of H afforded by some irreducible constituent $W = W_\theta$ of $\mathrm{Res}_H^G(V)$. By Lemma 4.1a, W is self-dual if and only if θ and $\bar{\theta}$ are Galois conjugate over K_D. (Use that Γ is abelian.) If $\tau \in \Gamma \setminus \Gamma_D$, then $W^\tau = W_{\theta^\tau}$ affords $\mathrm{Tr}_{K_D(\theta)|K_D}(\theta^\tau)$ and is an irreducible $\mathbb{F}_p H$-module not isomorphic to W. But $\delta_{W^\tau} = \delta_W$ (see Sec. 2.7). Moreover, W^τ is self-dual if and only if W is self-dual.

Now consider first all irreducible constituents θ of $\mathrm{Res}_H^G(\chi)$ for which $\langle \chi, \theta \rangle_H \geq 2$. Then $\mathrm{Res}_H^G(V)$ contains a submodule $U_\theta = W_\theta \oplus W_\theta$, and $\delta_{U_\theta} = \delta_{W_\theta}^2$. Consider next those θ for which both θ and θ^τ are irreducible constituents of $\mathrm{Res}_H^G(\chi)$ for some $\tau \in \Gamma \setminus \Gamma_D$. Then $\mathrm{Res}_H^G(V)$ contains a submodule $U_{\theta,\tau} = W_\theta \oplus W_\theta^\tau$, and again $\delta_{U_{\theta,\tau}} = \delta_{W_\theta}^2$. Let U_0 be the sum of all these submodules U_θ, $U_{\theta,\tau}$ of $\mathrm{Res}_H^G(V)$. Then there is a rational-valued generalized (Knörr) character ψ_0 of H such that $\psi_0^2 = \delta_{U_0}$.

If U_0 is a faithful $\mathbb{F}_p H$-module, we are done. Otherwise we may search for irreducible constituents θ of $\mathrm{Res}_H^G(\chi)$ for which θ and $\bar{\theta}$ are conjugate over K_D. In this case $U_\theta = W_\theta$ is self-dual, and Theorems 4.1b, 4.6a and 4.6b apply. If we add all these modules to U_0, the resulting $\mathbb{F}_p H$-module U is a submodule of $\mathrm{Res}_H^G(V)$, and it can be faithful, or not.

Proposition 5.2c. *Let V be a coprime, faithful, irreducible $\mathbb{F}_p G$-module affording $\mathrm{Tr}_{K_D(\chi)|K_D}(\chi)$ for some (absolutely) irreducible character χ of G. Let $H = C_G(v)$ for some v in V. Suppose there are irreducible constituents θ_i of $\mathrm{Res}_H^G(\chi)$ which are pairwise not algebraically conjugate and satisfy one of the following: $\langle \theta_i, \chi \rangle_H \geq 2$, $\langle \theta_i^\tau, \chi \rangle_H \geq 1$ for some $\tau \in \Gamma \setminus \Gamma_D$, or $\bar{\theta}_i$ is conjugate to θ_i over K_D. If then $\sum_i \theta_i$ is a faithful character of H, the assumptions in Theorem 3.3d are fulfilled (for appropriate U and ψ determined by the θ_i).*

Proof. Clear in view of the above discussion. □

5.3. Search for Real Vectors

Let now V be an arbitrary coprime FG-module, not necessarily faithful. It is obvious that if the direct summands of V admit (strongly) real vectors, then so does the module itself. Our objective is to ensure that the search for (strongly) real vectors is compatible with Clifford reduction.

Proposition 5.3a.. *Let $V = \operatorname{Ind}_H^G(U)$ be induced from the FH-module U.*

(i) *If there is a (strongly) real vector in U for H, there is a (strongly) real vector in V for G.*

(ii) *If there is a regular vector in V for $N = \operatorname{Core}_G(H)$, there is a real vector in V for G.*

Proof. Let $\{t_i\}$ be a right transversal to H in G (so that $N = \bigcap_i H^{t_i}$).

(i) Let $u \in U$ be real for H, and let $v = \sum_i u t_i$. Let W be a self-dual $FC_H(u)$-submodule of U with $C_H(W) \cap C_H(u) = C_H(U)$, and let $V_0 = \bigoplus_i W t_i$. Then V_0 is a subspace of V. Now $C_G(v) \cap H^{t_i}$ centralizes $u t_i$ and so $W t_i$ is self-dual as an $F[C_G(v) \cap H^{t_i}]$-module. Inducing up this self-dual module to $C_G(v)$ yields a self-dual $FC_G(v)$-module by Lemma 4.1a. By Mackey decomposition (1.2d) V_0 is a direct sum of such modules (taken over a set of representatives for the orbits of $C_G(v)$ on the cosets $H t_i$). It remains to show that $C_G(V_0) \cap C_G(v) = C_G(V)$. Now the centralizer in $C_G(v)$ of V_0 preserves each coset $H t_i$ and so lies in N. It follows that $C_G(V_0) \cap C_G(v) \subseteq N \cap C_H(W) \cap C_H(u) = C_N(U)$. Similarly $C_G(V_0) \cap C_G(v) \subseteq C_N(U t_i)$ for each i, and $\bigcap_i C_N(U t_i) = C_G(V)$. The statement for strongly real vectors is treated similarly.

(ii) Let $v = \sum_i u t_i$ be a regular vector in V for N. We may assume that all $v_i = u t_i \neq 0$. Let $H t_j C_G(v)$ be the different double cosets of G modulo $(H, C_G(v))$. By Mackey decomposition once again

$$\operatorname{Res}_{C_G(v)}^G(V) = \bigoplus_j \operatorname{Ind}_{C_G(v) \cap H^{t_j}}^{C_G(v)}(U t_j).$$

Let $V_0 = \bigoplus_j \operatorname{Ind}_{C_G(v) \cap H^{t_j}}^{C_G(v)}(F v_j)$. Then V_0 is a permutation module for $C_G(v)$ over F, hence self-dual. Since each $F v_j$ has a $C_G(v) \cap H^{t_j}$-invariant complement in $U t_j$ (Maschke), and since module induction respects direct sums, V_0 is a direct summand of V. As before $C_G(V_0) \cap C_G(v) \subseteq N$. Hence $C_G(V_0) \cap C_G(v) \subseteq N \cap C_G(v) = C_N(v) = C_G(V)$, as desired. □

Proposition 5.3b. *Suppose $V = U \otimes_F W$ for some (coprime) FG-modules U, W with $2 \le d = \dim_F U \le \dim_F W$ (say).*

(i) *If there are (strongly) real vectors in U and in W for G, there is a (strongly) real vector in V for G. Similar statement for regular vectors.*

(ii) *Let $\{u_i\}$, $\{w_j\}$ be F-bases of U and W, respectively. Let $v = \sum_{i=1}^{d} u_i \otimes w_i$, and let W_0 be the subspace of W generated by $w_1, \cdots, w_d$. Then $V_0 = U \otimes_F W_0$ is a self-dual $FC_G(v)$-module.*

(iii) *Assume that there is a $(d-1)$-dimensional subspace $\widetilde{W}$ of W such that $C_G(\widetilde{W})/C_G(W)$ has a regular orbit on W. If G induces all scalar multiplications on W, there is a real vector in V for G.*

Proof. (i) We have $C_G(U) \cap C_G(W) \subseteq C_G(V)$, the elements in $C_G(V)$ being those elements of G inducing on U and W scalar multiplications which are inverse to each other. For if $u \in U$, $w \in W$ are nonzero vectors, $v = u \otimes w$ and $g \in C_G(v)$, then $v = vg = ug \otimes wg$ implies that $ug = cu$ and $wg = c^{-1}w$ for some unique scalar $c = c_g(v) \in F^\star$. If u, w are regular vectors for G, that is, $C_G(u) = C_G(U)$ and $C_G(w) = C_G(W)$, then $C_G(v) = C_G(V)$.

Let $u \in U$ and $w \in W$ be real vectors for G. Let U_0 and W_0 be self-dual submodules of U and W for $C_G(u)$ and $C_G(w)$, respectively, with $C_G(u) \cap C_G(U_0) = C_G(U)$ and $C_G(w) \cap C_G(W_0) = C_G(W)$. We may and do assume that $u \in U_0$ and $w \in W_0$. (Otherwise replace U_0 by $U_0 \oplus Fu$; W_0 by $W_0 \oplus Fw$.) Let $v = u \otimes w$ and $V_0 = U_0 \otimes_F W_0$. If $g \in C_G(v)$ then, for some $c \in F^\star$, $c^{-1}g$ fixes u and cg fixes w and so g acts on V_0. If g centralizes V_0, then g acts as scalar c on U_0 and as the scalar c^{-1} on W_0 (as $u \in U_0$ and $v \in V_0$). Hence $C_G(v) \cap C_G(V_0) = C_G(V)$. (For the statement in parentheses take $U_0 = U$ and $W_0 = W$.)

(ii) Let $g \in C_G(v)$. Let (a_{ij}) be the matrix of g on U with respect to the given basis. Then

$$v = vg = \sum_i u_i g \otimes w_i g = \sum_i (\sum_j a_{ij} u_j) \otimes w_i g = \sum_j u_j \otimes (\sum_i a_{ij} w_i) g.$$

Thus $w_j g^{-1} = \sum_{i=1}^{d} a_{ij} w_i$ for each $j = 1, \cdots, d$. It follows that g acts on W_0 through the inverse transpose matrix $(a_{ij})^{-t}$. This shows that W_0 and V_0 are $FC_G(v)$-modules. Lifting the eigenvalues of g on U and on W_0 to characteristic 0, we see that if φ is the Brauer character of $C_G(v)$ on U, then $\varphi\bar{\varphi}$ is that on $U \otimes_F W_0$. By Lemma 4.1a, $V_0 = U \otimes_F W_0$ is a self-dual $FC_G(v)$-module.

(iii) Choose notation such that $\widetilde{W}$ is generated by $w_1, \cdots, w_{d-1}$. By hypothesis we find $w = w_d$ in W outside $\widetilde{W}$ such that $C_G(W_0) = C_G(W)$ for the subspace W_0 of W generated by $\widetilde{W}$ and w. Define v and φ as before, so that $V_0 = U \otimes_F W_0$ is a self-dual $FC_G(v)$-module affording $\varphi\bar\varphi = |\varphi|^2$. It remains to show that $C_G(v)/C_G(V)$ is faithful on V_0. Let $g \in C_G(v)$ act trivially on V_0. Then $|\varphi(g)|^2 = d^2$ and so $|\varphi(g)| = d$. Thus g acts on U by multiplication with some scalar c, whence on W_0 via c^{-1}. By hypothesis there is $g_0 \in G$ acting as scalar multiplication with c^{-1} on W. Then $gg_0^{-1} \in C_G(W_0) = C_G(W)$ and so $g = g_0$ on W and $g \in C_G(V)$.   □

Remark. Replacing G by $G \times Z$ for some suitable subgroup Z of $Z(\mathrm{GL}(W))$ (acting trivially on U) one can achieve that G induces all scalar transfomations on W. Every vector in V which is real for $G \times Z$ is real for G.

Proposition 5.3c. *Suppose* $V = \mathrm{Ten}_H^G(W)$ *for some (coprime) FH-module W with $\dim_F W \geq 2$. If there is a (strongly) real vector in W for H, then there is also a (strongly) real vector in V for G.*

Proof. We prove the lemma for the case of real vectors, the proof for strongly real vectors being similar (and easier). Let $\{t_i\}_{i=1}^n$ be a right transversal to H in G (with $t_1 = 1$). For $x \in G$ let $i \mapsto ix$ be the permutation (on the indices) induced by x, as in Sec. 1.2. Let $w_0 \in W$ be real for H, and let W_0 be a subspace of W which is a self-dual $FC_H(w_0)$-module satisfying $C_H(w_0) \cap C_H(W_0) = C_H(W)$. Without loss of generality we assume that $w_0 \neq 0$ and, replacing W_0 by $W_0 \oplus Fw_0$ if necessary, that $w_0 \in W_0$. Let $v_0 = w_0t_1 \otimes \cdots \otimes w_0t_n$ and $V_0 = W_0t_1 \otimes_F \cdots \otimes_F W_0t_n$. Then V_0 is a subspace of $V = Wt_1 \otimes_F \cdots \otimes_F Wt_n$. We assert that V_0 is a self-dual $FC_G(v_0)$-module satisfying $C_G(v_0) \cap C_G(V_0) = C_G(V)$.

Let $x \in C_G(v_0)$. Then there are scalars $c_i \in F$ such that $\prod_{i=1}^n c_i = 1$ and $w_0t_ix = c_iw_0t_{ix}$ for each i. Hence x acts on V_0, and $c_i^{-1}t_ixt_{ix}^{-1} \in C_H(w_0)$. Let $Ht_j\langle x \rangle$ be the distinct double cosets of G modulo $(H, \langle x \rangle)$. Suppose the $\langle x \rangle$-orbit of Ht_j has size n_j, and let the scalar b_j be the product of the scalars c_i belonging to this orbit. Then $\sum_j n_j = n$, $\prod_j b_j = 1$, $t_jx^{n_j}t_j^{-1} \in H$ and $b_j^{-1}t_jx^{n_j}t_j^{-1} \in C_H(w_0)$ for each j. By formula (1.2e) the Brauer character χ of G afforded by V takes the value

$$\chi(x) = \prod_j \theta(t_jx^{n_j}t_j^{-1}) = \prod_j \theta(b_j^{-1}t_jx^{n_j}t_j),$$

where θ is the Brauer character of H afforded by W. Passing from H, W to $C_H(w_0), W_0$ (character θ_0) we get the character χ_0 of $C_G(v_0)$ afforded by V_0. Thus $\chi_0(x) = \prod_j \theta_0(b_j^{-1} t_j x^{n_j} t_j)$. Apply Lemma 4.1a.

Suppose $x \in C_G(v_0)$ acts trivially on V_0. Then $i \mapsto ix$ is be the identity permutation, whence $x \in \mathrm{Core}_G(H)$. Also, for each i then $c_i^{-1} t_i x t_i^{-1}$ acts as a scalar on W_0, and centralizes $w_0 \in W_0$. Hence $c_i^{-1} t_i x t_i^{-1} \in C_H(W_0) \cap C_H(w_0) = C_H(W)$. It follows that x acts on the ith component of $V = \mathrm{Ten}_H^G(W)$ as the scalar c_i. Using that $\prod_i c_i = 1$ we get that $x \in C_G(V)$, as desired. $\qquad\square$

5.4. Clifford Reduction

Let V be a faithful, irreducible, coprime FG-module in characteristic p. Assume there is no (strongly) real vector in V for G but that (G, V) is a minimal counterexample in the following sense:

- Whenever G_0 is a central extension of a subgroup of G by a p'-group and V_0 is a $F_0 G_0$-module for which $\mathrm{char}(F_0) = p$ and $\dim_{F_0} V_0 < \dim_F V$, then there is a (strongly) real vector in V_0 for G_0. •

Theorem 5.4. *Suppose (G, V) is a minimal counterexample in the above sense (for real or strongly real vectors). Then G has a unique minimal nonabelian normal subgroup, say E, and this is either quasisimple or of extraspecial type. Moreover E is absolutely irreducible on V, and all abelian normal subgroups of G are cyclic and central.*

Proof. We concentrate on real vectors, the argumentation for strongly real vectors being similar. We only make use of results holding for both kinds of vectors. It is clear that G is not abelian, because otherwise there were a regular G-orbit on V by (3.4b), and $d = \dim_F V \geq 2$ as G is faithful on V. We proceed in several steps.

(1) *V is an absolutely irreducible FG-module:*

For otherwise embed F (properly) into the field $F_0 = \mathrm{End}_{FG}(V)$, and let $\Gamma = \mathrm{Gal}(F_0|F)$. Then $F_0 \otimes_F V = \bigoplus_{\sigma \in \Gamma} V_0^\sigma$ for some absolutely irreducible $F_0 G$-module V_0. Then V_0 is faithful and $\dim_{F_0} V_0 < d$. Thus V_0 contains a real vector v_0 for G, that is, $\mathrm{Res}_{C_G(v_0)}^G(V_0)$ has a faithful self-dual submodule U, say. Let Γ_0 be the stabilizer of U in Γ, and let $W = \sum_{\sigma \in \Gamma/\Gamma_0} U^\sigma$ and $v = \sum_{\sigma \in \Gamma/\Gamma_0} u^\sigma$. V_0 and V are isomorphic as G-sets (see Sec. 2.7). It follows that $C_G(v) = C_G(v_0)$ and, since field and group automorphisms commute, that W is a self-dual $FC_G(v)$-submodule of V. This contradicts our choice of V.

(2) V *is a primitive FG-module:*

Otherwise $V = \mathrm{Ind}_H^G(U)$ for some proper subgroup H of G and some FH-module U. Since $\dim{}_F U < d$ there is a real vector in U for H by the choice of V. But then by part (i) of Proposition 5.3a there is a real vector in V for G, against our assumption.

(3) *The irreducible constituents of* $\mathrm{Res}_N^G(V)$ *are absolutely irreducible for all normal subgroups N of G:*

Otherwise choose N maximal such that the assertion is false. Then $N \neq G$ by (1), and there is an irreducible submodule W of $\mathrm{Res}_N^G(V)$ such that $F_0 = \mathrm{End}_{FN}(W)$ is a proper extension field of F. Let $\Gamma = \mathrm{Gal}(F_0|F)$, and let $F_0 \otimes_F W = \bigoplus_{\sigma \in \Gamma} U^\sigma$ for some absolutely irreducible F_0N-module U. By (2) W is G-invariant. Hence to every $x \in G$ there exists a unique $\sigma = \sigma_x \in \Gamma$ such that $(Ux)^\sigma \cong U$, and the assignment $x \mapsto \sigma_x$ is a homomorphism making F_0 into a "G-field". The kernel of this Galois action of G is a normal subgroup N_0 containing N. Let Γ_0 be the image in Γ of G (so that $G/N_0 \cong \Gamma_0$). Then $W_0 = \bigoplus_{\sigma \in \Gamma_0} U^\sigma$ is an irreducible FN_0-module with $\mathrm{End}_{FN_0}(W_0) \cong F_0$. Hence $N_0 = N$ by the choice of N and so $G/N \cong \Gamma$. It follows that

$$F_0 \otimes_F V \cong \mathrm{Ind}_N^G(U)$$

and that $\mathrm{Res}_N^G(V) = W$ is irreducible. We have $\dim{}_{F_0} U < d$. By the choice of V there is a real vector in U for N, and thus there is a real vector in $F_0 \otimes_F V$ for G by part (i) of Proposition 5.3a. As in (1) we get a real vector in V for G.

From (1), (2), (3) it follows that every abelian normal subgroup of G is cyclic and central in G (acting by scalar multiplications). Without loss of generality we may assume that G induces all scalar multiplications on V. The generalized Fitting subgroup of G is nonabelian for otherwise G were cyclic and so had a regular orbit on V.

(4) $\mathrm{Res}_N^G(V)$ *is absolutely irreducible for all nonabelian normal subgroups N of G:*

Otherwise, in view of (3), there is a nonabelian normal subgroup N of G such that the restriction of V to N is a proper multiple eU of some absolutely irreducible FN-module U ($e \geq 2$). This U is faithful. Let θ be the Brauer character of N afforded by U, and let $G(\theta)$ be the extended representation group of θ. By Theorem 1.9c there is an $FG(\theta)$-module $\widehat{U}$ extending U (in the usual sense), and there is an $FG(\theta)$-module W such that $V \cong \widehat{U} \otimes_F W$ (viewed as an $FG(\theta)$-module). Thus $\dim_F \widehat{U} < d$ and $\dim_F W = e < d$. By the choice of V there are real vectors for $G(\theta)$ in $\widehat{U}$ and in W. But then by part (i) of Proposition 5.3b there is a real vector in V for G.

(5) *Conclusion:*

Let E be a minimal nonabelian normal subgroup of G. By (4) $\mathrm{Res}_E^G(V) = W$ is absolutely irreducible. In view of (2), (3) $C_G(E) = Z = Z(G)$ and EZ is the generalized Fitting subgroup of G.

Suppose first that E is solvable. Then it is a q-group of "symplectic type" for some prime $q \neq p$. Either q is odd and $E = \Omega_1(EZ)$ is of exponent q, or E is a 2-group of extraspecial type. Also, E is the unique minimal nonabelian normal subgroup of G. This is clear when q is odd or $q = 2$ and $|Z(E)| = 4$ (E of type 0). Suppose $E \cong 2_{\pm}^{1+2m}$ for some m. Note that G is irreducible on $E/Z(E)$ as each proper G-invariant subgroup of E is cyclic and central in G. Thus E cannot be dihedral of order 8, and E is unique except possibly when $|Z|$ is divisible by 4. But then E is the unique minimal nonabelian G-invariant subgroup of $E \circ Z_4$.

Let E be nonsolvable. Then E is the central product of the distinct G-conjugates of some quasisimple group E_0, and W is the tensor product of $n = |G : N_G(E_0)|$ distinct G-conjugates of the unique *absolutely* irreducible constituent W_0 of $\mathrm{Res}_{E_0}^E(W)$. Assume $E \neq E_0$ ($n > 1$). Let θ, θ_0 be the Brauer characters of N, N_0 afforded by W, W_0, respectively, and let $G(\theta)$, $G_0(\theta_0)$ be the extended representation groups. So $G_0(\theta_0)$ is a central extension of G_0 by a cyclic p'-group. By Theorem 1.9c there is an $FG_0(\theta_0)$-module $\widehat{W_0}$ extending W_0 (in the usual sense). Since $\dim_F \widehat{W_0} = \dim_F W_0 = (\dim_F W)^{\frac{1}{n}} = \sqrt[n]{d} < d$, there is a real vector in $\widehat{W_0}$ for $G(\theta_0)$. Since $\mathrm{Res}_N^G(V) = W$, we have $G(\theta) = G$. By Theorem 1.9d there is a finite extension $\widehat{G}$ of G by an abelian p'-group, containing a subgroup $\widehat{G}_0$ mapping onto $G_0(\theta)$, such that

$$V = \mathrm{Ten}_{\widehat{G}_0}^{\widehat{G}}(\widehat{W_0})$$

(We may arrange matters such that the Clifford correspondent of θ is trivial.) By Proposition 5.3c there is a real vector in V for G, a contradiction. Hence $E = E_0$ is quasisimple. $\qquad\square$

5.5. Reduced Pairs

The group G is said to be *reduced* if all abelian normal subgroups of G are central and if G has a unique minimal nonabelian normal subgroup, E, called the core of G, which is either of extraspecial type or is quasisimple. If G is reduced with core E and V is a faithful, coprime FG-module, then (G, V) is a *reduced pair* (over F) provided E is absolutely irreducible on V. Then $C_G(E) = Z(G)$ acts as a group of scalar multiplications on V and $Z(G)E$ is the generalized Fitting subgroup of G. We say that the pair (G, V) is "large" if $Z(G) \cong F^\star$, that is, if G induces all scalar transformations on V.

The pair (G, V) is *nonreal reduced* if it is reduced and if there is no real vector in V for G. The minimal counterexample (G, V) described in Theorem 5.4 is a nonreal reduced pair. The objective of the next two chapters is to give a complete classification of such pairs. In the quasisimple case we even shall describe all reduced pairs (up to isomorphism) admitting no regular vectors.

We say that two reduced pairs (G, V) and $(\widetilde{G}, \widetilde{V})$ over F are *isomorphic* if there is a group isomorphism $\alpha : G \to \widetilde{G}$ making $\widetilde{V}$ into an FG-module isomorphic to V. In other words, identifying $G = \widetilde{G}$ the Brauer characters afforded by V and $\widetilde{V}$ are conjugate under an automorphism α of G.

It is immediate that this notion of "isomorphism" for reduced pairs preserves (regular) orbits, (strongly) real vectors, and abelian vectors.

5.6. Counting Methods

The approach to the classification theorems generally is in two steps: First we use counting arguments in order to reduce the discussion to "small" groups, often Atlas groups. Then we proceed by a case-by-case analysis of the remaining groups. Sometimes we use the computer, and then we refer to [Groups, Algorithms, and Programming, 2006]. This will be briefly cited as [GAP].

Let us describe the basic methods. Suppose (G, V) is a reduced pair over $F = \mathbb{F}_r$, r being a power of the prime p not dividing $|G|$, and let $d = \dim{}_F V$. Embed G into $\mathrm{GL}(V)$, and let $G_0 = N_{\mathrm{GL}(V)}(E)$. Then $Z = Z(G_0) = C_{G_0}(E) \cong F^\star$ is cyclic of order $r - 1$. Assume (G, V) is large (replacing G by GZ, if necessary). Let $\bar{G}_0 = G_0/Z$ and $\bar{G} = G/Z$, which are subgroups of $\mathrm{Aut}(E)$ (as $Z \cap E = Z(E)$). For each $v \in V^{\sharp}$, $C_G(v) \cap Z = 1$ and $C_G(v)Z/Z = C_{\bar{G}}(Zv)$. Hence there is a 1-1 correspondence between the G-orbits on $V^{\sharp}$ and the $\bar{G}$-orbits on the projective 1-space $P_1(V)$, with corresponding point stabilizers.

Define the *bottom* $\beta(G)$ of G (and similarly for each finite group) as the set of all noncentral subgroups of G of prime order. If there is $v \in V$ such that $C_G(v) \neq 1$, the stabilizer contains at least one subgroup in $\beta(G)$ (Sylow). Hence if $\bigcup_{\gamma \in \beta(G)} C_V(\gamma) \neq V$, there is a regular vector in V for G. Usually we show that $\sum_{\gamma \in \beta(G)} |C_V(\gamma)| < |V|$, ignoring zero spaces or possible intersections. But sometimes this will be improved using common diagonalization, or arguing "projectively", because if

$$(5.6a) \qquad \sum_{\bar{\gamma} \in \beta(\bar{G})} |C_{P_1(V)}(\bar{\gamma})| < |P_1(V)|,$$

then there is a regular G-orbit on V as well. Arguing in this manner, for any given $\bar{\gamma} \in \beta(\bar{G})$, choose a subgroup γ of G of smallest possible order mapping onto $\bar{\gamma}$. Then γ is a cyclic group of prime power order. If $G_0 = X \circ Z$ is a central product over $Z(E)$, clearly we may pick γ as a subgroup of X, and then γ either has prime order or $\gamma \cap Z(E) \neq 1$.

Let $\beta^*(G)$ denote the set of all noncentral subgroups of G which are cyclic of order 4 or of odd prime order. Then

$$(5.6b) \qquad \bigcup_{\gamma \in \beta^*(G)} C_V(\gamma) \neq V$$

implies that there is some $v \in V$ such that $C_G(v)$ is an elementary abelian 2-group. Then v is a strongly real vector for G (and $C_G(v)$ has a regular orbit on V by Eq. (3.4b)). For a set ω of primes we denote by $\beta^*_\omega(G)$ the set of ω-subgroups in $\beta^*(G)$; define $\beta_\omega(G)$ similarly. If for instance $\bigcup_{\gamma \in \beta_\omega(G)} C_V(\gamma) \neq V$, there is $v \in V$ such that $C_G(v)$ is a ω'-group.

Let $g \in G_0$. We define the *fixed point ratio* by

$$(5.6c) \qquad f(g) = f(g, V) = \dim {}_F C_V(g) / \dim {}_F V.$$

Of course $f(g) = f(\langle g \rangle)$ depends only on (the generators of) the group $\langle g \rangle$. For $z \in Z = F^\star$ the fixed space $C_V(z^{-1}g) = V_z(g)$ is the eigenspace of g on V to the eigenvalue z, or it is zero. (It is for instance zero if the order $o(z)$ of z does not divide $o(g)$.) We have $\bigoplus_{z \in Z} V_z(g) \subseteq V$, with equality if and only if g is a p'-element and F is large enough (containing all eigenvalues of g). Note that g is faithful on V. There are at most $d = \dim_F V$ distinct eigenspaces of g on V. Hence if $f(zg) \leq f$ for all $z \in Z$ then $|\bigcup_{z \in Z} C_V(zg)| \leq d \cdot r^{df}$.

The following estimate will be applied very often. Let $g \in G_0$. Suppose again that $f(zg) \leq f$ for all $z \in Z$, and assume that $\frac{1}{2} \leq f < 1$. Then if $z_i \in Z$ are the distinct eigenvalues of g on V,

$$(5.6d) \qquad \sum_i |C_V(z_i^{-1} g)| \leq r^{\lfloor df \rfloor} + r^{d - \lfloor df \rfloor} \leq 2 r^{df}.$$

In order to verify this, without loss of generality we may assume that g is a p'-element and F is large enough, so g has the (distinct) eigenvalues z_i on V with the multiplicities $d_i \geq 1$, $1 \leq i \leq n$, such that $\sum_{i=1}^{n} d_i = d$. Arrange these eigenvalues such that $d_1 \geq d_2 \geq \cdots \geq d_n$. By assumption $d > df \geq d/2$ and $df \geq \lfloor df \rfloor \geq d_1$. Suppose d_j' is another such decreasing sequence of positive integers for $j = 1, \cdots, n'$, satisfying $\sum_{j=1}^{n'} d_j' = d$ and $df \geq d_1'$. Then $\sum_{i=1}^{n} r^{d_i} \leq \sum_{j=1}^{n'} r^{d_j'}$ if and only if $(d_1, d_2, \cdots) \leq (d_1', d_2', \cdots)$ in lexicographical ordering. Consequently $\sum_{i=1}^{n} r^{d_i} \leq r^{\lfloor df \rfloor} + r^{d - \lfloor df \rfloor}$, which is at most $2 r^{\lfloor df \rfloor}$ if $f > \frac{1}{2}$, and at most $2 r^{df}$ if $f = \frac{1}{2}$.

Notation. Suppose g has on V the eigenvalues z_i with multiplicities d_i, $1 \leq i \leq n$. Arrange these such that $d_1 \geq d_2 \geq \cdots \geq d_n$. Then we indicate the *spectral pattern* by writing

$$g_V = [z_1^{(d_1)}, \cdots, z_n^{(d_n)}] = [d_1, \cdots, d_n],$$

the latter (weak form) in the case when only the dimensions of the eigenspaces are of interest. Thus $\sum_{i=1}^{n} d_i \leq d = \dim_F V$, and we have equality if and only if g is a p'-element of order dividing $r - 1$. It is obvious but important in applications that each element of G_0 in the conjugacy class of g has the same spectral pattern of the first kind, and each element in the coset Zg has the same weak pattern. In particular, $|C_{P_1(V)}(Zg)| = \frac{1}{r-1} \sum_{i=1}^{n} (r^{d_i} - 1)$.

Now we turn to character theory. Let χ be the Brauer character of G_0 afforded by V, and assume we know χ (to some extent). Let $g \in G_0$ be a p'-element of order s, say. Then V is a projective $F\langle g \rangle$-module, $\mathrm{Res}_{\langle g \rangle}^{G_0}(\chi)$ an ordinary character and so (Sec. 1.3)

$$(5.6e) \qquad \dim_F C_V(g) = \langle \chi, 1 \rangle_{\langle g \rangle} = \frac{1}{s} \sum_{i=1}^{s} \chi(g^i).$$

Hence the fixed point ratio $f(g)$ is just the average over the ratios $\chi(g^i)/\chi(1)$ for $i = 1, \cdots, s$. Recall from Sec. 1.5 that the $\chi(g^i)$ for the generators g^i of $\langle g \rangle$ are the algebraic conjugates of $\chi(g)$. If for instance s is a prime, therefore the strong pattern g_V is determined by $\chi(g)$ since the sth roots of unity $\neq 1$ are linear independent over $\mathbb{Z}$ (and their sum equals -1). We can compute the spectral pattern from the character table (in terms of $F = \mathbb{F}_r$).

5.7. Two Examples

To illustrate the counting methods we discuss in some detail two reduced pairs, one of extraspecial type and one of quasisimple type.

Example 5.7a. Let (G, V) be a large reduced pair over $F = \mathbb{F}_r$ with core $E \cong 5^{1+2}_+$. Since E is faithful and absolutely irreducible on V, we have $\text{char}(F) = p \neq 5$, $d = \dim_F V = 5$ and $5 \mid r - 1$. Hence $r \geq 11$. Embed G into $\text{GL}(V) = \text{GL}_5(r)$, and consider $G_0 = N_{\text{GL}(V)}(E)$. From Theorem 4.3c and Proposition 4.3a it follows that

$$G_0 = X \circ Z$$

is a central product over $Z(E)$, where X is the standard holomorph of E and where $Z = Z(G_0) \cong F^*$. We also know that $X = E : S$ is a semidirect product where the complement $S \cong \text{Sp}_2(5)$ is determined up to conjugacy (under E). We are going to show that there is a regular vector in V for G.

By definition of a reduced pair G is irreducible on $U = E/Z(E) = EZ/Z$. If $p = 2$ or 3, then G/EZ has order 3 or is a quaternion group of order 8, respectively, and it is easy to show that there is a regular G-orbit on V. So let $p > 5$. Then G_0 is a p'-group, and it is enough to consider $G = G_0$. (Usually we are treating this "worst" case ignoring whether G_0 is a p'-group or not, counting scalar products with the 1-character "symbolically" as dimensions of fixed spaces.) Let χ be the Brauer character of X (and of G) afforded by V. This χ is one of four faithful irreducible characters of degree 5 of the holomorph X which, however, are conjugate under automorphisms of X by Theorem 4.3c, so lead to isomorphic reduced pairs. Therefore we shall sometimes speak of "the" character associated to X. By Theorem 4.5a, $\text{Res}^X_S(\chi) = \xi$ is "the" generic Weil character of S. In the terminology of the [Atlas, p. 2], $\xi = \chi_2 + \chi_7$ or $\chi_3 + \chi_6$.

We may identify $\bar{X} = X/Z(E)$ with G/Z. Each prime order element $Z(E)x$ of $\bar{X}$ is represented either by a (noncentral) element of E (of order

5) or lies in a coset Ex for some element $x \in S$ of prime order. There are $(5^2 - 1)/4 = 6$ subgroups of order 5 in $U = E/Z(E)$, their generators having weak spectral pattern $[1^{(5)}]$ as χ vanishes on noncentral elements of E. Thus at most $\frac{6}{r-1} \cdot 5(r-1) = 30$ points in $P_1(V)$, or $6(5r) = 30r$ vectors in V, are fixed by these subgroups. There are two conjugacy classes $5A_0B_0$ of elements in S of order 5, and if y belongs to $5A_0$, then $\xi(y) = \sqrt{5}$ (and y^{-1} belongs to $5B_0$ with $\xi(y^{-1}) = -\sqrt{5}$). There are $\frac{1}{2}|5A_0B_0| = 6$ (conjugate) subgroups in S of order 5. By Theorem 4.4, y is good for U and $|E : C_E(y)| = 5$. We compute that $\dim {}_F C_V(y) = 1$ and

$$y_V = [\varepsilon_1^{(2)}, \varepsilon_2^{(2)}, 1^{(1)}] = [2, 2, 1]$$

for some primitive 5th roots of unity $\varepsilon_1 \neq \varepsilon_2$ which are inverse to each other ($5 \mid r - 1$ and $\varepsilon_1 + \varepsilon_2$ is a square root of 5 in F). By Theorem 4.4 the coset Ey contains just $|U : C_U(y)| = 5$ good cosets of $Z(E)$ (which are conjugate under E). So there are $5^2 - 5 = 20$ bad cosets of $Z(E)$ in Ey. If an element y_0 in Ey is bad, then $\chi(y_0) = 0$ and so $o(y_0) = 5$ and $(y_0)_V = [1^{(5)}]$ (weak pattern). We therefore have

$$\sum_{\gamma \in \beta_5(G)} |C_V(\gamma)| \leq 30r + 6\big(5(2r^2 + r) + 20(5r)\big).$$

Let $\bar{x} \in \bar{X}$ have order 3, and let $x \in S$ be such that $\bar{x}$ is contained in Ex. Then x belongs to the unique conjugacy class $3A_0$ of elements of order 3 in S. It determines the coset Ex in X, which in turn determines the coset ZEx in G. There are $|S : C_S(x)| = |3A_0| = 2 \cdot 10$ cosets of E in X conjugate to Ex under S. By Theorem 4.4 the coset $Z(E)x$ is good for U, and the good cosets lying in Ex are conjugate under E (and so have the same weak spectral pattern). From $\chi(x) = -1$ we deduce that $|C_U(x)| = \chi(x)^2 = 1$, hence $|U : C_U(x)| = |U| = 5^2$, and that

$$x_V = [z_1^{(2)}, z_2^{(2)}, 1] = [2, 2, 1]$$

if $3 \mid r - 1$, where $z_1 \neq z_2$ are primitive 3rd roots of unity, and $x_V = [1]$ otherwise ($\dim {}_F C_V(x) = 1$). It follows that in Ex there are just $5^2 - 5 = 20$ bad cosets $Z(E)x'$. Since $\chi(x') = 0$, the bad element x' cannot have (prime) order 3. But each coset of $Z(E)$ in X of order 3 can be represented by an element of order 3. Hence we may ignore the bad cosets. We conclude that

$$\sum_{\gamma \in \beta_3(G)} |C_V(\gamma)| \leq 10 \cdot 5^2(2r^2 + r).$$

if $3 \mid r - 1$, and we may replace $2r^2 + r$ by r otherwise.

Let $j \in S$ represent the unique (central) involution in S (belonging to $1A_1$). We have $|C_U(j)| = \chi(1)^2 = (-1)^2 = 1$ and so $|U : C_U(j)| = 5^2$. Arguing as before we get $\sum_{\gamma \in \beta_2(G)} |C_V(\gamma)| \leq 5^2(r^3 + r^2)$. Consequently $\sum_{\gamma \in \beta(G)} |C_V(\gamma)| \leq 25r^3 + 585r^2 + 940r$. This is less than $|V| = r^5$ for $r \geq 10$. Hence there is a regular G-orbit.

Example 5.7b. Suppose (G, V) is a large reduced pair over $F = \mathbb{F}_r$, with core $E = 2.A_6$, and $d = \dim_F V = 4$. Then $p \geq 7$ by coprimeness, and V affords, as an FE-module, one of the characters χ_8, χ_9 as a Brauer character (Atlas, p. 5). These characters are the Weil characters ξ_1, ξ_1° of $\mathrm{Sp}_2(9)$ in the terminology of Theorem 4.5a. So they are rational-valued and fuse in $2.A_6.2_2$. But we are in an exceptional situation: The characters extend to $2.A_6.2_1$, requiring $\sqrt{\pm 3}$. More precisely, χ_8 extends to $2^+ S_6$ requiring $\sqrt{3}$ and to $2^- S_6$ requiring $\sqrt{-3}$, and for χ_9 it is vice versa. Recall that $2^+ S_6$ is that covering group of S_6 in which transpositions lift to involutions (and for which the characters are given in the Atlas).

In the following we assume that $\chi = \chi_9$ on E, for convenience. Exchanging the conjugacy classes $3A, 3B$ leads from one character to the other. Similarly, exchanging the conjugacy classes $6A, 6B$ leads from one isoclinic variant of $2.A_6.2_1$ to the other. So the arguments will carry over.

Embed G into $\mathrm{GL}(V)$, and let $G_0 = N_{\mathrm{GL}(V)}(E)$. Then $Z = Z(G) = C_{G_0}(E) \cong F^\star$. In the counting argument we assume (implicitly) that F contains a square root of 3 or -3 (or both, which happens when $4 \mid r - 1$). Let us consider the (worse) case that $\sqrt{-3} \in F$ and that

$$G = G_0 = X \circ Z$$

where $X = 2^+ S_6$. We ask whether there is a regular G-orbit on V, or a real vector in V for G, at least.

Let $x \in E$ be an element of order 3 belonging to $3A_0$ (mapping onto 3-cycles in S_6). Then $\chi(x) = 1$ and $x_V = [1^{(2)}, z^{(2)}]$ for some primitive 3rd root of unity z (viewed as an element of Z). In particular $W = C_V(x)$ has dimension 2. If an element in E fixes a nonzero vector in V, then it is conjugate to x [Atlas]. Let $y \in C_E(x)$ be an element of order 3 in the class $3B_0$ (mapping onto double 3-cycles). Then $\chi(y) = -2$ and so $y_V = [z^{(2)}, \bar{z}^{(2)}]$. Of course W is stable under y, and we have

$$y_W = [z^{(1)}, \bar{z}^{(1)}].$$

In order to see this, we have to exclude that y acts as a scalar multiplication on W. In that case, xy and xy^2 would have nontrivial fixed points on V by common diagonalization. But xy and xy^2 do belong to the class $3B_0$. We have $|3A_0| = |3B_0| = 40$. Since these are the unique conjugacy classes of order 3 in E and in G, we have $\sum_{\gamma \in \beta_3(G)} |C_V(\gamma)| \leq \frac{40}{2}(r^2 + 2r) + \frac{40}{2}(2r^2) = 60r^2 + 40r$.

On the two conjugacy classes $5A_0B_0$ of order 5 in E the character $\chi = \chi_9$ takes the value -1 and so the weak pattern is $[1, 1, 1, 1]$ if $5 \mid r - 1$ and empty otherwise. From $|5A_0| = |5B_0| = 72$ we see that there are just $\frac{1}{2}|5A_05B_0| = 72/((5-1)/2) = 36$ subgroups of order 5 in E. Hence $\sum_{\gamma \in \beta_5(G)} |C_V(\gamma)| \leq 36(4r)$.

Let $g \in X$ be an element of order 6 in the class $6A_0$ such that $g^2 = x$. Then $\chi(g) = \pm\sqrt{-3}$, and $j = g^3$ is an involution in the class $2B_0$ (mapping onto transpositions), which satisfies $\chi(j) = 0$. If an element in $X \smallsetminus E$ fixes a nonzero vector in V, then it is conjugate to j [Atlas]. Hence $j_V = [1^{(2)}, -1^{(2)}]$ and

$$g_W = j_W = [-1^{(2)}],$$

because $g = jx^{-1} = x^{-1}j$ has no fixed points on $V^\sharp$. The conjugacy classes $2A$ and $2C$ in S_6 lift to classes of elements of order 4 in X, and these have to be considered when $4 \mid r - 1$. Using that χ vanishes also on $2A_0$ and $2C_0$, we then get the contribution

$$(|2B| + |2A| + |2C|)(2r^2) = (15 + 45 + 15)(2r^2) = 150r^2.$$

There is a regular G-orbit on V provided $(60r^2 + 40r) + 144r + 150r^2 < r^4 = |V|$. This is true when $r \geq 16$. Hence $r = 7, 11$ or 13. We cannot have $r = 11$ since $3 \mid r - 1$; we can exclude $r = 11$ also when assuming that $\sqrt{3} \in F$ using that then $4 \nmid r - 1$, replacing the summand $150r^2$ by $30r^2$. One can also show that there is a regular G-orbit on V when $r = 13$. Either one argues as in the $r = 7$ case below (where there is no regular orbit, however), using that

$$\dim {}_F W = \dim {}_F C_V(x) = \langle \chi, 1 \rangle_{\langle x \rangle}$$

is independent of F, or applying the method to be developed in Proposition 7.3c below.

So let $r = 7$ in what follows. For each nonzero vector $v \in V$ we have $C_E(v) = \langle x \rangle$ if $v \in W = C_V(x)$ and $C_E(v) = 1$ otherwise (see above). So we have E-orbits on $V^\sharp$ of size $|E : \langle x \rangle| = 240$ and regular orbits. Let Δ be an E-orbit of size 240. Then $\Delta \cap W \neq \varnothing$ and $|\Delta \cap W| = |N_E(W) : \langle x \rangle|$. Since $N_E(W) \supseteq N_E(\langle x \rangle)$ and $|N_E(\langle x \rangle) : \langle x \rangle| = 12$, we can infer that $N_E(W) = N_E(\langle x \rangle)$. Moreover, from $|W^\sharp| = 7^2 - 1 = 48$ we deduce that there are exactly four E-orbits Δ_i of size 240, each satisfying $|\Delta_i \cap W| = 12$ ($0 \leq i \leq 3$). From $|V^\sharp| = 7^4 - 1 = 2.400$ we conclude that there are just two regular E-orbits Ω_1, Ω_2.

Write $-j$ for the product of j with the generator of $Z(X)$. We know that $-j$ centralizes W and acts as -1 on V/W. Hence each Δ_i remains an X-orbit, with point stabilizer $\langle x, -j \rangle \cong Z_6$. The element $x(-j)$ belongs to the class $6A_0$ in X and is conjugate to g. Since $\chi(g) = \pm\sqrt{-3}$ and g is diagonal on V, from Lemma 4.1c it follows that no vector in the Δ_i is real for X (as 6 does not divide $7 + 1 = 8$). On the other hand, Ω_1, Ω_2 fuse in X, because $-j$ does not fix any vector in Ω_1 or Ω_2 and $X = \langle E, -j \rangle$. So $\Omega = \Omega_1 \cup \Omega_2$ is a regular orbit for X.

From $y_W = [z, \bar{z}]$ we see that there are just $6 + 6 = 12$ (nonzero) eigenvectors of y in W. The group $\langle Z(X), y \rangle \cong Z_6$ acts semiregularly on $W^\sharp$, and if $\Delta_i \cap W$ contains an eigenvector of y for some i, all six eigenvectors of y to the given eigenvalue belong to $\Delta_i \cap U$. Hence there exists Δ_i such that $\Delta_i \cap U$ does not contain an eigenvector of y.

Consider the action of $\langle yz \rangle$ on $V^\sharp$. Clearly Ω_1, Ω_2 and Ω are fixed by yz. Hence there is $v_j \in \Omega_j$ such that $C_{EZ}(v_j) = \langle yz \rangle$. From Lemma 4.1c it follows that no vector in Ω is real for EZ, hence not real for $G = XZ$. Note again that yz acts diagonally on V and that 3 does not divide $7 + 1$. Also, $\langle yz \rangle$ acts on the E-orbits Δ_i, and such a Δ_i remains an EZ-orbit if and only if there is $w_i \in \Delta_i \cap W$ such that $C_{EZ}(w_i) = \langle x, yz \rangle$ (by conjugacy of y, y^{-1} in E). From the observation in the preceding paragraph we infer that just one orbit, say Δ_0, is preserved by $\langle yz \rangle$, and the other ones are fused. So $\Delta = \Delta_1 \cup \Delta_2 \cup \Delta_3$ is an EZ-orbit with point stabilizer $\langle x \rangle$, and Δ_0 has the stabilizer $\langle x, yz \rangle$. The vectors in Δ are real for EZ, those in Δ_0 are not.

We conclude that Ω, Δ and Δ_0 are the distinct G-orbits on $V^\sharp$, with point stabilizers conjugate to $\langle yz \rangle \cong Z_3$, $\langle x(-j) \rangle \cong Z_6$ and $\langle x(-j), yz \rangle \cong Z_3 \times S_3$ ($= Z_3 \mathrm{wr}\, S_2$), respectively. No vector in V is real for G.

Chapter 6

Reduced Pairs of Extraspecial Type

In this chapter we classify the nonreal reduced pairs of extraspecial type, up to isomorphism. In terms of their cores we have just three types (Q_8), (2^5_-) and (3^3_+), the pairs being defined over certain prime fields (of order 3, 5, 7 or 13).

6.1. Nonreal Reduced Pairs

Throughout (G, V) is a reduced pair over a finite field $F = \mathbb{F}_r$ of characteristic p where the core $E \cong q^{1+2m}_{\pm,0}$ is of extraspecial type (Sec. 5.5). So either E is an extraspecial q-group of odd exponent $q \neq p$ and order q^{1+2m}, or $q = 2 \neq p$ and E is extraspecial of $+$ or $-$ type and order 2^{1+2m}, or E is the central product of such a 2-group with a cyclic group of order 4 (type 0). By the character theory of E we have $d = \dim{}_F V = q^m$. Furthermore $|Z(E)|$ is a divisor of $r - 1$. Without loss of generality we assume that $E \cong 2^{1+2m}_0$ when $q = 2$ and F contains the 4th roots of unity $(4 \mid r - 1)$.

We now give a detailed description of certain nonreal reduced pairs, including orbit structures and point stabilizers (which will be of relevance). With one exception the pairs are large. The target will be to show that there are no further nonreal reduced pairs of extraspecial type.

Type (Q_8) : Let $r = p$ be one of the primes $5, 7, 11$ or 23. In Sec. 3.2 we have seen that the quaternion group $E \cong Q_8$ embeds into $\mathrm{GL}_2(p)$ such that $G = N_{\mathrm{GL}_2(p)}(E)$ acts transitively on the nonzero vectors in $V = \mathbb{F}_p^{(2)}$. This G is a p'-group. The stabilizer in G of any nonzero vector is cyclic of order $4, 3, 2$ or 1, correspondingly. Complete reducibility and Lemma 4.1c imply that (G, V) is a nonreal reduced pair when $p = 5$ or $p = 7$.

For $p = 5$ this G is the standard holomorph of $2^{1+2}_0 \cong E \circ Z_4$ (and a 5-complement in $\mathrm{GL}_2(5)$). For $p = 7$ we have $G = X \circ Z_6 = X \times Z_3$ where $X \cong 2^- S_4$ is the standard holomorph of $E \cong 2^{1+2}_-$ with value field $\mathbb{Q}(\sqrt{2})$ (since 2 is a square mod 7). This G has a unique subgroup $G_1 \cong \mathrm{Sp}_2(3) \times Z_3$ of index 2, which evidently gives rise to a further nonreal reduced pair (G_1, V).

Let $r = p = 13$. Embed E into $\mathrm{GL}_2(13)$. The normalizer $G = N_{\mathrm{GL}_2(13)}(E)$ has two orbits on nonzero vectors in $V = \mathbb{F}_{13}^{(2)}$, with cyclic stabilizers of order 4 and 3. Again (G, V) is a nonreal reduced pair.

Type (2_-^5) : Let GV be the largest Bucht group studied in Sec. 3.2. Here G is faithful on V and transitive on $V^\sharp$, any point stabilizer H being cyclic order 8. Also, $F = \mathbb{F}_3$, $\dim_F V = 4$, and G contains $E \cong 2_-^{1+4}$ as a normal subgroup, which is absolutely irreducible on V ($G \cong E.(Z_5 : Z_4)$). Complete reducibility and Lemma 4.1c imply that (G, V) is a nonreal reduced pair (as $8 \mid 3^2 - 1$ but $8 \nmid 3^1 + 1$).

Let $r = p = 7$. Embed $E \cong 2_-^{1+4}$ into $\mathrm{GL}_4(7)$. Then $G = N_{\mathrm{GL}_4(7)}(E)$ is isomorphic to $X \circ Z_6$ where X is the standard holomorph of E with value field $\mathbb{Q}(\sqrt{2})$. Here G is a $7'$-group and has two orbits on nonzero vectors in $V = \mathbb{F}_7^{(4)}$, with point stabilizers Z_6 resp. $\mathrm{SL}_2(3)$. No vector in V is real for G (cf. Prop. 6.6b below). Similar statement for the unique subgroup $G_1 = Y \circ Z_6$ of index 2 in G, where $Y/E \cong \Omega_4^-(2)$.

Type (3_+^3) : Let $X = E : \mathrm{Sp}_2(3)$ be the standard holomorph of $E = 3_+^{1+2}$, and let χ be one of the faithful irreducible characters of X of degree 3, which by Theorem 4.3c are conjugate under automorphisms of X. In view of the discussions in Theorem 4.5a there is, up to conjugacy in X, a unique complement $S \cong \mathrm{Sp}_2(3)$ to E in X such that $\mathrm{Res}_S^X(\chi) = 1_S + \xi_2$ contains the 1-character. Here ξ_2 is an irreducible Weil character of degree 2, with $\mathbb{Q}(\xi_2) = \mathbb{Q}(\sqrt{-3})$.

Let $r = p = 7$ or 13. Let V be a (coprime) FX-module affording χ as a Brauer character. Embed E and X into $\mathrm{GL}(V)$ through χ, and let $G = N_{\mathrm{GL}(V)}(E) = X \circ Z_{r-1}$ (Proposition 4.3a). For $r = 7$ there are three G-orbits on $V^\sharp$ with point stabilizers $S.1$, $Z_3.2 = Z_6$ and $Z_3^{(2)}.2 = Z_3^{(2)} \times Z_2$ (where $A.B$ indicates that A is the stabilizer taken in X). It follows that both (X, V) and (G, V) are nonreal reduced pairs. For $r = 13$ we have six G-orbits on $V^\sharp$, with point stabilizers $S.1$ (four X-orbits fusing in G), $Z_3.2 = Z_6$ (3 times), $Z_3^{(2)}.2 = Z_3^{(2)} \times Z_2$ and $Z_1.4 = Z_4$ (one regular X-orbit). Just the pair (G, V) is nonreal reduced in this $r = 13$ case.

Theorem 6.1. *Up to isomorphism there are just* 9 *nonreal reduced pairs of extraspecial type, and these are described above. All other reduced pairs (G, V) of extraspecial type admit a strongly real vector $v \in V$ for G such that $C_G(v)$ has a regular orbit on V.*

This will be established in the course of this chapter.

6.2. Fixed Point Ratios

Let (G, V) be reduced of extraspecial type, as above. We embed the core E, and G, into $\mathrm{GL}(V)$ and let $G_0 = N_{\mathrm{GL}(V)}(E)$. If r is odd and $4 \nmid r - 1$, the field F contains a square root either of 2 or of -2. Hence from Theorems 4.3c, 4.3d and 4.3e we deduce, in view of Proposition 4.3a, that

$$\text{(6.2a)} \qquad\qquad G_0 = X \circ Z$$

is a unique central product over $Z(E)$ where X is that standard holomorph of E fitting into $\mathrm{GL}(V)$. (This refers only to $E \cong 2_\pm^{1+2m}$, where we have two standard holomorphs determined by the value field $\mathbb{Q}(\sqrt{2})$ or $\mathbb{Q}(\sqrt{-2})$.) There is an ordinary irreducible character χ of G_0 which agrees on p'-elements with the Brauer character of G_0 afforded by V. This is an obvious extension to the central product of "the" character associated to X, that is, of one of the faithful irreducible characters of X of degree q^m (conjugate under automorphisms of X and so leading to isomorphic reduced pairs).

$$\text{(6.2b)} \qquad\qquad U = E/Z(E) = EZ/Z$$

carries in the natural way the structure of a symplectic resp. orthogonal $\mathbb{F}_q G_0$-module. In fact, $\bar{X} = X/E = G_0/ZE = \bar{G}_0$ may be identified with $\mathrm{Sp}_{2m}(q)$ for odd q and for $q = 2$ and $r \equiv 1 \pmod{4}$, and with $\mathrm{O}_{2m}^\pm(2)$ otherwise (the sign depending on E). By definition $\bar{G} = GZ/ZE$ acts *irreducibly* on U, because E is the unique minimal nonabelian normal subgroup of G and all abelian normal subgroups are cyclic and central (contained in $Z(G) = Z \cap G$). The concept of "good" elements for U applies to all (noncentral) elements of G_0; in particular (4.4) applies.

Theorem 6.2c. *Suppose $g \in G_0$ is a noncentral p'-element of order s where s is a prime or $s = 4$.*

(i) *Suppose $q \mid s$. Then $\dim {}_F C_V(g) \leq 2q^{m-1}$ when $s = q$ is odd, and $\dim {}_F C_V(g) \leq \frac{1}{s}(2^m + (s-1) \cdot 2^{m-1})$ otherwise. If g is not good for U, then even $\dim {}_F C_V(g) \leq q^{m-1}$.*

(ii) *Suppose $q \nmid s$. Then $\dim {}_F C_V(g) \leq \frac{1}{s}(q^m + (s-1)q^{m-1})$. If s is odd and $s \geq q + 1$, then even $\dim {}_F C_V(g) \leq \frac{1}{s}(q^m + (s-1)q^{m-2})$.*

Proof. Of course we use that $\dim {}_F C_V(g) = \langle \chi, 1 \rangle_{\langle g \rangle}$ by Eq. (5.6e).

(i) If g is not good for U, then $\chi(g) = 0$ by Theorem 4.4. Then χ vanishes on each generator of $\langle g \rangle$ and so $\langle \chi, 1 \rangle_{\langle g \rangle} = \frac{q^m}{s} = q^{m-1}$ when $s = q$ is a prime. For $s = 4$ (and $q = 2$) observe that $\chi(g^2) = -q^m$ if $g^2 \in Z$, and otherwise $\chi(g^2)$ is an integer of absolute value at most q^{m-1} by Theorem 4.4. The result follows.

Suppose g is good for U. Then $|\chi(g)|^2 = |C_U(g)|$ by Theorem 4.4. Since g is not central in G_0, $|\chi(g)|^2 = q^n$ for some $n \leq 2m - 1$. Consider first the case $q = 2$. If $s = 2$, then $\chi(g)$ is an integer, hence n is even and $\chi(g) = \pm 2^{n/2}$. It follows that $\langle \chi, 1 \rangle_{\langle g \rangle} = \frac{1}{2}(2^m \pm 2^{n/2}) \leq \frac{1}{2}(2^m + 2^{m-1})$. Suppose next that g has order $s = 4$. Then $\chi(g) \in \mathbb{Z}[i]$ $(i^2 = -1)$. If n is odd, there are signs ε_i such that $\chi(g) = 2^{\frac{n-1}{2}}(\varepsilon_1 + i\varepsilon_2)$. Then $\chi(g^{-1}) = 2^{\frac{n-1}{2}}(\varepsilon_1 - i\varepsilon_2)$. Either the integer $\chi(g^2) = 0$ or $= -2^{\frac{n-1}{2}}$ if $g^2 \in E$, and $|\chi(g^2)| \leq 2^{2m-1}$ otherwise (Theorem 4.4). It follows that $\langle \chi, 1 \rangle_{\langle g \rangle} \leq \frac{1}{4}(2^m + 3 \cdot 2^{m-1})$, as desired. If n is even, then $\frac{n}{2} \leq m - 1$ and $\chi(g^2) \leq 2^{m-1}$, and we obtain the same estimate.

Suppose that $s = q \geq 3$. Then $\chi(g) \in \mathbb{Z}[\varepsilon_q]$ $(\varepsilon_q = e^{2\pi i/q})$. Let $\mathfrak{q}$ be the unique prime ideal in $\mathbb{Z}[\varepsilon_q]$ above q (totally ramified). Since $\mathfrak{q} = \bar{\mathfrak{q}}$ and $\chi(g)\overline{\chi(g)} = |\chi(g)|^2 = q^n$, and since $\mathbb{Q}\left(\sqrt{(\frac{-1}{q})q}\,\right)$ is the (unique) quadratic number field contained in $\mathbb{Q}(\varepsilon_q)$, we infer that $\chi(g)/\left((\frac{-1}{q})q\right)^{\frac{n}{2}}$ is a $\mathfrak{q}$-adic integer. It follows that this is even an integer in $\mathbb{Z}[\varepsilon_q]$. From $|\chi(g)/\left((\frac{-1}{q})q\right)^{\frac{n}{2}}| = 1$, and using that the conjugates over $\mathbb{Q}$ have absolute value 1 likewise, we obtain that $\chi(g)/\left((\frac{-1}{q})q\right)^{\frac{n}{2}} = \varepsilon$ is a root of unity (with $\varepsilon^{2q} = 1$). If $n = 2a$ is an even integer, then $\chi(g) = \pm \varepsilon q^a$, and from $a \leq m - 1$ it follows that $\langle \chi, 1 \rangle_{\langle g \rangle} \leq \frac{1}{s}(q^m + (s-1)q^{m-1})$, which is not greater than $2q^{m-1}$. So let $n = 2a + 1$ be odd. Using that $\sqrt{(\frac{-1}{q})q} = \sum_{k=1}^{q-1} (\frac{k}{q})\varepsilon_q^k$ (Gauss) we obtain that

$$\chi(g) = \pm \varepsilon q^a \sqrt{(\frac{-1}{q})q} = \pm \varepsilon q^a \sum_{k=1}^{q-1} (\frac{k}{q})\varepsilon_q^k.$$

Adding or subtracting $\varepsilon q^a \sum_{k=1}^{q-1} \varepsilon_q^k = 0$ we see that the maximum possible multiplicity of an eigenvalue of g on V is $2q^a$. Hence $\dim {}_F C_V(g) \leq 2q^a \leq 2q^{\frac{n-1}{2}} \leq 2q^{m-1}$.

(ii) In this coprime case, g is good for U and $|\chi(g)|^2 = |C_U(g)|$ (Theorem 4.4). Also, $E_0 = C_E(g)$ is a proper extraspecial subgroup of E, or of extraspecial type (mapping onto $C_U(g)$). Let $|E_0/Z(E_0)| = q^{2n}$, and let θ be the unique irreducible constituent of $\mathrm{Res}^G_{E_0}(\chi)$. Then $\theta(1) = q^n$ ($n \leq m - 1$). By Theorem 1.6c there is a sign $\pm$ and a linear character μ of $\langle g \rangle$ such that $\chi(y) = \pm\mu(y)\theta(1)$ for all generators y of $\langle g \rangle$. Moreover, the sign is such that

$$q^{m-n} = \langle \chi, \theta \rangle_{E_0} \equiv \pm 1 \ (\mathrm{mod}\, s)$$

when s is an odd prime. In this case

$$\langle \chi, 1 \rangle_{\langle g \rangle} \leq \frac{1}{s}(q^m + (s-1)\theta(1)),$$

and the result follows. For $s = 4$ use again that $\chi(g^2) = -q^m$ when $g^2 \in Z$, and apply Theorem 4.4 otherwise.

Assume in addition that s is odd and $s \geq q + 1$. If $\theta(1) \leq q^{m-2}$, the result follows. So let $\theta(1) = q^{m-1}$ ($n = m - 1$). Then the sign must be negative, $s = q + 1$, and $\chi(y) = -\mu(y)\theta(1)$ for all generators y. Indeed $s = 3$ and $q = 2$. If μ is the trivial character of $\langle g \rangle$, even $\langle \chi, 1 \rangle_{\langle g \rangle} = \frac{1}{s}(q^m - (s-1)q^{m-1}) = 0$. If μ is not trivial, $\langle \chi, 1 \rangle_{\langle g \rangle} = \frac{1}{3}(2^m + 2^{m-1})$ as the sum over the 3rd roots of unity is zero. $\qquad\square$

Remark. The bounds given in Theorem 6.2c apply also to the eigenspaces of g on V, replacing g by zg for the elements $z \in Z$ of order dividing $s = o(g)$. For all $\gamma \in \beta^*(G)$ (Sec. 5.5) we have the following "generic" upper bounds for the fixed point ratios (depending only on q): $f(\gamma) \leq \frac{1}{3}(1 + \frac{2}{q})$ when $q > 3$, $f(\gamma) \leq \frac{2}{3}$ when $q = 3$, and $f(\gamma) \leq \frac{5}{8}$ for $q = 2$. Observe that $\frac{2}{q} \leq \frac{1}{4}(1 + \frac{3}{q}) \leq \frac{1}{3}(1 + \frac{2}{q})$ for $q > 3$. If g is an involution, then $f(g) \leq \frac{q+1}{2q}$.

6.3. Point Stabilizers of Exponent 2

Recall that $\bar{G}$ is an irreducible p'-subgroup of $\bar{G}_0 = \mathrm{Sp}_{2m}(q)$ when q is odd or $q = 2$ and $4 \mid r - 1$, and of $\bar{G}_0 = \mathrm{O}^\pm_{2m}(2)$ otherwise. For small r this usually implies that $\bar{G}$ is a *proper* subgroup of $\bar{G}_0$. So we require some information on the maximal subgroups of these classical groups, but only for $m \leq 7$ and $q \leq 3$. The reader is referred to [Aschbacher, 1984] and [Liebeck, 1985] for thorough studies of this topic (see also [Kleidman and Liebeck, 1990]). We quote from [Dickson, 1901, pp. 94, 206]:

$$|\mathrm{Sp}_{2m}(q)| = q^{m^2}(q^{2m} - 1)(q^{2(m-1)} - 1)\cdots(q^2 - 1),$$

$$|\mathrm{O}^\pm_{2m}(q)| = 2(q^m \mp 1)q^{m(m-1)}(q^{2(m-1)} - 1)\cdots(q^2 - 1).$$

This yields the estimates $|\mathrm{Sp}_{2m}(q)| < q^{2m^2+m}$ and $|\mathrm{O}^\pm_{2m}(q)| < q^{2m^2-m+1}$.

Lemma 6.3a. *Assume* $\bar{G} \neq \bar{G}_0$. *Then* $|\bar{G}| < q^{4m+4}$ *or* $\bar{G}$ *is isomorphic to a subgroup of* $\bar{G}_0$ *from a certain distinguished finite list, the proper irreducible subgroups of* $\bar{G}_0$ *of maximal order being known.*

This refers to Theorems 4.2, 5.2, 5.4 and 5.5 in [Liebeck, 1985]. The corresponding lists of (nonparabolic) maximal irreducible subgroups can be found in Tables 3.5.C, 3.5.E and 3.5.F in [Kleidman–Liebeck, 1990]. For example, if $\bar{G}_0 \cong \mathrm{Sp}_{2m}(2)$ the list consists of $\mathrm{Sp}_{2k}(2) \mathrm{wr}\, \mathrm{S}_\ell$ where $m = k\ell$ with $\ell > 1$, $\mathrm{Sp}_{2k}(2^l).Z_\ell$ where $m = k\ell$ with ℓ prime, $\mathrm{O}_{2m}^{\pm}(2)$, and of S_{2m+2} when m is even, and $\mathrm{O}_{2m}^{-}(2)$ is "the" proper irreducible subgroup of maximal order $(m \geq 3)$.

Theorem 6.3b. *There exist* $v \in V$ *such that* $C_G(v)$ *is an elementary abelian 2-group except possibly when* $q = 2$ *and* $m \leq 5$, *or when* $q = 3$ *and* $m \leq 3$.

Proof. Suppose $V = \bigcup_{\gamma \in \beta^*(G)} C_V(\gamma)$. In view of the estimate (5.6b) it suffices to show that then necessarily $q = 2$ and $m \leq 5$, or $q = 3$ and $m \leq 3$. Our estimates will be rather crude, at first. Of course, it suffices to consider only those γ where the $C_V(\gamma)$ are different. We use that each $\gamma \in \beta^*(G)$ has at least two generators.

Let first $q > 3$. Then $f(\gamma) \leq \frac{1}{3}(1 + \frac{2}{q})$ by Theorem 6.2c. Using that there are at most $\dim{}_F V = q^m$ distinct eigenvalues on V for each $g \in G$ we get $2|V| = 2r^{q^m} \leq r^{(q^m + 2q^{m-1})/3} q^m q^{2m} |\bar{G}|$. Since $|\bar{G}| = |GZ/EZ| \leq |\mathrm{Sp}_{2m}(q)| < q^{2m^2 + m}$, this yields that

$$r \leq \left(\frac{q^{3m}}{2} |\bar{G}| \right)^{\frac{3}{2(q^m - q^{m-1})}} < \left(\frac{q^{m^2 + 2m}}{\sqrt{2}} \right)^{\frac{3}{q^m - q^{m-1}}}.$$

The function $m \mapsto \frac{m^2 + 2m}{q^m - 1}$ is decreasing for $m \geq 1$ (and $q > 2$). So it suffices to consider the case $m = 1$ in order to see that we have $r < 8$ for $q \geq 11$. But $|Z(E)| = q$ is a divisor of $|F^*| = r - 1$, so that we indeed have $q \leq 7$. For $q = 7$ we get $r < 16$ for $m = 1$, and $r < 3$ for $m = 2$. Since r is a prime power congruent 1 mod 7, the only possibility is $m = 1$, $r = 8$. But in this case $\bar{G}$ is an irreducible subgroup of $\mathrm{Sp}_2(7)$ of (odd) order 21, and we get a contradiction (namely $r < 8$).

For $q = 5$ we get $r < 29$ for $m = 1$ and $r < 7$ for $m = 2$, and so the only possibilities are $m = 1$ and $r = 16$ or $r = 11$. We have seen in Example 5.7a that for $m = 1$ there is always a vector $v \in V$ such that $C_G(v)$ is an elementary abelian 2-group (even with $C_G(v) = 1$). So we are reduced to the cases where $q = 2, 3$.

Let $q = 3$. Then $f(\gamma) \leq f = 2/3$ for each $\gamma \in \beta^*(G)$ by Theorem 6.2c. Using (5.6d) we get $2|V| = 2r^{3^m} \leq (r^{2 \cdot 3^{m-1}} + r^{3^{m-1}})3^{2m}|\bar{G}|$, which gives

$$r^{3^{m-1}} \leq \frac{3^{2m}}{2}\left(1 + \frac{1}{r^{3^{m-1}}}\right)|\bar{G}|.$$

From $|\bar{G}| \leq |\mathrm{Sp}_{2m}(3)| < 3^{2m^2+m}$ we obtain that $m \leq 5$, and for $m = 5$ we must have $r = 2$. But the case $q = 3$, $r = 2$ cannot happen.

It remains to rule out the case $m = 4$ ($q = 3$). The above estimate yields that we must have $r \leq 5$. Then $r = 4$ (as $q = 3 \mid r - 1$). Now $\bar{G}$ is a proper irreducible subgroup of $\mathrm{Sp}_8(3)$ (of odd order) and so $|\bar{G}| \leq |\mathrm{Sp}_4(9).2)| = 2^{10} \cdot 3^8 \cdot 5^2 \cdot 41$ by [Liebeck, 1985, Theorem 5.2]. This bound suffices to get a contradiction.

Let $q = 2$. Then $f(\gamma) \leq f = \frac{5}{8}$ for each $\gamma \in \beta^*(G)$ by Theorem 6.2c. Note that $f2^m$ is an integer for $m \geq 3$. (For the time being we only discuss the cases where $m \geq 6$.) Applying (5.6d) we get $2|V| = 2r^{2^m} \leq (r^{5 \cdot 2^{m-3}} + r^{3 \cdot 2^{m-3}})2^{2m}|\bar{G}|$ and thus

$$r^{3 \cdot 2^{m-3}} \leq 2^{2m-1}\left(1 + \frac{1}{r^{2^{m-2}}}\right)|\bar{G}|.$$

This yields at once that we must have $m \leq 8$. For $m = 8$ we obtain that $r = 3$ (since r is odd). In this case $\bar{G}_0 \cong \mathrm{O}_{16}^{\pm}(2)$. Knowing that $\bar{G}$ is a $3'$-group and that 3^7 is a divisor of $|\mathrm{O}_{16}^{\pm}(2)|$ this case is ruled out.

For $m = 7$ we obtain that $r \leq 7$. Again the case $r = 7$ is ruled out using the upper bound for the order of $\mathrm{O}_{14}^{\pm}(2)$ in place of that for $\mathrm{Sp}_{14}(2)$. For $r = 5$ we also have $|\bar{G}| \leq |\mathrm{O}_{14}^-(2)|$, by the comment after (6.3a), and this suffices to get a contradiction. For $r = 3$ we use that the order of a proper irreducible ($3'$-) subgroup of $\mathrm{O}_{14}^{\pm}(2)$ is bounded above by $|\mathrm{GU}_7(2)| = 2^{21} \cdot 3^9 \cdot 5 \cdot 7 \cdot 11 \cdot 43$ [Liebeck, 1985, Theorems 5.4, 5.5]. We obtain a contradiction.

Let $m = 6$. The usual estimate gives that $r^{24} \leq 2^{89}$, hence $r \leq 13$. Since $|\mathrm{Sp}_{12}(2)|$ is divisible by each (odd) prime $p \leq 13$, we have $|\bar{G}| \leq |\mathrm{O}_{12}^-(2)|$ by Lemma 6.3a, and this leads to $r \leq 5$. Consider $r = 5$. By Lemma 6.3a we are led to a study of the irreducible $5'$-subgroups of $\mathrm{O}_{12}^{\pm}(2)$ and of S_{14} (viewing U as shortened S_{14}-permutation module over $\mathbb{F}_2$). In the first case we get $|\bar{G}| \leq |\mathrm{GU}_6(2)| = 2^{15} \cdot 3^8 \cdot 5 \cdot 7 \cdot 11$ by the theorems of

Liebeck quoted above, and this bound is good enough. There is no problem with the bound $|\bar{G}| \leq |S_{14}|$.

So let $r = 3$. We get the desired contradiction (to the above inequality) if we can show that that the order of an irreducible $3'$-subgroup of $O_{12}^{\pm}(2)$ is bounded above by 2^{27}. Observe that $4m + 4 = 28$. We argue by inspection of the list of maximal subgroups of the orthogonal groups. The maximal subgroups of $O_{12}^{+}(2)$ which are irreducible are either isomorphic to $O_6^{+}(2)\text{wr}\,S_2$, $O_4^{+}(2)\text{wr}\,S_3$ or to almost simple groups for which the $3'$-part of the order is less than 2^{27}, like for $GU_6(2)$ or $O_6^{+}(4).2$ (cf. Table 3.5.E in [Kleidman–Liebeck, 1985]). The maximal subgroups of $O_{12}^{-}(2)$ which are irreducible are either isomorphic to $O_4^{-}(2)\text{wr}\,S_3$ or to $O_6^{-}(4).2$, or are certain almost simple groups for which the $3'$-part of the order is less than 2^{27}, like for S_{13} (cf. Table 3.5.F in [Kleidman–Liebeck, 1990]). From this last table one infers that the $3'$-subgroup of $O_6^{-}(4).2$ have the desired small order. Hence the result. $\qquad\square$

Comments 6.3c. Theorem 6.3b reduces the discussion to some few cases. One can show that then there are point stabilizers of exponent 2 (or 1) when r is large enough:

Suppose $q = 3$ and $m = 3$. The crude bound $\sum_{\gamma \in \beta^*(G)} |C_V(\gamma)| \leq \frac{1}{2}(r^{\frac{2}{3}3^m} + r^{\frac{1}{3}3^m}) \cdot 3^{2m} \cdot |\bar{G}|$ used in the above proof is sufficient to show, taking the upper bound $|\bar{G}| \leq |Sp_6(3)| = 2^{10} \cdot 3^9 \cdot 5 \cdot 7 \cdot 13$, that there is $v \in V$ such that $C_G(v)$ is an elementary abelian 2-group provided $r > 19$. For $m = 2$ the corresponding holds provided $r > 157$, and for $m = 1$ whenever $r > 211$, but it is easy to improve these estimates on the basis of Theorems 4.4, 4.5a. See Sec. 6.5 below.

Let $q = 2$. If $m = 5$, the analogous estimate, taking the upper bound $|\bar{G}| \leq |Sp_{10}(2)| = 2^{25} \cdot 3^6 \cdot 5^2 \cdot 7 \cdot 11 \cdot 17 \cdot 31$, yields that there is $v \in V$ such that $C_G(v)$ is an elementary abelian 2-group provided $r > 37$. In the orthogonal case, taking the upper bound $|\bar{G}| \leq |O_{10}^{-}(2)| = 2^{21} \cdot 3^6 \cdot 5^2 \cdot 7 \cdot 11 \cdot 17$, this holds for $r > 23$. For $m \leq 4$ we argue a bit more carefully:

$m = 4$: In $Sp_8(2)$ there are just $a = 346.832.896$ elements of order at most 4 [Atlas, pp. 124, 125]. By Theorem 6.2c, $|C_V(\gamma)| \leq r^{10}$, r^8, r^9 for $\gamma \in \beta_3(G)$, $\beta_4^*(G)$ or $\beta_s(G)$ for a prime $s \geq 5$, respectively. If there is no vector $v \in V$ such that $C_G(v)$ is an elementary abelian 2-group, then in view of the estimate (5.6d)

$$2 \cdot r^{16} \leq 2^8 a(r^{10} + r^6) + 2^8 \cdot (|Sp_8(2)| - a)(r^9 + r^7).$$

Using that $|\mathrm{Sp}_8(2)| = 2^{16} \cdot 3^7 \cdot 5^2 \cdot 7 \cdot 17$ this forces that $r \leq 71$.

$m = 3$: In $\mathrm{Sp}_6(2)$ there are just $a = 80.707$ elements of order at most 4 [Atlas, p. 47]. By Theorem 6.2c, $|C_V(\gamma)| \leq r^4$, r^5, r^4 for $\gamma \in \beta_3(G)$, $\beta_4^*(G)$ or $\beta_s(G)$ for a prime $s \geq 5$, respectively. From $2 \cdot r^8 \leq 2^6 a(r^5 + r^3) + 2^6(|\mathrm{Sp}_6(2)| - a)(2r^4)$ we get $r \leq 147$. Hence there is $v \in V$ such that $C_G(v)$ is an elementary abelian 2-group provided $r > 139$.

$m = 2$: In $\mathrm{Sp}_4(2) \cong S_6$ there are just $b = 480$ elements of 2-power order or of prime order. Hence at most 480 cosets of G_0 mod Z contain noncentral elements of order 4 or of odd prime order. From $2 \cdot r^4 \leq 2^4 \cdot 480(r^2 + r^2)$ we get $r \leq 83$.

$m = 1$: From $2 \cdot r^2 \leq |G_0/Z|(2r) = 48r$ we get $r \leq 23$.

The corresponding bounds for the orthogonal cases are somewhat better. We emphasize that a more careful computation yields much better estimates which, however, do not suffice for our purposes.

This makes the following case-by-case analysis necessary. We attempt to avoid computer calculations, and this is possible in the present situation. We shall use both (elementary) counting methods and theoretical arguments.

6.4. Characteristic 2

Theorem 6.4. *Let $p = 2$. Then there is a strongly real vector $v \in V$ for G such that $C_G(v)$ has a regular orbit on V.*

Proof. Here G has odd order (by coprimeness) and so is solvable by the Feit–Thompson theorem. By virtue of Theorem 6.3b we can assume that $q = 3$ and that $m \leq 3$ (and dim $_F V = 3^m$).

Of course, F is a field of order r where r is a 2-power and $r - 1$ is divisible by 3. By Theorem 4.3c there is a subgroup $S \cong \mathrm{Sp}_{2m}(3)$ which is a complement to E in X. Let H be the subgroup of S mapping onto $\bar{G} = GZ/EZ$. We know that H is faithful and irreducible on $U = EZ/Z$, whence $O_3(H) = 1$. Also, H has odd order. For $m = 1$ this forces that $H = 1$. Hence it remains to examine the cases $m = 2, 3$. By Theorem 4.5a $\mathrm{Res}_S^X(\chi) = \xi$ is a Weil character of S and $\xi = \xi_1 + \xi_2 = 2\xi_1 + 1_S$ on $2'$-elements of S, where the ξ_i are irreducible characters of S. Thus there is v in $V^\sharp$ such that $C_S(v) \supseteq H$ (by coprimeness). We have $N_E(Fv) \neq E$ as E is irreducible on V. Moreover, $N_E(Fv)$ is H-invariant and so $N_E(Fv) = Z(E)$ as H is irreducible on U. It follows that

$$N_{HE}(Fv) = H \times Z(E).$$

Hence $C_{GZ}(v) = C_{HE}(v) = H$ by Lemma 5.1b, and v is strongly real for G provided ξ takes only real values on H. Recall from Theorem 4.5a that ξ takes rational values on the $3'$-elements of H.

Let $m = 2$. Then $\xi = \chi_2 + \chi_{21}$ (resp. $\bar{\xi} = \chi_3 + \chi_{22}$) in the Atlas notation [Atlas, p. 26]. Checking the list of (irreducible) maximal subgroups of $\mathrm{Sp}_4(3)$ we see that H must be in a group of the form $2.(2^4 : A_5)$ or of the form $2.S_6$. In both cases necessarily $|H| = 5$ and $\xi(x) = -1$ for each generator x of H (belonging to the conjugacy class $5A$).

Let $m = 3$. The irreducible odd order group H can only be contained in (almost maximal) subgroups of $\mathrm{Sp}_6(3)$ of the following types: $2.A_5$, $\mathrm{Sp}_2(13)$, $\mathrm{SL}_3(3)$, $\mathrm{U}_3(3).2$, $\mathrm{Sp}_2(3)\mathrm{wr}\,S_3$ and $\mathrm{Sp}_2(27) : 3$ [Atlas, p. 113]. Considering the maximal (and second maximal) subgroups of these groups we get that either H is cyclic of order 7 or is a Frobenius group of order 21. Note that H cannot be of order 13 or of type $13 : 3$, because 7 is the unique Zsigmondy prime divisor of $3^6 - 1$ which divides $3^3 + 1 = 28$, the order of a Singer cycle in $\mathrm{Sp}_6(3)$. This implies that H is contained in groups of types $\mathrm{Sp}_2(13)$ or $\mathrm{PGL}_2(7)$ (embedded in $\mathrm{U}_3(3).2$), which are real groups, or in a group of type $\mathrm{Sp}_2(27) : 3$ (but not in $\mathrm{Sp}_2(27)$ unless $|H| = 7$). Also, H is uniquely determined in $\mathrm{Sp}_6(3)$ up to conjugacy. By the character table of $\mathrm{Sp}_6(3)$ the elements $h \in H$ of order 7 have $\xi(h) = -1$ and $\dim {}_F C_V(h) = 3$.

We may assume without loss that H is Frobenius of type $7 : 3$. We have to show that ξ takes only real values on the elements of order 3, and that there is a regular H-orbit on V. Inspection of the character tables of $\mathrm{Sp}_2(13)$, $\mathrm{U}_3(3).2 \cong G_2(2)$ and $\mathrm{Sp}_2(27) : 3$ yields that we must have $\xi(y) = 0, 3$ or -9 for each element $y \in H$ of order 3, as desired. Since the fixed spaces of h or y on V have dimension at most 11, and since $21 \cdot r^{11} < r^{27}$ for $r \geq 4$, there is a regular H-orbit on V. $\qquad\square$

Remark. Theorem 6.4 solves the $k(GV)$ problem in characteristic 2, in view of our previous results. The Robinson–Thompson Theorem 5.2b tells us that it suffices to find a (strongly) real vector in V for G. Arguing by induction on $\dim {}_F V$ by Theorem 5.4 we are led to the situation that (G, V) is a reduced pair, and then Theorem 6.4 applies. A different approach (for groups G of odd order) has been given in [Gluck, 1984].

6.5. Extraspecial 3-Groups

Theorem 6.5. *Suppose p and q are odd. Then there exists a strongly real vector $v \in V$ such that $C_G(v)$ has a regular orbit on V, except when (G, V) is isomorphic to one of the three nonreal reduced pairs of type (3^3_+) for $r = p = 7, 13$.*

Proof. By Theorem 6.3b we may assume that $q = 3$ and $m \leq 3$. Thus $p \geq 5$ and $3 \mid r - 1$, whence $r \geq 7$. By Theorem 4.3c, $X = E : S$ is a semidirect product of $E \cong 3^{1+2m}_+$ with $S \cong \mathrm{Sp}_{2m}(3)$, the complements being conjugate unless $m = 1$ (in which case $S \cdot Z(E)/Z(E)$ is determined up to conjugacy in $X/Z(E)$). Also, each element of $S \cdot Z(E)$ is good for $U = E/Z(E)$. Recall also that $\mathrm{Res}^X_S(\chi) = \xi$ is a Weil character of S (Theorem 4.5a). Let j be the central involution of S, which inverts the elements of U.

Let $m = 1$. Then G_0 is a p'-group. Without loss of generality let $G = G_0$. The character table of $S \cong \mathrm{Sp}_2(3)$ can be found in [I, p. 288]. Using the notation used there $\xi = \chi_2 + \chi_7$ (resp. $\bar{\xi} = \chi_3 + \chi_6$). We apply the counting techniques described in Sec. 5.6. There is one conjugacy class of elements x in S of order 4 with $|S : C_S(x)| = 2 \cdot 3$. We have $\xi(x) = 1$ and hence $x_V = [1, i, -i]$ if $4 \mid r - 1$ and $x_V = [1]$ otherwise, and $|U : C_U(x)| = 3^2$. There are two conjugacy classes of elements of order 3 in S, represented by y and y^{-1} where $\xi(y) = i\sqrt{3}$ say. So $y_V = [z^{(2)}, 1]$ for some primitive 3rd root of unity z, and $|E : C_E(y)| = 3$ (and $|S : C_S(y)| = 4$). By Theorem 4.4, E permutes the good cosets of $Z(E)$ in Ey transitively by conjugation, and so there are just $3^2 - 3 = 6$ cosets of $Z(E)$ in Ey which are not good for U. Also, χ vanishes on the elements of order 3 in these bad cosets (so that each eigenvalue has multiplicity 1). Similar statement for the noncentral elements of E (of order 3). Since there are four subgroups of order 3 in $U = E/Z(E)$, we get

$$\sum_{\gamma \in \beta^*(G)} |C_V(\gamma)| \leq 3 \cdot 3^2 (3r) + 4 \cdot 3(r^2 + r) + (4 \cdot 6 + 4)(3r).$$

This is less than $|V| = r^3$ provided $r > 20$. For $r = 19$ we may replace the first summand on the right by $\sum_{\gamma \in \beta^*_4(G)} |C_V(\gamma)| \leq 3 \cdot 3^2 r$ (as $4 \nmid 19 - 1$). This gives the result for $r = 19$. The remaining cases are $r = 13$ and $r = 7$, which lead to the nonreal reduced pairs of type (3^3_+).

Let $m = 2$. Now $S \cong \mathrm{Sp}_4(3)$ has order $2^7 \cdot 3^4 \cdot 5$. For $p \neq 5$ we may assume that $G = G_0$, and otherwise $\bar{G}$ is an irreducible (solvable)

$5'$-subgroup of $\mathrm{Sp}_4(3)$. We have $\xi = \chi_2 + \chi_{21}$ (resp. $\bar{\xi} = \chi_3 + \chi_{22}$) in the Atlas notation [Atlas, p. 27].

The usual counting method yields that $\sum_{\gamma \in \beta^*(G)} |C_V(\gamma)| < |V| = r^9$ for $r \geq 10$, but not for $r = 7$. An even more precise estimate gives this inequality also for $r = 7$, but this requires the computer [GAP]. So we only sketch the estimate, and then proceed by giving the (theoretical) argument proposed in [Gluck–Magaard, 2002a] for $m = 2$ and $m = 3$.

There three conjugacy classes $2B_0$ and $4A_0, 4A_1$ of elements of order 4 in $S \cong \mathrm{Sp}_4(3)$ lead, on the basis of (4.4), to the estimate

$$\sum_{\gamma \in \beta_4^*(G)} |C_V(\gamma)| \leq 3^4 \cdot 135(r^3 + 3r^2) + 3^2 \cdot 270(3r^3) + 3^4 \cdot 270(2r^3 + r^2 + r).$$

Similarly we obtain $\sum_{\gamma \in \beta_5(G)} |C_V(\gamma)| \leq 1.296 \cdot 3^4(4r^2 + r)$. There are four conjugacy classes $3A_0B_0$ and $3C_0$, $3D_0$ of elements of order 3 in S, and there are 40 subgroups of order 3 in U. This yields the upper bound $\sum_{\gamma \in \beta_3(G)} |C_V(\gamma)| \leq 3 \cdot 40(r^6 + r^3) + 3^2 \cdot 120(2r^4 + r) + 3^2 \cdot 240(r^5 + 2r^2) + (40 + 40(3^4 - 3) + 120(3^4 - 3^2) + 240(3^4 - 3^2))(3r^3)$. Now sum up.

In what follows let either $m = 2$ or $m = 3$. Let P be the stabilizer in S of a maximal totally isotropic subspace W of U. Thus $P = R : L$ is a maximal parabolic subgroup of S where W is the standard module for $L \cong \mathrm{GL}_m(3)$ and where the unipotent radical $R \cong \mathrm{Sym}^2(W)$ is an irreducible L-module and L'-module (Appendix B and Appendix A7). Also $U = W \oplus W^*$ where the dual L-module W^* is not isomorphic to W by Lemma 4.1c. For $m = 3$ we know that W and W^* are not even isomorphic as modules for $L' \cong \mathrm{SL}_3(3) \cong \mathrm{PSL}_3(3)$ ($L = L' \times \langle j \rangle$). For $m = 2$ the modules $W = \langle e_1, e_2 \rangle$ and $W^* = \langle e_1^*, e_2^* \rangle$ are isomorphic L'-modules, but the unique further (L-conjugate) irreducible L'-submodules $\langle e_1 + e_2^*, e_2 - e_1^* \rangle$ and $\langle e_2 + e_1^*, e_2^* - e_1 \rangle$ of U are nondegenerate. In terms of this hyperbolic basis, let $\tau \in S$ be the linear (symplectic) map on U sending e_i to e_i^* and e_i^* to $-e_i$ (as in Theorem 4.5b). Then $\tau^2 = j$ and, at any rate, $N_S(L') = N_S(L) = \langle L, \tau \rangle$. Of course τ interchanges W and W^*. Each element of L is conjugate to its inverse within $N_S(L)$. Hence $\chi = \xi$ is real-valued on L.

In view of (A2) $\mathrm{H}^1(L, W) = 0 = \mathrm{H}^1(L, W^*)$ (as $j \in L$). Identifying $W^* = \mathrm{Hom}(W, Z(E))$, and $W = (W^*)^*$, therefore by (A7) there exist L-invariant elementary abelian subgroups A, A^* of E mapping isomorphically onto W, W^*, respectively. Both $C_V(A)$ and $C_V(A^*)$ are 1-dimensional, and distinct. Let $\widetilde{V} = C_V(A) \oplus C_V(A^*)$. Since both direct summands of $\widetilde{V}$ are

L-invariant, L' acts trivially on $\widetilde{V}$. Also, τ interchanges $A = [A \times Z(E), L]$ and $A^* = [A^* \times Z(E), L]$, hence also $C_V(A)$ and $C_V(A^*)$. There is $v \in \widetilde{V}$ outside $C_V(A)$ and $C_V(A^*)$ which is not an eigenvector of τ (as $r \geq 7$).

Let $H = C_{G_0}(v)$. We assert that $H \cap ZE = 1$. Otherwise, since $C_Z(v) = 1$, $H \cap ZE$ is a nontrivial L'-invariant elementary abelian subgroup of ZE whose image in $U = ZE/Z$ is nonzero and totally singular and so must be W or W^*. Since H contains $[H \cap ZE, L]$, which is A or A^*, it follows that A or A^* centralize v. This contradicts the choice of v. Hence the assertion.

We know that $N_{G_0}(L') = N_G(L) = \langle L, \tau \rangle \times Z$. Assume $H \subseteq N_{G_0}(L')$. Using that $H \cap ZE = 1$ and that $\langle L, \tau \rangle / L'$ is elementary abelian (of order 4) then even $H \subseteq L \times \langle -1 \rangle$. In this case v is a strongly real vector for G. Also, $C_G(v) = H \cap G$ has a regular orbit on V. This is immediate for $m = 2$ (as $r \geq 7$). For $m = 3$ and $p \neq 13$ the group $L = \mathrm{GL}_3(3)$ is a p'-group ($p \geq 5$) and so has a regular orbit on each faithful irreducible L-submodule of V (which exists) by Theorem 7.2a below (see also the remark after this theorem). When $p = 13$ use a simple counting argument, noting that every involution in L' has r^{15} fixed points on V, and that every element of order 3 at most r^{12} (character table), and that $\dim {}_F C_V(j) = 14$.

So the proof is accomplished if we can show that $H \subseteq N_{G_0}(L')$ or, at least, that v can be chosen in V_0 such that this holds for its stabilizer. At any rate we have $H' \subseteq S$. To see this use that $\tau^2 = j$. We have $j \in L'$, even $j \in L'' \cong Q_8$ when $m = 2$, whereas $j \notin L'$ for $m = 3$ (in which case $L = L' \times \langle j \rangle$). Since τ interchanges $C_V(A)$ and $C_V(A^*)$, we get that j acts on V_0 as 1 or -1. Hence j normalizes Fv and so $[H, j] \subseteq H$. But j maps onto the central involution in $G_0/ZE \cong \mathrm{Sp}_{2m}(3)$ and so $[H, j] \subseteq H \cap ZE = 1$. It follows that indeed $H' \subseteq C_{G_0}(j)' = (SZ)' = S' = S$.

Recall that $L' \subseteq H$ (but $\tau \notin H$). Since $H_0 = H'L'$ centralizes v and ξ does not contain the 1-character, H_0 is a proper subgroup of S. Suppose that H_0 normalizes W. Then $\Lambda = \mathrm{Hom}(W, Z(E))$ is an irreducible H_0-module which agrees with $W^* = W^\tau$ on L'. We have $\mathrm{H}^1(L', \Lambda) = 0$, either by (A2) since $j \in L'$ ($m = 2$), or by (A9), (A7) since $\Lambda \cong W^\tau$ ($m = 3$). We claim that H_0 normalizes A. For $m = 2$ we know that $j \in H_0$ and so $\mathrm{Ext}_{H_0}(W, Z(E)) = 0$ by (A2) and (A7). For $m = 3$ we have either $H_0 \subseteq L$ or H_0 is a subgroup of P containing the (irreducible) unipotent radical $R = O_3(P)$. In the latter case Λ and R are nonisomorphic irreducible H_0-modules. Application of the exact inflation–restriction sequence (Appendix

A2) yields that $\mathrm{H}^1(H_0/R, \Lambda) \cong \mathrm{H}^1(H_0, \Lambda)$. In fact, the sequence

$$0 \to \mathrm{H}^1(H_0/R, \Lambda) \to \mathrm{H}^1(H_0, \Lambda) \to \mathrm{Hom}_{H_0}(R, \Lambda) = 0$$

is exact. Now $H_0/R \cong L$ or L' and $\Lambda \cong W^*$ as L-module. We conclude that $\mathrm{Ext}_{H_0}(W, Z(E)) = \mathrm{H}^1(H_0, \Lambda) = 0$, as before. It follows that there is an H_0-invariant subgroup of E mapping isomorphically onto W, and this must be $A = [A \times Z(E), L']$, as claimed. Hence H_0 normalizes $C_V(A)$ and H_0' centralizes $V_0 = C_V(A) \oplus Fv$, whence $C_V(A^*)$. Clearly $N_E(C_V(A^*)) \supseteq A^*Z(E)$, and we must have equality since $|E : N_E(C_V(A^*))| \geq \dim {}_F V = |E : A^*Z(E)|$ and E is irreducible on V. Therefore H_0' normalizes W^* likewise, which implies that $H_0' \subseteq L$. We conclude that $H_0 \subseteq N_S(L) = N_S(L')$ and $H \subseteq N_{G_0}(L')$, which gives the result. A similar argument works when H' normalizes W^*.

Let $m = 3$. Then $N_S(L) = L\langle\tau\rangle$ is a maximal subgroup of $S \cong \mathrm{Sp}_6(3)$ [Atlas, p. 113]. From $\tau \notin H$ we get that $H' \supseteq L'$ is contained in a maximal parabolic subgroup stabilizing either W or W^*. We are done in this case.

Thus it remains to consider the case $m = 2$ (and $r = 7$). We inspect the overgroups of L' in S which do not normalize L' [Atlas, p. 26]. Let y be an element in L' of order 3. Since $\xi(y)$ is real and y centralizes V_0, y belongs to the class $3D_0$ of S (see above). It follows that $L'/\langle j \rangle$ cannot be in a maximal subgroup of type $2^4 : A_5$ of $S/\langle j \rangle \cong \mathrm{PSp}_4(3)$, because each element of order 3 in A_5 centralizes a V_4-subgroup in the shortened permutation module 2^4 over $\mathbb{F}_2$, whereas $|C_S(y)| = 2 \cdot 54$. Also, L' cannot map into a maximal subgroup of $\mathrm{PSp}_4(3)$ of type $3^{1+2}_+ : \mathrm{Sp}_2(3)$ as $j \in L'$. So L' must be contained in maximal subgroups $X \cong \mathrm{Sp}_2(3)\mathrm{wr}\,Z_2$ or $Y \cong 2.S_6$ of S.

X permutes the two nondegenerate irreducible L'-submodules of U, so is the unique overgroup of L' of this kind. Further, L' must be diagonally embedded in the base group $B = L_1 \times L_2$ of X ($L' \cong L_i \cong \mathrm{Sp}_2(3)$). We claim that $H_0 \supseteq O_2(X) \cong Q_8 \times Q_8$. Otherwise $H_0 \cap O_2(L_i) \subseteq Z(L_i)$ for each i, because L' acts irreducibly on $O_2(L_i)/Z(L_i)$. A Sylow 3-subgroup of H_0 acts on $O_2(H_0) \cap B \subseteq O_2(L')Z(X) \cong Q_8 \times Z_2$, inducing at most an automorphism group of order 3. We conclude that $|H_0|_3 = 3$, because otherwise there is an element of order 3 in $H_0 \smallsetminus L'$ centralizing $L' = \langle y, O_2(L') \rangle$. We infer that L' has index 2 or 1 in $H_0 \cap B$, which is impossible by assumption. Now

$$\mathrm{Res}^S_B(\xi) = \xi_1 \otimes \xi_2$$

where ξ_i is an (appropriate) Weil characters of L_i (corresponding to a central decomposition $E \cong 3_+^{1+2} \circ 3_+^{1+2}$; cf. Lemma 7.7b for more details). Since each ξ_i is the sum of two irreducible Weil characters of degree 1 and 2, which remain irreducible on $O_2(L_i)$, we obtain that $O_2(B) = O_2(X)$ fixes exactly one point in $P_1(V)$, hence at most one point in $P_1(\widetilde{V})$.

Suppose $L' \subseteq Y$. Then Y is the unique $2.S_6$ subgroup of S containing L'. This may be seen by using that $\mathrm{PSp}_4(3) \cong \Omega_5(3)$ and noting that $Y/\langle j \rangle \cong O_4^-(3)$ is the stabilizer of a minus point in the orthogonal space [Atlas]. Also, since L' embeds into X, its central quotient group centralizes a 1-dimensional plus type subspace [Atlas]. If L' were contained in two different $2.S_6$ subgroups of S, the fixed space of $L'/\langle j \rangle \cong A_4$ on the orthogonal space is (at least) 3-dimensional and so A_4 has a faithful irreducible module over $\mathbb{F}_3$ of dimension $5 - 3 = 2$, which is not true. Now any two A_4 subgroups of A_6 are conjugate under $\mathrm{Aut}(A_6)$, and each A_4 subgroup is contained in exactly two A_5 subgroups of A_6 [Atlas, p. 4]. We conclude that L' is contained in exactly two $2.S_5$ subgroups of Y, each of which can fix at most one point in $P_1(\widetilde{V})$. Each solvable subgroup of Y above L' normalizes L'.

Consequently, if we remove from $P_1(\widetilde{V})$, besides $C_V(A)$, $C_V(A^*)$ and the eigenspaces of τ, eventually further $3 = 1 + 2$ points ($|P_1(\widetilde{V})| = r + 1 \geq 8$), there remains a 1-dimensional subspace Fv for which $C_{G_0}(v) \subseteq L \times \langle -1 \rangle$. This completes the proof. $\qquad\square$

6.6. Extraspecial 2-Groups of Small Order

Proposition 6.6a. *Let $E \cong 2_{\pm,0}^{1+2}$. There exists $v \in V$ such that $C_G(v)$ is an elementary abelian 2-group unless (G, V) is isomorphic to one of the four nonreal reduced pairs of type (Q_8) for $p = r = 5, 7, 13$.*

Proof. We cannot have $E \cong 2_+^{1+2}$, because then X either is dihedral ($r \equiv 7 \pmod 8$) or semidihedral ($r \equiv 3 \pmod 8$) of order 16 and cannot be irreducible on $U = E/Z(E)$. So $E \cong Q_8$ or $E \cong Q_8 \circ Z_4$ (in case $4 \mid r - 1$). In both cases $|G_0| = 24(r - 1)$ and $G_0/Z \cong S_4$. Since $\bar{G}$ is an irreducible p'-subgroup of $\bar{G}_0 \cong S_3$, we have $p \neq 3$, and we may take $G = G_0$.

There are at most six noncentral cyclic subgroups γ in G of order 4 (and at most three if $4 \nmid r - 1$), each having a fixed space on V of dimension at most 1, because $\dim{}_F C_V(\gamma) \leq \frac{1}{4}(2 + 3 \cdot 1) \leq \frac{5}{4}$ by Theorem 6.2c. There are at most eight noncentral subgroups of order 3 in G (and at most four such subgroups if $3 \nmid r - 1$), each having a fixed point space of dimension at most 1 by Theorem 6.2c (with $s = 3$, $q = 2$). Hence

$$\sum_{\gamma \in \beta^*(G)} |C_V(\gamma)| \leq 14r < r^2 = |V|$$

provided $r > 14$. For $r = 11$ the sum on the left is at most $7r$. (For $r = 11$ we know from Sec. 3.2 that $G = G_0 \cong \mathrm{GL}_2(3) \times Z_5$ is transitive on $V^\sharp$ and that $|C_G(v)| = 2$ for each non-zero vector v.) For $r = 5, 7, 13$ we get the nonreal reduced pairs of type (Q_8) described in Sec. 6.1. $\qquad\square$

Proposition 6.6b. *Let* $E \cong 2^{1+4}_{\pm,0}$. *There exists* $v \in V$ *such that* $C_G(v)$ *has a regular orbit on* V, *and this vector* v *is strongly real for* G *unless* (G, V) *is isomorphic to one of the three nonreal reduced pairs of type* (2^5_-).

Proof. We treat the three types for E separately.

(i) Let $E \cong 2^{1+4}_+$. Then $\bar{G}$ is an irreducible p'-subgroup of $\bar{G}_0 \cong O_4^+(2) \cong S_3 \mathrm{wr} S_2 \cong 3^2 : D_8$. Therefore we have $p \neq 3$, and we may assume that $G = G_0$. Further $r - 1$ is not divisible by 4 and so $r \geq 7$. For $r = 7$ by [GAP] computation (or directly) one finds $v \in V$ such that $C_G(v) \cong D_{12}$, which is a real group and has a regular orbit on V. So let $r > 7$ (hence $r \geq 11$).

Let S be a Sylow 3-subgroup of X (mapping onto $O_3(\bar{X})$) and let $Y = N_X(S)$. By the Frattini argument $YE = X$. Hence Y is irreducible on $U = E/Z(E)$ and so $Y \cap E = Z(E)$. There are two Y-orbits on $S^\sharp$ of sizes 4. Let g_1, g_2 in S represent these orbits, and choose notation such that $|C_U(g_1)| = 1$ and $|C_U(g_2)| = 2^2$. In fact, g_1 permutes the 6 nonsingular vectors in U in two orbits of size 3 and g_2 fixes 3 of them. (Consider a central decomposition $E = Q_8 \circ Q_8$.) From Theorem 4.4 it follows that $\chi(g_1) = \pm 1$ and $\chi(g_2) = \pm 2$. Computation of $\dim {}_F C_V(g_i) = \frac{1}{3}(4 + 2\chi(g_i))$ shows that $\chi(g_1) = 1$ and $\chi(g_2) = -2$. We have $(g_1)_V = [1^{(2)}, z_1, z_2]$ and $(g_2)_V = [z_1^{(2)}, z_2^{(2)}]$, provided $3 \mid r - 1$, $z_1 \neq z_2$ being the primitive 3rd roots of unity. For each noncentral involution x in E we have $x_V = [1^{(2)}, -1^{(2)}]$. There are $9 = 15 - 6$ singular points in U, hence $2 \cdot 9$ noncentral involutions in E. We have

$$18r^2 + \sum_{\gamma \in \beta_3(G)} |C_V(g)| \leq 18r^2 + 2 \cdot 2^4(r^2 + 2r) + 2 \cdot 3^2(2r^2) < r^4 = |V|$$

since $r \geq 11$. Hence there is a vector $v \in V$ such that $H = C_G(v)$ is a 2-group satisfying $H \cap ZE = 1$. It follows that H is isomorphic to a subgroup of a dihedral group D_8. Hence χ takes only rational values on H, because $\exp(H) \mid 4$ and $4 \nmid r - 1$. It is obvious that H has a regular orbit on V.

(ii) Let $E \cong 2_{-}^{1+4}$. By (A9) there is a subgroup T of X such that $T \cdot Z(E) = X$ and $T \cap E = Z(E)$, and this is determined up to conjugacy. (Here $U = E/Z(E)$ is the projective Steinberg module for $\bar{X}' \cong \Omega_4^-(2)$.) Assume $T' \cong A_5$. Then inspection of the character table, on the basis of Theorem 4.4, shows that $\operatorname{Res}_{T'}^X(\chi) = \chi_1 + \chi_2$ or $\chi_1 + \chi_3$ or χ_4 in the Atlas notation [Atlas, p. 2]. However, in these cases the character χ takes the value 1 on the conjugacy class $3A$ of elements of order 3 in A_5 and so these elements have trivial fixed space on U by Theorem 4.4. But the elements of order 3 in $O_4^-(2)$ stabilize singular or nonsingular points in U. Consequently $T \cong 2.S_5$ and

$$\operatorname{Res}_{T'}^X(\chi) = \chi_6 + \chi_7,$$

where χ_6, χ_7 are algebraically conjugate but fuse to a rational-valued character of T. (The characters χ_8 of degree 4 are ruled out by their values $\chi_8(6A_0) = \pm\sqrt{-3}$.) Let y represent the unique conjugacy class of elements in T of order 3. Then $|T : C_T(y)| = 2 \cdot 10$ and $\chi(y) = -2$. Hence $\dim_F C_V(y) = \frac{1}{3}(4 - 2 - 2) = 0$ and, when $3 \mid r - 1$, then $y_V = [z_1^{(2)}, z_2^{(2)}]$ where $z_1 \neq z_2$ are the primitive 3rd roots of unity. In view of Theorem 4.4, y is good for U and $|E : C_E(y)| = 2^2$. It follows that $\sum_{\gamma \in \beta_3(G)} |C_V(\gamma)| \leq 2^2 \cdot 10(2r^2) = 80r^2$ if $3 \mid r - 1$, and $\bigcup_{\gamma \in \beta_3(G)} C_V(\gamma) = 0$ otherwise. There are two conjugacy classes $5A_0B_0$ in T of elements of order 5, represented by g, g^{-1} say. Then $|T : C_T(g)| = 6$ and $\chi(g) = -1$. It follows that $|E : C_E(g)| = 2^4$, by (4.4), and that $C_V(g) = 0$. Hence each nonidentity eigenvalue of g on V occurs with multiplicity 1 (if $5 \mid r - 1$). Thus $\sum_{\gamma \in \beta_5(G)} |C_V(\gamma)| \leq 2^4 \cdot 6(4r) = 384r$. There are five singular points in U, hence $5 \cdot 2$ noncentral involutions in ZE, each having a 2-dimensional fixed space on V. Letting $\hat{\beta}(G)$ be the union of $\beta_3(G)$, $\beta_5(G)$ and the subgroups generated by these involutions, we therefore have the estimate

$$\sum_{\gamma \in \hat{\beta}(G)} |C_V(\gamma)| \leq 90r^2 + 384r.$$

This is less than $|V| = r^4$ if $r > 11$. For $r = 11$ the right hand side may be replaced by $10r^2 + 384r$. Hence there is $v \in V$ outside $\bigcup_{\gamma \in \hat{\beta}(G)} C_V(\gamma)$ when $r \geq 11$. Then $H = C_{G_0}(v)$ is a 2-group and $H \cap ZE = 1$. So H is isomorphic to a Sylow 2-subgroup of S_5, which is dihedral of order 8. As in (i) v is strongly real for G, and $C_G(v)$ has a regular orbit on V.

It remains to examine the cases $r = 3$ and $r = 7$. In the former case G is a $3'$-group and so $\bar{G}$ maps into $O_2^-(4) \cong 5 : 4$, but $O_5(\bar{G}) \neq 1$ as it is irreducible. If GV is not the largest Bucht group, $C_G(v)$ has order $1, 2$ or 4 for each $v \in V^\sharp$. Then v is strongly real for G by Lemma 4.1c. Let $r = 7$. There is a subgroup $S \cong SL_2(3)$ of T, containing the element y say. Then χ splits up on $S\Delta_{Z_3} Z_3$ the 1-character. If $G \supseteq X' \times Z_3$, we obtain a vector $v \in V^\sharp$ such that $C_G(v) \cong SL_2(3)$, and v is not real for G. In this case there is only one further orbit on $V^\sharp$, with point stabilizer Z_3 or Z_6 (depending on whether $G = X' \times Z_3$ or $X \times Z_3$), and Lemma 4.1c applies. Exclude now these two nonreal reduced pairs of type (2^5_-) just described. Then either $G \subseteq X$, or $\bar{G}$ must map again into $O_2^-(4)$. The above estimate applies ignoring the elements of order 3 and 5.

(iii) Let $E \cong 2_0^{1+4}$ (so that $4 \mid r - 1$). By (A9) and the character table of $\bar{G}_0 \cong Sp_4(2) \cong S_6$ there is a subgroup $T \cong 2.A_6$ of X such that $Z(T) = T \cap Z$, and $\operatorname{Res}_T^X(\chi) = \chi_8$ or χ_9 in the Atlas notation [Atlas, p. 5].

Consider first the case $p = 3$. Then $\bar{G}$ is an irreducible $3'$-subgroup of $\bar{G}_0$. Inspection of the maximal and second maximal subgroups of S_6 [Atlas, pp. 4, 2] shows that $\bar{G}$ is isomorphic to a subgroup of $O_2^-(4).2 \cong 5 : 4$. Without loss of generality we may assume that $\bar{G} \cong O_2^-(4).2$, in which case the elements of order 2 (and 5) in $\bar{G}$ belong to $O_2^-(4) \subseteq A_6$. Let $x \in T$ be an element of order 5 whose coset generates $O_5(\bar{G})$. Then $\chi(x) = -1$ and $C_V(x) = 0$ [Atlas] and $|E : C_E(x)| = 2^4$ by Theorem 4.4. It follows that $\sum_{\gamma \in \beta_5(G)} |C_V(\gamma)| \leq 2^4(4r)$ (assuming the worse case $5 \mid r - 1$). Let $\tilde{\beta}_4(G)$ be the set of cyclic subgroups of G of order 4 or 2 mapping onto the five subgroups of order 2 in $\bar{G}$. Let $\bar{y}$ be an involution in $\bar{G}$. Then $U = E/Z(E)$ is the regular $\mathbb{F}_4\langle \bar{y} \rangle$-module. There is $y_0 \in T$ mapping onto $\bar{y}$, and this y_0 has order 4 and satisfies $y_0^2 \in Z(E)$ and $\chi(y_0) = 0$ [Atlas]. Thus y_0 is bad for U, $C_V(y_0) = 0$ and $(y_0)_V = [i^{(2)}, -i^{(2)}]$. Let $\langle y \rangle \in \tilde{\beta}_4(G)$ be such that y maps onto $\bar{y}$. If $Z(E)y = Z(E)y_0$, then this coset describes one of the $|C_U(\bar{y})| = 4$ complements to U in $\langle U, Z(E)y_0 \rangle$, which are all conjugate to $\langle Z(E)y_0 \rangle$. So assume that $Z(E)y \neq Z(E)y_0$. Then y cannot be an involution and $Z(E)y$ has order 4. If y is bad for U, then y^2 is an involution in $E \smallsetminus Z(E)$ and so $\chi(y) = \chi(y^2) = 0$ and $y_V = [i, -i, 1, -1]$. If y represents one of the four good cosets, then $|\chi(y)|^2 = 4$ and $y_V = [\pm i^{(2)}, 1, -1]$ or $[\pm 1^{(2)}, i, -i]$. Hence

$$\sum_{\gamma \in \tilde{\beta}_4(G)} |C_V(\gamma)| \leq 5\left(4(2r^2) + \frac{1}{2}8(4r) + \frac{1}{2}4(r^2 + 2r)\right).$$

It follows that $\sum_{\gamma \in \beta_5(G) \cup \tilde{\beta}_4(G)} |C_V(\gamma)| \leq 50r^2 + 164r$, which is less than $r^4 = |V|$ since $r \geq 9$ (as $p = 3$ and $4 \mid r - 1$). Hence there is $v \in V$ such that $C_G(v) = C_{EZ}(v)$ is an elementary abelian 2-group.

Let next $p = 5$. Then $\bar{G}$ is an irreducible $5'$-subgroup of $\mathrm{Sp}_4(2) \cong S_6$, and we may assume that $\bar{G} \cong \mathrm{O}_4^+(2) \cong 3^2 : D_8$ [Atlas]. For $r = 5$ we find, using a random search, $v \in V$ such that $H = C_G(v)$ is a S_4 group. Then the nontrivial H-submodule W of V is the deleted permutation module or its tensor product with the sign character, and H has a regular orbit on W. So let $r \geq 25$. We may argue as for (i), but using the character table (for T). We find $v \in V$ such that $C_G(v)$ maps isomorphically into a Sylow 2-subgroup of $\bar{G}$. Using that the elements in $\bar{G}$ of order 4 belong to A_6 and that χ vanishes on the elements of order 4 in $T \cong 2.A_6$ we see that v is strongly real for G. Also, $C_G(v)$ has a regular orbit on V.

Let finally $p > 5$ ($r \geq 13$), and let $G = G_0$. Write $E = E_0 \circ Z_4$ with $E_0 \cong 2_+^{1+4}$. Let $X_0 = N_G(E_0)$. Then $\bar{X}_0 = X_0/ZE \cong \mathrm{O}_4^+(2)$. As in the second paragraph of the proof for (i) let S be a Sylow 3-subgroup of X_0, which maps isomorphically onto $\bar{S} = O_3(\bar{X}_0)$, and let $Y = N_{X_0}(S)$. Then $Y \cdot ZE = X_0$ and $Y \cap ZE = Z$. Let P be the stabilizer in $\bar{X}_0$ of a maximal totally singular subspace W_1 of $U_0 = E_0/Z(E_0)$. Thus $P \cong \Lambda^2(W_1) : \bar{L}$ where W_1 is the standard module for $\bar{L} \cong \mathrm{GL}(W_1) \cong S_3$ (Appendix B). Also $U_0 = W_1 \oplus W_2 = W_1 \oplus W_3$ where the W_i are all the distinct, but isomorphic, irreducible $\bar{L}$-submodules of $U_0 \cong U$ (noting that $|\mathrm{Hom}_{\bar{L}}(U_0/W_1, W_2)| = 2$). Since $\mathrm{H}^n(\bar{L}, W_i) = 0$ for all $n \geq 1$, in view of (A2), there exist a subgroup L of $X_0 = N_{\mathrm{GL}(V)}(E_0)$ mapping onto $\bar{L}$ and L-invariant elementary abelian subgroups A_i of E mapping onto the W_i. By Appendix (C1) we may choose L such that $L \cong \mathrm{GL}(A_i) \cong S_3$. The fixed spaces $C_V(A_i)$ are 1-dimensional, pairwise distinct, and centralized by L'. Since an involution t in L acts as ± 1 on these spaces, we may replace t by tz for the generator z of $Z(E_0)$, if necessary (and L correspondingly), so that L centralizes at least two of these fixed spaces, say $C_V(A_1)$ and $C_V(A_2)$. We may choose $v \in C_V(A_1) \oplus C_V(A_2)$ outside all $C_V(A_i)$.

We claim that $N_{ZE}(Fv) = Z$. For otherwise $N_{ZE}(Fv)/Z$ would be an irreducible L-submodule of $U \cong U_0$ and hence one of the W_i. It follows that $LA_i \subseteq N_{G_0}(Fv)$ and that A_i centralizes v, because L centralizes v and $[A_i, L] = A_i$. Hence the claim. In particular $C_{ZE}(v) = 1$. Let $H = C_G(v)$, and let $\bar{H}$ be its isomorphic image in $\bar{G}$.

From the character table we infer that $\bar{H} \neq \bar{G}$. If $H \cong A_6, S_5$ or A_5, then H is a real group, and H has a regular orbit on V by Theorem 7.2a below. H is a real group in the cases where $\bar{H}$ is contained in a maximal subgroup of S_6 of type $S_4 \times Z_2$ or in a solvable subgroup of S_5 (as $H \supseteq L$), and then H has a regular orbit on V (as $r \geq 13$). It remains to investigate the case where $\bar{H}$ is contained in a maximal subgroup of type $O_4^+(2) \cong \bar{X}_0$ of S_6. Then $\bar{H} \subseteq \bar{X}_0$ since $\bar{X}_0$ is the unique 3-Sylow normalizer in $\bar{G}_0$ containing $\bar{L}$. We assert that $H \subseteq L \times Z_2$, which will complete the proof.

Otherwise $O_3(\bar{H}) = O_3(\bar{X}_0)$ and $O_3(H) \cong S$. Changing notation, if necessary, we may let $S = O_3(H)$. Then $H \subseteq Y = N_{X_0}(S)$ and $Y/ZS \cong D_8$. Since S centralizes v and is faithful on V, by Proposition 1.6a we have a proper decomposition

$$V = C_V(S) \oplus [V, S]$$

of FY-modules ($p \neq 3$). This is impossible since Y is irreducible on V, as follows from Theorem 1.8b. For $\mathrm{Res}_S^Y(V)$ is the sum of four nonisomorphic 1-dimensional FS-modules corresponding to an Y-orbit on $S^\sharp$ when $3 \mid r-1$, and of two 2-dimensional modules otherwise. $\qquad\square$

Proposition 6.6c. *Let $E \cong 2_-^{1+6}$. There exists a strongly real vector $v \in V$ for G such that $C_G(v)$ has a regular orbit on V.*

Proof. $O_6^-(2)$ contains no irreducible $3'$-subgroup [Atlas, p. 26]. Hence $p > 3$ and $r \geq 7$ (as $4 \nmid r - 1$). For $r = 7$ by computation with [GAP] (or directly) one finds $v \in V$ such that $C_{G_0}(v) \cong S_4$, which is a real group and which obviously has a regular orbit on V ($\dim_F V = 8$). So let $r \geq 11$ in what follows.

Let $\bar{Y} \cong 3_+^{1+2} : \mathrm{GL}_2(3)$ be a maximal subgroup of $\bar{X} \cong O_6^-(2)$ which is irreducible. Let $S \cong 3_+^{1+2}$ be a subgroup of X such that $\bar{S} = SE/E = O_3(\bar{Y})$, and let $Y = N_X(S)$. By the Frattini argument Y covers $\bar{Y}$ and so is irreducible on $U = E/Z(E)$. Each faithful irreducible representation of S has degree 3 or 6 depending on whether the underlying field of scalars contains a primitive 3rd root of unity or not. Since S is faithful on U (and $Z(S)$ fixed point free), therefore even S is irreducible on U. Note that Y inverts the elements of $Z(S)$. It follows that $Y \cap E = Z(E)$ and that $C_X(Z(S)) = Y'Z(E)$ has index 2 in Y, mapping onto the maximal subgroup $\bar{Y}' \cong \mathrm{GU}_3(2)$ of $\Omega_6^-(2)$. Using that $\mathrm{GL}_2(3)$ has trivial Schur multiplier we see that $Y'Z(E) = Y' \times Z(E)$ and that Y' is a weak holomorph of $S \cong 3_+^{1+2}$ (as $\mathrm{M}(\mathrm{Sp}_2(3)) = 1$).

We show that Y' is the standard holomorph of S, that is, Y' splits over S. In view of (A2) we find a subgroup L_0 of Y' such that $L_0 S = Y'$ and $L_0 \cap S = Z(S)$. We have to exclude that L_0/L_0' is cyclic (of order 9). In this case L_0, hence $\Omega_6^-(2)$, had an element of order 18, which is false [Atlas, p. 27]. Hence there is a subgroup $L \cong \mathrm{Sp}_2(3)$ in Y', and $Y' = LS$.

Since S is faithful on V and Y inverts the elements of $Z(S)$, there is a faithful irreducible FY-submodule V_1 of V of dimension 6. In fact $V_1 = [V, Z(S)]$ and the restriction to $C_X(Z(S)) = Y' \times Z(E)$ of V_1 affords the sum of two distinct (absolutely) irreducible characters φ, $\bar{\varphi}$ of degree 3, which are faithful and which are interchanged by Y. Let $V_0 = C_V(Z(S))$, so that

$$V = V_0 \oplus V_1$$

is a decomposition of FY-modules (Proposition 1.6a). Since Y is transitive on $(S/S')^\sharp$ (and $\dim {}_F V_0 = 2$), it follows from Theorem 1.8b that $V_0 = C_V(S)$. We assert that L is faithful and irreducible on V_0. Assume there is an element $x \in L$ of order 4 which is trivial on V_0. We have $\varphi(x) = 1 = \bar{\varphi}(x)$ as $\mathrm{Res}_L^{Y'}(\varphi)$ is the sum of two irreducible characters of degrees 1 and 2. It thus follows that $\chi(x) = 1+1+2 = 4$, hence $|C_U(x)| = 16 = 4^2$ by Theorem 4.4. Now the image $\bar{x}$ of x in $O_6^-(2)$ has order 4 and preserves an $\mathbb{F}_4$-structure on U, because $\bar{x} \in \mathrm{SU}_3(2)$. Since $\bar{x}$ is a unipotent transformation on $U \cong \mathbb{F}_4^{(3)}$, it must have two Jordan blocks on U of sizes 1 and 2. But this forces that $\bar{x}^2 = 1$, a contradiction. Thus x does not centralize V_0. Since $L/Z(L) \cong A_4$ has no faithful irreducible F-representation of degree 2, we see that L is faithful and irreducible on V_0, as asserted.

We infer that L maps injectively into $Y/C_Y(V_0)$. Since both $Z(E)$ and $Z(L)$ act as -1 on V_0, it follows that $|C_Y(V_0) : S| = 2$. Moreover $C_Y(V_0)' = S$ as $Z(E)$ centralizes S but $Z(L)$ inverts the elements of $S/Z(S)$.

At most 8 points in $P_1(V_0)$ are fixed by some nonidentity element in $L/Z(L) \cong A_4$, because the elements of order 4 in L do not fix any point (as $4 \nmid r - 1$). We conclude that there is $v \in V_0$ such that $N_L(Fv) = Z(L)$ (as $r \geq 11$). Since $S \subseteq C_Y(V_0)$ is irreducible on $U = E/Z(E)$ and E is irreducible on V, we have $N_E(Fv) = Z(E)$. It follows that the index $|C_Y(v) : C_Y(V_0)| \leq 2$ and that $C_Y(v)' = S$. Let $H = C_X(v)$. Since $H \supseteq S$ acts irreducibly on U, we have $H \cap E = 1$. Thus H maps isomorphically onto an irreducible subgroup $\bar{H}$ of $\bar{X}$.

By (A9), (A10) $N_X(Fv)$ maps onto a proper (irreducible) subgroup of $\bar{X} \cong O_6^-(2)$ and so its Sylow 3-subgroups have order $|S| = 3^3$ or 3^4.

In the latter case $S \subset S_0$ for some Sylow 3-subgroup S_0 of $N_X(Fv)$. But then $S_0 \subseteq N_X(S) = Y$, even $S_0 \subseteq Y' = LS$. Since S centralizes V_0, it follows that $N_L(Fv)$ contains an element of order 3, contradicting the fact that $N_L(Fv) = Z(L)$. Consequently $N_X(Fv)/C_X(v)$ is a (cyclic) 2-group (of order dividing $r - 1$). Now $|N_X(Fv) : C_X(v)| = 2$ as $4 \nmid r - 1$, and $N_X(Fv) = C_X(v) \times Z(E)$. From Lemma 5.1b we infer that $C_{G_0}(v) = C_X(v) = H$. We shall prove that $H \subseteq Y$ and so H/S has order 2 or 4.

Assume $H \nsubseteq Y$. If $\bar{H}$ is contained in a conjugate of $\bar{Y}$, the Sylow 2-subgroups of H' are cyclic, because the Sylow 2-subgroups of $\bar{Y}/\bar{S} \cong \mathrm{GL}_2(3)$ are semidihedral and S is a Sylow 3-subgroup of H. By a simple transfer argument H' has a normal 2-complement. But then $H \subseteq N_X(S) = Y$. So $\bar{H}$, being irreducible, is contained in a maximal subgroup $P_0 = 3^3 : (S_4 \times Z_2)$ of $\bar{X} \cong O_6^-(2) \cong \mathrm{PSp}_4(3)$ [Atlas, p. 26]. Here $S_4 = \mathrm{PGL}_2(3)$ is irreducible on $R = 3^3$, because $P = R : \mathrm{GL}_2(3)$ is a maximal parabolic of $\mathrm{Sp}_4(3)$ stabilizing an isotropic line W, and $R \cong \mathrm{Sym}^2(W)$. Now $\bar{S}$ is a Sylow 3-subgroup of $\bar{H}$ of order 3^3, which by assumption is not normal in $\bar{H}$. We may assume that $|R \cap \bar{S}| = 3^2$ (Sylow). No element of order 3 in P_0/R centralizes $R \cap \bar{S} = R \cap \bar{H}$ and so $|C_{\bar{H}}(R \cap \bar{H}) : R \cap \bar{H}| \leq 2$. It follows that $\bar{H}/C_{\bar{H}}(R \cap \bar{H})$ is a subgroup of $\mathrm{GL}_2(3)$ having no normal Sylow 3-subgroups, hence containing $\mathrm{SL}_2(3)$. On the other hand, this is a homomorphic image of a subgroup of $S_4 \times Z_2$, which is impossible.

$\mathrm{Res}_H^X(\chi)$ is rational-valued, because on X the values of χ lie in $\mathbb{Q}(\sqrt{2})$ or $\mathbb{Q}(\sqrt{-2})$ by Theorem 4.3e, and the exponent of H is 6 or 12. Hence it remains to show that H has a regular orbit on V. There are at most $3 \cdot 3^3 = 81$ involutions in H, each having at most r^6 fixed points on V. If y is one of the $2 \cdot 13$ elements of order 3 in H, then $\dim{}_F C_V(y) \leq 4$ by Theorem 6.2c. As $r \geq 11$, $\sum_{\gamma \in \beta(H)} |C_V(\gamma)| \leq 81r^6 + 13r^4 < r^8 = |V|$. $\square$

6.7. The Remaining Cases

We often shall use that certain groups cannot appear as subgroups of X, e.g., by the nonsplitting properties stated in (A9), (A10). Other groups will be ruled out on the basis of Theorem 4.4 and the character table. For example, let $m = 5$ and let X be the holomorph to any of $2_{\pm,0}^{1+10}$. Assume $Y = A_9$ occurs as a subgroup of X. Then there must be a nonnegative integer combination $\mathrm{Res}_Y^X(\chi) = \sum_{i=1}^6 n_i \chi_i$ of the first six irreducible characters of A_9 [Atlas, p. 37] such that $\sum_i n_i \chi_i(1) = 2^5$ and $|\sum_i n_i \chi_i(y)|^2 = 0$ or a 2-power ($y \in Y$). This is impossible.

Proposition 6.7a. *Let $E \cong 2^{1+2m}_+$ for $m \geq 3$. There exists a strongly real vector $v \in V$ for G such that $C_G(v)$ has a regular orbit on V.*

Proof. By Theorem 6.3b it suffices to consider the cases $m = 3, 4, 5$. By Theorem 4.5b there is a subgroup $L \cong L_m(2)$ of X normalizing two elementary abelian subgroups A, A^* of E yielding a decomposition $U = W \oplus W^*$ into totally singular subspaces. Here $A \cong W$ is the standard module of L and $A^* \cong W^*$ its dual. Recall also that $\operatorname{Res}^X_L(\chi) = 2 \cdot 1_L + \xi$ for some rational-valued irreducible (Weil) character ξ. We further know that $N_X(L) = LZ(E)\langle \tau \rangle$ where τ interchanges A and A^* and induces the inverse transpose automorphism on L ($\tau^2 \in Z(E)$). Now $C_V(A)$ and $C_V(A^*)$ are distinct 1-dimensional subspaces of V, and $L = L'$ centralizes $\tilde{V} = C_V(A) \oplus C_V(A^*)$. Also, τ interchanges $C_V(A)$ and $C_V(A^*)$ and fixes no point in $P_1(\tilde{V})$ if $\tau^2 \neq 1$ (as $4 \nmid r - 1$) and two points otherwise. We choose $v \in \tilde{V}$ outside $C_V(A)$ and $C_V(A^*)$ and, if possible, such that v is no eigenvector of τ. Thus if τ stabilizes Fv, then $r = 3$ and τ is an involution. Let $H = C_{G_0}(v)$.

Assume $H \cap ZE \neq 1$. Since $H \cap Z = 1$, $H \cap ZE$ then is an L-invariant elementary abelian subgroup of ZE whose image in $U = ZE/Z$ is nonzero and totally singular and so must be W or W^*, because $U = W \oplus W^*$ and W, W^* are irreducible and nonisomorphic L-modules (Lemma 4.1c). Since H contains $[H \cap EZ, L]$, which is A or A^*, it follows that A or A^* centralize v. This contradicts the choice of v. Hence $H \supseteq L$ map injectively onto subgroups $\bar{H} \supseteq \bar{L}$ of $\bar{G}_0 \cong O^+_{2m}(2)$.

Let $\bar{\tau} = \tau ZE$. Suppose $\bar{H}$ contains $N_{\bar{G}_0}(\bar{L}) = \langle \bar{L}, \bar{\tau} \rangle$. Then there is an element $\tau_0 \in H$ with $\tau_0 ZE = \bar{\tau}$, and then $L = L^{\tau_0}$ must agree (within H). Thus τ_0 and τ act on L in the same manner. We deduce that $\tau \tau_0^{-1} \in C_{ZE}(L) = Z$. As τ_0 centralizes v, τ fixes Fv. Thus $r = 3$ and τ is an involution.

By (A9), (A10) $\bar{H}$ is neither $\bar{G}_0 = O^+_{2m}(2)$ nor $\Omega^+_{2m}(2)$, nor is any subgroup of odd index in these groups by (A4). If $\bar{H}$ is contained in a maximal parabolic P of $\Omega^+_{2m}(2)$ stabilizing a maximal totally singular subspace of U, this subspace must be W or $W^* = W^\tau$, by the action of $\bar{L}$ on U, and necessarily $H = L$ as the Levi complement $\bar{L}$ is irreducible on the unipotent radical $R \cong \Lambda^2(W)$ resp. $\Lambda^2(W^*)$ of P (Appendix B and Appendix A7).

Let $m = 3$. Then $H = L$ or $r = 3$ and $H = L\langle \tau \rangle \cong \operatorname{PGL}_2(7)$, the further possibilities $H \cong A_7, S_7$ being ruled out by Theorem 4.4 and the character tables. The extensions of ξ to $\operatorname{PGL}_2(7)$ take only real values

[Atlas, p. 3]. Also,

$$\sum_{\gamma \in \beta(H)} |C_V(\gamma)| \leq 21r^6 + 28r^5 + 28r^4 < r^8 = |V|$$

for $r > 3$. For $r = 3$, $H \cap G$ is contained in groups of type D_{16} (not in L) or $7 : 2$ (not in L). An involution of H in L has r^6 fixed points on V, an involution outside r^5. Use that $5r^6 + 4r^5 < r^8$ respectively $7r^5 + r^2 < r^8$.

Let $m = 4$. We assert that then $H = L$. We inspect the maximal subgroups of $O_8^+(2)$ [Atlas, p. 85]. The possibilities $H \cong A_9, S_9$ are ruled out by the character tables. $\bar{H}$ cannot be in a maximal subgroup of $O_8^+(2)$ of type $Sp_6(2) \times Z_2$, the stabilizer of a nonsingular point (as $\bar{H} \supseteq \bar{L}$). If $\bar{H} \neq \bar{L}$, then $\bar{H}$ must be contained in an (irreducible) $Sp_6(2)$ group [Atlas, p. 46], which in turn contains an S_8 subgroup. Thus if $H \neq L$, then $\bar{\tau} \in \bar{H}$, $r = 3$ and either $H = L\langle \tau \rangle \cong S_8$ or $H \cong Sp_6(2)$. But in this case χ restricts to an S_8 subgroup as $2\chi_1 + \chi_3$ (Atlas notation) where, by Theorem 4.4, $\chi_3(2C) = -4$ on the involutions $2C$ outside A_8. But then $\chi_3(10A) = +1$ and $\chi(10A) = 3$, contradicting Theorem 4.4. Hence the assertion. The counting method yields a regular orbit on V for $H \cap G$ for $r > 3$ (hence $r \geq 7$). For $r = 3$ use that $H \cap G$ is contained in groups of type $2^4 : 2^2$, $2^3 : 7$, $5 : 4$ or $5 : 2^2$ [Atlas], and that $2^6 r^{12} < r^{16}$.

Let $m = 5$. Then $H = L$ or $r = 3$ and $H = L\langle \tau \rangle \cong L_5(2) : 2$ by inspection of the maximal subgroups of $O_{10}^+(2)$ [Atlas, p. 146]. Observe that ξ extends to real-valued characters of $L_5(2) : 2$ [Atlas, p. 70]. As before we get a regular orbit for $H \cap G$. $\qquad \square$

Proposition 6.7b. *Let $E \cong 2_0^{1+2m}$ for $m \geq 3$. There exists a strongly real vector $v \in V$ for G such that $C_G(v)$ has a regular orbit on V.*

Proof. By Theorem 6.3b we only have to examine the cases $m = 3, 4, 5$. Write $E = E_0 \circ Z_4$ where $E_0 \cong 2_+^{1+2m}$. Let $Y = N_X(E_0)$. Then $\bar{Y} = Y/E \cong O_{2m}^+(2)$ is a maximal subgroup of $\bar{X} \cong Sp_{2m}(2)$. As before pick a subgroup $L \cong L_m(2)$ of Y and L-invariant elementary abelian subgroups A, A^* of E_0 of order 2^m. Again $\mathrm{Res}_L^X(\chi) = 2 \cdot 1_L + \xi$ for some rational-valued irreducible (Weil) character ξ. We also find $\tau \in X$ inducing the inverse transpose automorphism on L such that $N_{G_0}(L) = LZ\langle \tau \rangle$ and $\tau^2 \in Z(E)$ (see the remark to Theorem 4.5b). This τ interchanges $A = [\Omega_1(AZ(E)), L]$ and $A^* = [\Omega_1(A^*Z(E)), L]$. Now $L = L'$ centralizes $\tilde{V} = C_V(A) \oplus C_V(A^*)$, and τ interchanges the 1-dimensional fixed spaces $C_V(A)$ and $C_V(A^*)$. Multiply τ by a scalar in Z, if necessary, such that its order is as small as possible.

Let $m = 3$. Choose $w \in \tilde{V}$ outside $C_V(A) =$ and $C_V(A^*)$ and, if possible, such that τ centralizes w. Let $H = C_{\bar{G}_0}(w)$. As before one shows that $H \cap ZE = 1$. So by (A10) $H \supseteq L$ map injectively onto proper subgroups $\bar{H} \supseteq \bar{L}$ of $\bar{G}_0 \cong \mathrm{Sp}_6(2)$. We cannot have $H \cong A_7, S_7$ by the character tables and (4.4). Similarly H is not isomorphic to $\Omega_6^+(2) \cong A_8$. It remains to study the situation where $\bar{H}$ is contained in a maximal $G_2(2)$ subgroup of $\bar{G}_0$ [Atlas, p. 46]. We cannot have $H \cong G_2(2)$ by the character table and (4.4). Suppose $H \cong G_2(2)'$. Then $H = H' \subseteq G_0' = X'$ and [Atlas, p. 14] $\mathrm{Res}_H^X(\chi) = \chi_1 + \chi_4$ or $\chi_1 + \chi_5$, where χ_4 and χ_5 are algebraically conjugate and not real-valued (but fuse in $G_2(2)$; $\mathrm{Res}_H^X(\chi) = \chi_1 + \chi_3$ is excluded by considering the value on the conjugacy class $6A$). There is a unique $G_2(2)'$ overgroup of $\bar{L}$ in $\bar{G}_0$, namely that normalized by $\bar{\tau} = \tau ZE$. (To see this, recall that $N_{\bar{G}_0}(\bar{L}) = \bar{L}\langle\bar{\tau}\rangle$ by Theorem 4.5b, and use that $\mathrm{Sp}_6(2)$ contains a unique conjugacy class of $G_2(2)'$ subgroups and that each $G_2(2)'$ group has a unique conjugacy class of (maximal) $L_3(2)$ subgroups.) Hence both H and H^τ contain L and map (isomorphically) onto $\bar{H}$ ($\bar{\tau} \notin \bar{H}$).

Now U is the unique (absolutely) irreducible $\mathbb{F}_2 H$-module of dimension 6 (arising naturally from the action of $G_2(2)$ on the $\mathbb{F}_2$ Cayley algebra [Atlas, p. 14], [B-Atlas, p. 23]). One knows that $\mathrm{H}^1(H, U)$ has order 2 [Sin, 1996]. Using that $H \cong G_2(2)'$ is perfect and has trivial Schur multiplier we see that there are just 2 conjugacy classes of complements to E in HE. If two such complements, say H and H^τ, are conjugate under E, say $H^\tau = H^{y^{-1}}$ ($y \in E$), then $L^y = \{x[x,y] \,|\, x \in L\} \subseteq H$ and $[L,y] \subseteq H \cap E = 1$, that is, $y \in Z(E) \subseteq Z$. We conclude that either $H = H^\tau$, or $H \neq H^\tau$ are the unique overgroups of L mapping onto $\bar{H}$.

Clearly $\langle H, A \rangle \supseteq E$ as $\bar{H}$ is irreducible on $U = E/Z(E)$. Hence $H = H'$ does not centralize $C_V(A)$. It follows that $C_{\tilde{V}}(H) = Fw$. If τ fixes Fw, then by the choices of τ and of w it centralizes w. But $\tau \notin H = C_{\bar{G}_0}(w)$. We conclude that τ does not fix Fw, whence $H \neq H^\tau$, and these two groups fix just the two points Fw and $Fw\tau$ in $P_1(\tilde{V})$. Note that $|P_1(\tilde{V})| \geq 6$. If we choose $v \in \tilde{V}$ such that Fv is different from these two points, different from $C_V(A)$ and $C_V(A^*)$, and if possible such that Fv is the eigenspace of τ on $\tilde{V}$ to the eigenvalue 1, then $H_0 = C_{\bar{G}_0}(v)$ either is L or τ is an involution and $H_0 = L\langle\tau\rangle \cong \mathrm{PGL}_2(7)$. This v is strongly real for G, and in the worst case we get the estimate

$$\sum_{\gamma \in \beta(H_0)} |C_V(\gamma)| \leq 21r^6 + 28r^5 + 28r^4.$$

This is less than $|V| = r^8$ except when $r = 5$. But in this exceptional case H_0, being a $5'$-group, has a regular orbit on V by Theorem 7.2a below.

For $m = 4, 5$ we choose $v \in \widetilde{V}$ outside $C_V(A)$ and $C_V(A^*)$ such that Fv is not fixed by τ. Then $H = C_{G_0}(v)$ agrees with L, and $L \cap G$ has a regular orbit on V. In the $m = 4$ case this follows by inspection of the maximal subgroups of $\mathrm{Sp}_8(2)$ [Atlas, p. 123]. In the $m = 5$ case we deduce from Lemma 6.3a (and Table 3.5.C in [Kleidman–Liebeck, 1990]) that the (isomorphic) image of H in $\mathrm{Sp}_{10}(2)$ must be properly contained in $\mathrm{O}_{10}^{\pm}(2)$ subgroups, and inspection of the [Atlas, pp. 146, 147] yields $H = L$. □

Proposition 6.7c. *Let* $E \cong 2_-^{1+2m}$ *with* $m \geq 4$. *There exists a strongly real vector* $v \in V$ *for* G *such that* $C_G(v)$ *has a regular orbit on* V.

Proof. By Theorem 6.3b we only have to consider the cases $m = 4, 5$. Decompose $E = E_0 \circ E_1$ where $E_0 \cong 2_+^{1+2(m-1)}$ and $E_1 \cong Q_8$. Let $X_0 = C_X(E_1)$ and $X_1 = C_X(E_0)$. Then X_0 is the standard holomorph of E_0 (in F) and X_1 that of E_1 (in F). Moreover, the central product $Y = X_0 \circ X_1$ (amalgamating $Z(E_0) = Z(E_1)$) is a maximal subgroup of X. As an Y-module we may decompose $V = V_0 \otimes_F V_1$ where V_0 is the natural module for X_0 and V_1 that for X_1. Recall that $X_1 \cong 2^+ S_4 = \mathrm{GL}_2(3)$ if $r \equiv 3 \pmod 8$ and $X_1 \cong 2^- S_4$ if $r \equiv 7 \pmod 8$ ($4 \nmid r - 1$). Let χ_0 and χ_1 denote the obvious characters of X_0 and X_1, respectively, so that $\mathrm{Res}_Y^X(\chi) = \chi_0 \otimes \chi_1$.

Let $L_0 \cong L_{m-1}(2)$, A_0, A_0^* and W_0, W_0^* be defined for X_0 as in Theorem 4.5b. Hence $U_0 = E_0 Z/Z = W_0 \oplus W_0^*$ where W_0 is the standard module for L_0 and W_0^* is its nonisomorphic dual. Here $\widetilde{V}_0 = C_{V_0}(A_0) \oplus C_{V_0}(A_0^*)$ is a 2-dimensional subspace of V_0 which is centralized by $L_0 = L_0'$. Further $N_{X_0}(L_0) = L_0 Z(E_0)\langle \tau_0 \rangle$ for some τ_0 which induces the inverse transpose automorphism on L_0 ($\tau_0^2 \in Z(E_0)$). We know that $\mathrm{Res}_{L_0}^{X_0}(\chi_0) = 2 \cdot 1_{L_0} + \xi_0$, where the Weil character ξ_0 is rational-valued.

For $r = 3$ pick any $v_1 \in V_1^\sharp$, so that $C_{X_1}(v_1) \cong S_3$, and choose $v_0 \in \widetilde{V}_0$ outside $C_{\widetilde{V}_0}(A_0) \cup C_{\widetilde{V}_0}(A_0^*)$, and let $v = v_0 \otimes v_1$. Otherwise $r \geq 7$, and we choose $v \in \widetilde{V} = \widetilde{V}_0 \otimes_F V_1$ which is not centralized by A_0, A_0^* nor by any element of prime order (2 or 3) in $\langle \tau_0 \rangle X_1 Z$. (We have at most 12 subgroups of order 3 and at most $2 + 2 \cdot 12$ involutions in $\langle \tau_0 \rangle X_1 Z$ outside Z, each with at most r^2 fixed points on $\widetilde{V}$, and $38 r^2 < r^4 = |\widetilde{V}|$.) Let $L = L_0 \times C_{X_1}(v_1)$ for $r = 3$ and $L = L_0$ otherwise. Of course χ takes only rational values on L. In view of the $m = 3$, $m = 4$ cases of the proof for Proposition 6.7a it is clear that there there is a regular $(L \cap G)$-orbit on V.

Let $H = C_{G_0}(v)$, so that $H \supseteq L$. We assert that $H \cap ZE = 1$. Otherwise $H \cap ZE$ is an L_0-invariant elementary abelian subgroup of ZE mapping onto an L_0-submodule $\neq 0$ of

$$U = U_0 \perp U_1 = (W_0 \oplus W_0^*) \perp U_1.$$

U is a completely reducible $\mathbb{F}_2 L_0$-module, W_0, W_0^* being irreducible and U_1 centralized by L_0, each point in U_1 being nonsingular. Hence the image of $H \cap ZE$ must be W or W^*. But this implies that one of A or A^* centralizes v, in contrast to the choice of v. Therefore $H \supseteq L$ map isomorphically onto (proper) subgroups $\bar{H} \supseteq \bar{L}$ of $\bar{G}_0 \cong O_{2m}^-(2)$.

The element $\bar{\tau}_0 = \tau_0 ZE$ of $\bar{X}_0 \cong O_{2(m-1)}^+(2)$ is not in $\bar{H}$ when $r > 3$. For otherwise H contains an element τ with $\tau ZE = \bar{\tau}_0$, both $L = L_0$ and L^τ being subgroups of H mapping onto $\bar{L} = \bar{L}^{\bar{\tau}_0}$. Thus $L_0 = L_0^\tau$ and both τ_0, τ induce the inverse transpose automorphism on L_0. We conclude that $\tau \cdot \tau_0^{-1} \in C_{ZE}(L_0) = ZE_1$ and so $\tau = \tau_0 x_1 z \in H$ for some $x_1 \in E_1$ and $z \in Z$, against our construction.

Let $m = 4$. Inspection of the maximal subgroups of $\bar{G}_0 \cong O_8^-(2)$ [Atlas, p. 89] shows that $\bar{H}$ must be contained in subgroups of the following types:

(i) A $\bar{G}_0$-conjugate of $\bar{Y} = YZ/ZE$, isomorphic to $O_6^+(2) \times O_2^-(2) \cong S_8 \times S_3$.

(ii) A maximal parabolic stabilizing a maximal totally singular subspace of U.

(iii) The stabilizer $\mathrm{Sp}_6(2) \times Z_2$ of some nonsingular point in U.

Here we already ruled out the possibility that $\bar{H}$ is contained in an irreducible $\mathrm{PGL}_2(7)$ subgroup of $O_8^-(2)$. In this case $\bar{L} \cong \mathrm{PSL}_2(7)$ would be normal in $\bar{H} \cong \mathrm{PGL}_2(7)$ and act on U with irreducible constituents of equal dimensions (Theorem 1.8). In (i) we have $\bar{H} \subseteq \bar{Y}$ by the action of $\bar{L}$ on U and, as in the proof for Proposition 6.7a, we get $H = L$ or $r = 3$ and $H \cong \mathrm{PGL}_2(7) \times S_3$. In (ii) the parabolic P is the stabilizer of W_0 or $W_0^* = W_0^{\bar{\tau}_0}$, and $P \supseteq \bar{L} = \bar{L}^{\bar{\tau}_0}$. So $P = R : \bar{L}$, where the unipotent radical R is a central extension of $W_0 \otimes U_1$ by $\Lambda^2(W_0)$ if $P = N_{\bar{G}_0}(W_0)$ (Appendix B2). $\bar{H}$ cannot contain $\Lambda^2(W_0)$ (resp. $\Lambda^2(W_0^*)$) for otherwise H contains an overgroup of L_0 mapping isomorphically onto a maximal parabolic of $\Omega_6^+(2)$, contradicting (A10). Using that L_0 is irreducible on $\Lambda^2(W_0)$ (resp. $\Lambda^2(W_0^*)$) we get $H = L$. In (iii) the nonsingular point must be in U_1, hence $r > 3$. Here $\bar{H}$ is contained in an $\mathrm{Sp}_6(2)$ group by the choice of v, because the direct factor Z_2 is such that it preserves U_1 and centralizes $U_0 = U_1^\perp$. We cannot have $\bar{H} \cong \mathrm{Sp}_6(2)$ or $\mathrm{O}_6^+(2)$, $\Omega_6^+(2)$ by (A10). By inspection of the maximal subgroups of $\mathrm{Sp}_6(2)$ and $G_2(2)$ [Atlas, pp. 46, 14], and using that $\bar{\tau}_0 \notin \bar{H}$, we obtain that either $H = L$ or $H \cong G_2(2)'$. In the latter case $H \subseteq X$ and

$$\mathrm{Res}_H^X(\chi) = 2\chi_1 + \chi_4 + \chi_5$$

in the notation of the Atlas (p. 14), where χ_4 and χ_5 are algebraically conjugate and fuse in $G_2(2)$ to a rational-valued character. Also $H \cap G$ has a regular orbit on V in this case, because

$$\sum_{\gamma \in \beta(H)} |C_V(\gamma)| \leq 63r^{12} + 336r^8 + (28 + 288)r^4 < r^{16} = |V|.$$

Let $m = 5$. Consider the maximal subgroups of $\bar{G}_0 \cong \mathrm{O}_{10}^-(2)$ [Atlas, p. 147]. Recall that A_9 does not occur as a subgroup of G_0. Furthermore, $\mathrm{O}_6^+(2) \times \mathrm{O}_4^-(2)$ and $\mathrm{O}_6^-(2) \times \mathrm{O}_4^+(2)$ cannot be overgroups of $\bar{H}$ (by their action on U), nor $M_{12} : 2$ (missing in the [Atlas]) and $\mathrm{GU}_5(2) : 2$ (as they do not have $L_4(2)$ subgroups). It remains to consider cases (i), (ii), (iii) defined as in the $m = 4$ case. This leads to $H = L$ or $r = 3$ and H is of type $S_8 \times S_3$ or $\mathrm{Sp}_6(2) \times S_3$, which are ruled out as in the $m = 4$ case of the proof for Proposition 6.7a. In (iii) $\bar{H}$ is contained in a stabilizer $\mathrm{Sp}_8(2) \times Z_2$ of a nonsingular point, necessarily in U_1, which implies that $r > 3$. It follows that either $H = L = L_0$ or $H \cong \mathrm{Sp}_2(6)$ maps onto on irreducible subgroup $\bar{H}$ of $\mathrm{O}_8^+(2)$ [Atlas]. However, then $\bar{H}$ contains $\bar{L}_0 \langle \bar{\tau}_0 \rangle \cong S_8$, which has also been ruled out. $\qquad\square$

This completes the proof for Theorem 6.1.

At this stage the $k(GV)$ conjecture is settled for solvable groups G and modules V in characteristic $p \neq 3, 5, 7, 13$ (by virtue of Theorems 5.2b and 5.4). Moreover, if G is a (solvable) group with a normal 2-complement, the conjecture holds. For the property of having a normal 2-complement is preserved when passing to subgroups, quotient groups and central extensions, so that Theorem 5.4 applies.

In particular, the $k(GV)$ conjecture holds for supersolvable groups [Knörr, 1984], which have a normal 2-complement, and for groups of odd order [Gluck, 1984].

Chapter 7

Reduced Pairs of Quasisimple Type

The main task of this chapter will be to determine the reduced pairs (G, V) of quasisimple type where there is no regular G-orbit on V. The classification of these pairs has been achieved by the efforts of [Liebeck, 1996], [Goodwin, 2000], [Riese, 2001] and [Köhler and Pahlings, 2001]. From this it is easy to classify the nonreal reduced pairs of quasisimple type.

7.1. Nonreal Reduced Pairs

Throughout (G, V) will be a reduced pair over $F = \mathbb{F}_r$ with quasisimple core, E. By coprimeness, and by the Feit–Thompson theorem, r is a power of some odd prime p. Embed E and G into $\mathrm{GL}(V)$ and let $G_0 = N_{\mathrm{GL}(V)}(E)$. Again we distinguish these pairs by their cores. It turns out that, in this sense, there are just 3 types of nonreal reduced pairs with quasisimple cores. Here all characters involved are Weil characters and, up to three exceptions in type $(\mathrm{Sp}_4(3))$, the pairs are large.

Type $(2.A_5)$: Let $r = p$ be one of the primes 11, 19, 29 or 59. We have seen in Sec. 3.2 that then $\mathrm{GL}_2(p)$ contains a subgroup $E \cong 2.A_5$ such that $G_0 = N_{\mathrm{GL}_2(p)}(E)$ acts transitively on the nonzero vectors in $V = \mathbb{F}_p^{(2)}$. Here $G = G_0 \cong E \circ Z_{p-1}$ is a p'-group, E is absolutely irreducible on V, and the centralizer in G of a nonzero vector is cyclic of order 5, 3, 2 or 1, respectively. We conclude, in view of Lemma 4.1c, that (G, V) is a nonreal reduced pair for $r = 11$ and for $r = 19$.

The Brauer character of E afforded by V is one of the two faithful irreducible Weil characters (of degree 2) of $\mathrm{SL}_2(5)$, which are algebraically conjugate and fuse in $2.A_5.2$. There are other primes $p \geq 11$ where $\mathrm{GL}_2(p)$ contains a copy E of $2.A_5 = \mathrm{SL}_2(5)$, because this only requires that 5 is a square mod p. We get a further nonreal reduced pair (G, V) for $p = 31$, where $G \cong 2.A_5 \times Z_{15}$ has two orbits on nonzero vectors with point stabilizers of order 3 and 5.

Type $(2.A_6)$: The group $E = 2.A_6$ has two faithful irreducible characters, χ_8 and χ_9, of degree 4 [Atlas, p. 5]. These are Weil characters for $E \cong$

$\mathrm{Sp}_2(9)$, and they fuse in $2.A_6.2_2$. There is a nonreal extension of χ_8 to 2^-S_6 (not given in the Atlas), and a nonreal extension of χ_9 to the isoclinic variant 2^+S_6, each requiring a square root of -3.

Let $r = p = 7$. Then $\sqrt{-3}$ is in $F = \mathbb{F}_7$ but $\sqrt{3}$ is not. Let V_i be an FE-module affording χ_i, embed E into $\mathrm{GL}(V_i)$ and let $G_i = N_{\mathrm{GL}(V_i)}(E)$ ($i = 8, 9$). Then G_i is isomorphic to $(2^-S_6) \times Z_3$ resp. to $(2^+S_6) \times Z_3$. There are three orbits of G_i on $V_i^\sharp$, the point stabilizers being of type Z_3, Z_6 and $Z_3 \mathrm{wr}\, Z_2 \cong Z_3 \times S_3$ (in both cases). This has been proved in Example 5.7b, where we also showed that no vector in V_i is real for G_i. We have two nonisomorphic nonreal reduced pairs (G_i, V_i) where, however, the underlying groups are isoclinic and the modules behave similarly.

Type ($\mathrm{Sp}_4(3)$) : The group $E = \mathrm{Sp}_4(3)$ has two faithful irreducible (Weil) characters $\chi = \chi_{21}$ or χ_{22} of degree 4 which are algebraically conjugate and which fuse in $\mathrm{Sp}_4(3).2$ [Atlas, p. 27]. So the resulting reduced pairs are isomorphic. There are unique conjugacy classes of subgroups $S \cong \mathrm{SL}_2(3)$ and $T \cong 3_+^{1+2} : Q_8$ of E on which χ splits off the 1-character, and no proper overgroups have this property [Atlas]; $Z(S)$ is generated by an element in class $2A$ and $Z(T)$ by one in classes $3A_0B_0$. By considering the elements of E in the classes $6E$, and their powers, one gets $\mathrm{Res}_S^E(\chi) = 1_S + \lambda + \xi_2$ where $\lambda \neq 1_S$ is linear and ξ_2 is an irreducible Weil character ($\mathbb{Q}(\xi_2) = \mathbb{Q}(\sqrt{-3})$).

Let $F = \mathbb{F}_r$ for $r = p = 7, 13$ or 19, and let V be a (coprime, faithful) FE-module affording χ as a Brauer character. Embed E into $\mathrm{GL}(V)$ through χ, and let $G = N_{\mathrm{GL}(V)}(E) = E \circ Z_{r-1}$. For $r = 7$ we have two G-orbits on $V^\sharp$ with point stabilizers $S.3 = S \times Z_3$ and $T.3 = T \times Z_3$ (where $A.B$ indicates that A is the point stabilizer taken in E). For $r = 13$ we have four G-orbits with stabilizers $S.3 = S \times Z_3$, $T.3 = 3_+^{1+2} : \mathrm{Sp}_2(3)$, $Q_8.6 = \mathrm{SL}_2(3) \circ Z_4$ and $Z_3.6 = Z_3 \mathrm{wr}\, Z_2$. For $r = 19$ there are five G-orbits with stabilizers $S.3 = S \times Z_3$, $T.3 = 3_+^{1+2} : \mathrm{Sp}_2(3)$, $Q_8.3 = \mathrm{SL}_2(3)$, $Z_3.3 = Z_3^{(2)}$ and $1.Z_9$. In each case (G, V) is nonreal reduced, as it is (E, V) for $r = 7$ and $(E \circ Z_6, V)$ for $r = 13, 19$.

Theorem 7.1. *Up to isomorphism there are exactly* 11 *nonreal reduced pairs of quasisimple type, and these are described above. All other reduced pairs (G, V) of quasisimple type admit a real vector $v \in V$ such that $C_G(v)$ has a regular orbit on V.*

This theorem will be derived from the classification of the reduced pairs of quasisimple type admitting no regular orbits.

7.2. Regular Orbits

We have seen in Example 5.1a that if (G, V) is a "permutation pair" over F of degree d and $r = p$ is equal to $d + 2$ or $d + 3$, then there is no regular G-orbit on V. These reduced pairs admit always strongly real vectors, and regular vectors if $r \geq d + 4$.

Theorem 7.2a. *Suppose the pair (G, V) does not admit a regular orbit. Then $G_0 = N_{\mathrm{GL}(V)}(E)$ is a p'-group. Up to a few isoclinisms, the isomorphism class of (G_0, V) is determined by G_0, $d = \dim_F V$, and possible $r = |F|$. Excluding the isomorphism classes of the permutation pairs these are listed below, together with the isomorphism type of a point stabilizer $H = C_{G_0}(v)$ of smallest possible order (depending on r):*

G_0	d	r	minimal stabilizer(s)
$A_5 \times Z$	3	11	Z_2
$2.A_5 \circ Z$	2	$11, 19, 29, 31, 41, 49, 61$	$Z_5, Z_3, Z_2, Z_3, Z_2, Z_2, Z_2$
$3.A_6 \circ Z$	3	$19, 31$	Z_2, Z_2
$2.S_6 \circ Z$	4	7	Z_3
$2.A_7 \circ Z$	4	11	Z_3
$L_2(7) \times Z$	3	11	Z_2
$\mathrm{Sp}_4(3) \circ Z$	4	$7, 13, 19, 31, 37$	H_1, H_2, Z_9, Z_3, Z_2
$\mathrm{PSp}_4(3) \times Z$	5	$7, 13, 19$	S_4, V_4, Z_2
$\mathrm{PSp}_4(3).2 \times Z$	6	$7, 11, 19$	D_{12}, V_4, Z_2
$\mathrm{Sp}_6(2) \times Z$	7	$11, 13, 17, 19$	$Z_2^{(3)}, V_4, Z_2, Z_2$
$(U_3(3) \times Z).2$	6	5	S_3
$U_3(3) \times Z$	7	5	Z_2
$U_3(3).2 \times Z$	7	5	Z_2
$6_1.U_4(3).2_2 \circ Z$	6	$13, 19, 31, 37$	$S_4 \times Z_2, S_3 \times Z_2, V_4, Z_2$
$(U_5(2) \times Z).2$	10	7	V_4
$2.O_8^+(2) \circ Z$	8	$11, 13, 17, 19, 23$	$S_4 \times Z_2, S_4, S_3, V_4, Z_2$
$2.J_2 \circ Z$	6	11	S_3

Remarks. There are six isoclinic but not isomorphic pairs in the above list ($2^{\pm}S_6$, the unitary groups $6_1.U_4(3).2_2 \circ Z$ for $r = 19, 31$, and the orthogonal groups for $r = 11, 19, 23$). For the groups $\mathrm{Sp}_4(3) \circ Z$ the point stabilizers $H_1 = \mathrm{SL}_2(3) \times Z_3$ ($r = 7$) and $H_2 = Z_3 \mathrm{wr} Z_2$ ($r = 13$) already appear in Sec. 7.1, and these are the only nonreal reduced pairs where there is no cyclic point stabilizer. It is easy to determine the possible subgroups G of G_0 for which there is no regular orbit, too.

Except for $G_0 = 2.A_5 \circ Z$, $d = 2$, $r = 49$ always $r = p$ is a prime. Here we deal with the two Weil characters of $E = 2.A_5$ of degree 2 , which require $\sqrt{5}$ and which fuse in $2.S_5$. Their sum is the Brauer character of an irreducible (but not absolutely irreducible) $\mathbb{F}_7 E$-module $\widetilde{V}$ which extends to $\widetilde{G} = N_{\mathrm{GL}(\widetilde{V})}(E) \cong (2.A_5 \circ Z_{48}).Z_2$. There is no regular $\widetilde{G}$-orbit on $\widetilde{V}$ (minimal stabilizer $H \cong V_4$).

There is only one further example of this kind: $E \cong L_2(7)$ has a faithful irreducible module $\widetilde{V}$ of degree 6 over $F = \mathbb{F}_5$ which, as before, is *not* absolutely irreducible (affording $\chi_2 + \chi_3$ in the Atlas notation, p. 3). This extends to $\widetilde{G} = N_{\mathrm{GL}(\widetilde{V})}(E) \cong (L_2(7) \times Z_{24}).Z_2$, and there is no regular $\widetilde{G}$-orbit on $\widetilde{V}$ (minimal stabilizer $H \cong Z_2$).

The proof for Theorem 7.2a will be carried out in Secs. 7.3 till 7.10. We now state some consequences.

Corollary 7.2b. *The reduced pairs described in Sec. 7.1 are the unique ones, up to isomorphism, which are nonreal of quasisimple type.*

Proof. It suffices to consider the reduced pairs listed in Theorem 7.2a. If the minimal stabilizer $H \cong C_{G_0}(v)$ is a real group, the result follows. It therefore remains to show that real vectors exist in the following two cases:

Case 1: $G = 2.A_7 \times Z_5$, $d = 4$, $F = \mathbb{F}_{11}$

Let $E = 2.A_7$. Note that $\chi = \chi_{10}$ resp. χ_{11} in the notation of the Atlas (p. 10). Straightforward computation shows that there is $v \in V$ such that $H = C_E(v)$ is generated by an element of order 3 in the conjugacy class $3B$, where χ takes the value 1. The normalizer $N_E(H)$ is a $5'$-group. Hence $C_G(v) = C_E(v)$ by (5.1b) and so v is a strongly real vector for G.

Case 2: $G = \mathrm{Sp}_4(3) \times Z_{15}$, $d = 4$, $F = \mathbb{F}_{31}$

We prove that there is a real vector $v \in V$ for G with $C_G(v) \cong Z_5$. (There is no *strongly* real vector in this case). Let $E = \mathrm{Sp}_4(3)$. Note that $\chi = \chi_{21}$ resp. χ_{22} in the notation of the [Atlas, p. 27]. We first show that there is a regular E-orbit on $V = V_\chi$. The elements of order 4 and 5 in E do not have nonzero fixed points on V. Each of the 45 noncentral involutions in E has r^2 fixed points, the $2 \cdot 40$ elements belonging to the classes $3A_0 B_0$ have r fixed points each, and the $2 \cdot 240$ elements belonging to $3D_0$ have r^2 fixed points on V. Now use that $45r^2 + 40r + 240r^2 < r^4 = |V|$ (as $r = 31$). So let $v \in V^\#$ belong to a regular E-orbit. Let $y \in E$ be an element in the (unique) conjugacy class $5A_0$ of E of elements of order 5, where χ takes the value -1. Then v, vy, vy^2, vy^3, vy^4 are linearly dependent ($d = 4$). Hence $vy = vz$ for some $z \in Z = Z_{30}$ of order 5. It follows that $H = C_G(v)$ contains $z^{-1}y$ and that $N_E(Fv)$ contains y. But $C_E(y) = \langle y \rangle$ and $N_E(Fv)$ is cyclic (isomorphic to a subgroup of $F^\star \cong Z_{30}$). We conclude that $N_E(Fv) = \langle y \rangle$ and, in view of Lemma 5.1b, that $H = \langle z^{-1}y \rangle$. We have $(z^{-1}y)_V = [1, z^2, z^3, z^4]$ and so $\mathrm{Res}_H^G(\chi)$ affords $1_H + \lambda^2 + \lambda^3 + \lambda^4$ for some linear character λ of order 5. The FH-module affording $1_H + \lambda^2 + \lambda^3$ is self-dual, faithful, and is a submodule of $\mathrm{Res}_H^G(V)$, as desired. $\square$

Corollary 7.2c. *Suppose* (G, V) *is a reduced pair of quasisimple type having no regular orbit on* V. *Then there is* $v \in V$ *such that* $C_G(v)$ *has a regular orbit on* V, *and* v *can be chosen to be real for* G *unless* (G, V) *is nonreal reduced. Whenever* H *is a subgroup of* G_0 *satisfying* $H \cap Z = 1$ *which is isomorphic to a minimal stabilizer as given in Theorem 7.2a, then* H *has a regular orbit on* $P_1(V)$.

Proof. From Eq. (3.4b) it follows that if H is abelian with $H \cap Z = 1$, then H has a regular orbit on $P_1(V)$. If (G, V) is a permutation pair (of degree d), by Example 5.1a there exists $v \in V$ such that $H = C_G(v) \cong S_d$ and $\mathrm{Res}_H^G(V)$ is the permutation module for S_d over F, which has a regular orbit on V. The assertion also holds for the two pairs studied in Corollary 7.2b. The minimal stabilizers H appearing in Theorem 7.2a which are real groups, are easily treated (with the help of the character tables).

   Hence it remains to examine the nonreal reduced pair of type $(\mathrm{Sp}_4(3))$ for $r = 7, 13$, where $H = H_1$ or H_2. So let $E = \mathrm{Sp}_4(3)$ and $\chi = \chi_{21}$ resp. χ_{22} [Atlas, p. 27]. Consider first the situation where $r = 7$ and $H \cong \mathrm{SL}_2(3) \times Z_3$. The unique central involution in H must map into the class $2A$ of $EZ/Z \cong U_4(2)$ and so fixes just $2(r + 1) = 16$ points in $P_1(V)$ (since $\chi(2A_0) = 0$). There are four conjugacy classes of elements of order 3 in E. The elements in the classes $3A_0B_0$ have spectral pattern $[\varepsilon^{(3)}, 1]$ for some primitive 3rd root of unity ε, hence fix $(r^3 - 1)/(r - 1) + 1 = 58$ points in $P_1(V)$. Elements in the class $3C_0$ fix $2(r + 1) = 16$ points (since $\chi(3C_0) = -2$), and those in $3D_0$ just $(r + 1) + 1 + 1 = 10$ points (as $\chi(3D_0) = 1$). We assume the worst case that all elements of order 3 in H belong to $3A_0B_0$. Let Y be a noncentral subgroup in H of order 3. Then the four H-conjugates of Y generate a subgroup $S \cong \mathrm{SL}_2(3)$ of H, and the subgroups in S of prime order fix at most $16 + 4 \cdot 58 = 248$ points in $P_1(V)$. Noting that $|P_1(V)| = (r^4 - 1)/(r - 1) = 400$, there are thus at least $400 - 248 = 152$ points which belong to regular orbits of S on $P_1(V)$. Hence there are at least 7 regular S-orbits on $P_1(V)$, including a total number of $7 \cdot |S| = 168$ points. There are three (central) subgroups (of order 3) in H outside S. By common diagonalization the intersections of the 3-dimensional eigenspaces of their generators on V have dimension at least 2. Hence the three subgroups of H outside S leave at most $58 + 2 \cdot 50 = 158$ points in $P_1(V)$ invariant. There is a point which is not fixed by any of these subgroups and belongs to a regular orbit of S, as desired.

In the $r = 13$ case $H \cong Z_3 \mathrm{wr} S_2$ has four subgroups of order 3 and three subgroups of order 2, each fixing at most $1 + (r^2 + r + 1)$ points in $P_1(V)$. The result follows. $\quad\square$

7.3. Covering Numbers, Projective Marks

For convenience we assume that the pair (G, V) is large $(Z \subseteq G)$. Recall that its core $E = E'$ is perfect, $L = E/Z(E)$ is nonabelian simple and that $Z(E)$ is part of the Schur multiplier $\mathrm{M}(L)$. So E is an epimorphic image of the unique (universal) covering group of L (Appendix A). Usually $\mathrm{M}(L)$ is "small", and is cyclic in most cases. At any rate, $Z(E)$ is a *cyclic* part of $\mathrm{M}(L)$. The groups $\bar{G} = G/Z$ and $\bar{G}_0 = G_0/Z$ are subgroups of $\mathrm{Aut}(L)$ containing L (as a normal subgroup). So $\bar{G}$ and $\bar{G}_0$ are almost simple groups, and we say that G, G_0 are *almost quasisimple* (just meaning that they are almost simple over their centre). The group of (outer) automorphisms $\mathrm{Out}(L)$ of L is known to be solvable (Schreier conjecture), and usually it is "small" as well. For example, if L is one of the 26 sporadic simple groups, then $\mathrm{M}(L)$ is cyclic and $\mathrm{Out}(L)$ has order 1 or 2, and if it has order 2, then it inverts the elements in $\mathrm{M}(L)$.

We associate to (G, V) the Brauer character χ of G afforded by V, which may be understood as an ordinary (absolutely) irreducible character of E extended to G. So we are interested in the faithful irreducible characters of E or, more generally, in the projective irreducible representations of L. The character χ may be regarded as afforded by a faithful projective representation of $\bar{G}$. By Schur every projective $\mathbb{C}$-representation of $\bar{G}$ can be lifted to an ordinary representation of its Schur cover.

In order to prove Theorem 7.2a we have to work through the classification of the finite simple groups. Usually we proceed in two steps: We use counting arguments as described in Sec. 5.6, as well as lower bounds for $d = \dim_F(V)$, to reduce the discussion to "small" simple groups L and "small" degrees d, and "small" r. It turns out that in most of the remaining cases $G = G_0 = X \circ Z$ where either $X = E.a$ is an Atlas groups or where χ is a Weil character. Whenever this does not suffice to finish a proof, we use computational methods following [Köhler–Pahlings, 2001].

Let $R_0(\bar{G})$ be the smallest integer $d_0 > 1$ such that there is a projective irreducible representation of $\bar{G}$ of degree d_0 in characteristic 0. Since $\bar{G}$ is a p'-group, $R_0(\bar{G}) = R_p(\bar{G})$ (similarly defined), hence $d \geq d_0$. Lower bounds for $R_0(\bar{G})$ are given in [Schur, 1911] and [Rasala, 1977] when $\bar{G}$ is an alternating or symmetric group, and in [Tiep–Zalesskii, 1996] when $\bar{G}$ is a simple classical group (Appendix C), for all simple groups of Lie type by [Seitz–Zalesskii, 1993]. For linear groups $\mathrm{GL}_m(q)$, symplectic groups $\mathrm{Sp}_{2m}(q)$, q odd, and unitary groups $\mathrm{SU}_m(q)$ this minimal degree is, with few exceptions, attained by the Weil characters (Appendix C and Theorems 4.5a, 4.5b). Being able to treat these Weil characters one can proceed to the characters of the second and third lowest degrees.

We usually begin by applying the following crude estimate, already used in [Liebeck, 1996]. For any nontrivial element x in the automorphism group of the simple nonabelian group L let us define the *covering number* $c(x)$ as the minimum number of L-conjugates of x required to generate the subgroup $\langle L, x \rangle$ of $\mathrm{Aut}(L)$. (This is not common terminology!) Let then

(7.3a) $$c(L) = \max\{c(x) : 1 \neq x \in \mathrm{Aut}(L)\}.$$

Upper bounds for $c(L)$ are available (see below).

Lemma 7.3b (Liebeck). *The fixed point ratio $f(g, V) \leq 1 - 1/c(Zg)$ for any $g \in G \setminus Z$. If there is no regular G-orbit on V, then*

$$r \leq (2|\bar{G}|)^{c(L)/d} \leq (2|\mathrm{Aut}(L)|)^{c(L)/R_0(L)}.$$

Proof. Write $c(g) = c(Zg)$. Since the $[V, g]^x = [V, g^x]$, $x \in E$, generate $V = [V, E]$, by definition $\dim_F[V, g] \geq d/c(g)$. By Proposition 1.6a, $V = [V, g] \oplus C_V(g)$. Hence $\dim_F C_V(g) \leq d - d/c(g)$. Of course $c(L) \geq c(g) \geq 2$. By (5.6d) $\sum_{z \in Z} |C_V(zg)| \leq 2r^{d-d/c(g)}$. If there is no regular G-orbit on V, therefore

$$r^d \leq 2|\bar{G}|r^{d-d/c(L)}.$$

This gives the final statement. $\square$

Now suppose $G = X \circ Z$ where $X = n.L.a$ is an Atlas group (and $Z(X) = Z \cap X$ has order n). Then we may identify $\bar{G}$ with the *almost simple* group $\bar{X} = X/Z(X)$. Choose a set $\{\bar{X}_i\}_{i=1}^s$ of representatives for the conjugacy classes of the subgroups of $\bar{X}$, and let X_i be any (small) subgroup of X mapping onto $\bar{X}_i$. Define m_{ij} as the number of fixed points of $\bar{X}_j$ in its action on the transitive $\bar{X}$-set $\bar{X}_i \backslash \bar{X}$ of right cosets mod $\bar{X}_i$. The isomorphism type of $\bar{X}_i \backslash \bar{X}$ is determined by the ith row vector of the *table of marks* (m_{ij}). Arranging matters such that $|\bar{X}_i| \leq |\bar{X}_j|$ for $i \leq j$ this is a lower diagonal matrix with $m_{ii} = |N_{\bar{X}}(\bar{X}_i) : \bar{X}_i|$ on the diagonal.

Proposition 7.3c. *Let t be any positive integer divisible by $n = |Z(X)|$ and dividing $\exp(X)$. For each i there exists a distinguished rational polynomial $\mu_{\chi,\bar{X}_i}^t$ of degree less than $d = \chi(1)$, with the following property: Whenever r is a power of a prime not dividing $|X|$ such that $t \mid r-1$ and χ can be written in $F = \mathbb{F}_r$, afforded by $V = V_\chi$, then $G_t = X \circ Z_t$ has exactly $\frac{r-1}{t} \cdot \mu_{\chi,\bar{X}_i}^t(r)$ orbits on $V^\sharp$ admitting a point stabilizer mapping onto $\bar{X}_i$.*

Proof. The crucial point will be the computation of these polynomials (their existence being somehow evident). Given r we usually regard the $\mu_{\chi,\bar{X}_i}^t$ as polynomials in r (omitting the argument). Of course, Z_t denotes the subgroup of order t of $Z = Z_{r-1}$ and G_t is the central product over $Z(X)$. If for instance $\mu_{\chi,1}^t$ vanishes (at r), there is no regular G_t-orbit on V, hence no regular orbit for $G = X \circ Z$. If $\mu_{\chi,1}^t$ does not vanish and t is a proper divisor of $r(X) = \gcd(r-1, \exp(X))$, we have to consider, finally, the polynomial $\mu_{\chi,1}^{r(X)}$. Sometimes it happens that $C_{G_t}(v) = C_G(v)$ for each v belonging to a regular orbit for G_t, that is, these orbits fuse to regular orbits for G. Then $\mu_{\chi,1}^t = \mu_{\chi,1}^{r(X)}$, so the polynomial only depends on t.

Let us construct the polynomials. Identify $\bar{G}_t = G_t/Z_t$ with $\bar{X}$. Recall that $C_{G_t}(v) \cap Z_t = 1$ and $C_{G_t}(v)Z_t/Z_t = C_{\bar{G}_t}(Z_t v)$ for each nonzero vector $v \in V$, so that the G_t-orbits on $V^\sharp$ are in 1-1 correspondence with the $\bar{X}$-orbits on $\Omega = \{Z_t v \mid v \in V^\sharp\}$. (We have $\Omega = P_1(V)$ when $Z_t = Z$.) As an $\bar{X}$-set, $\Omega \cong \bigcup_i \mu_i \cdot (\bar{X}_i \backslash \bar{X})$ for certain multiplicities μ_i, counting the number of $\bar{X}$-orbits with stabilizer conjugate to $\bar{X}_i$. Then $|C_\Omega(\bar{X}_j)| = \sum_i \mu_i \cdot m_{ij}$ for each j. It is easy to compute these multiplicities once the row vector $(|C_\Omega(\bar{X}_1)|, \cdots, |C_\Omega(\bar{X}_t)|)$ of marks of $\bar{X}$ on Ω is known, and this is given by

$$|C_\Omega(\bar{X}_j)| = |C_\Omega(X_j)| = \frac{1}{t} \sum_\lambda (r^{\langle \chi, \lambda \rangle x_j} - 1)$$

for all j. Here the sum is taken over all linear characters λ of X_j of order dividing t (and dividing $\exp(X)$). This is a polynomial in r of degree at most *d.* Solving (uniquely) the system of linear equations $(\mu_1, \cdots, \mu_s)(m_{ij}) = (|C_\Omega(X_1)|, \cdots, |C_\Omega(X_s)|)$ we obtain these multiplicities μ_i. Now, observing that $|C_{P_1(V)}(\bar{X}_i)| = \frac{t}{r-1} \cdot |C_\Omega(\bar{X}_i)|$, we define $\mu^t_{\chi, \bar{X}_i} = \frac{t}{r-1} \cdot \mu_i$ (as a polynomial in r). □

For almost all Atlas groups $\bar{X} = X/Z(X)$ of interest for us the table of marks is given in the [GAP] library, and these tables contain generators for the $\bar{X}_i$ which enable one to compute the character tables of the X_i and hence the *projective marks* $|C_{P_1(V)}(\bar{X}_i)|$. On this basis Köhler has implemented an algorithm for the computation of the polynomials [Köhler, 1999]. We shall also refer to the results given in [Köhler–Pahlings, 2001].

Example 7.3d. Let $L = A_5 = \bar{X}$. Then $\bar{X}_1 = 1, \bar{X}_2 \cong Z_2, \bar{X}_3 \cong Z_3, \bar{X}_4 \cong V_4, \bar{X}_5 \cong Z_5, \bar{X}_6 \cong S_3, \bar{X}_7 \cong D_{10}, \bar{X}_8 \cong A_4, \bar{X}_9 \cong A_5$ is a set of representatives for nonconjugate subgroups of $\bar{X}$. The table of marks is given by

$$
(m_{ij}) = \begin{pmatrix}
60 & & & & & & & & \\
30 & 2 & & & & & & & \\
20 & 0 & 2 & & & & & & \\
15 & 3 & 0 & 3 & & & & & \\
12 & 0 & 0 & 0 & 2 & & & & \\
10 & 2 & 1 & 0 & 0 & 1 & & & \\
6 & 2 & 0 & 0 & 1 & 0 & 1 & & \\
5 & 1 & 2 & 1 & 0 & 0 & 0 & 1 & \\
1 & 1 & 1 & 1 & 1 & 1 & 1 & 1 & 1
\end{pmatrix}.
$$

Let us consider one of the (conjugate) Weil characters, χ, of degree 2 for $E = 2.L = \mathrm{SL}_2(5) = X$ (characters χ_6, χ_7 in [Atlas, p. 2]). Here $\exp(X) = 60$ and $G = G_0 = X \circ Z$. Let $V = V_\chi$ over $F = \mathbb{F}_r$ (with r a power of a prime $p \geq 7$, and F containing a square root of 5).

Given any even, positive divisor t of $\exp(X)$ we compute the polynomials $\mu^t_{\chi, \bar{X}_i}$ by picking r such that $t = r(X) = \gcd(r - 1, \exp(X))$. Thus we consider the action of $\bar{G} = \bar{X} = A_5$ on $\Omega = P_1(V)$. For $r(X) = 2$, the corresponding mark vector is $(r + 1, 0, 0, 0, 0, 0, 0, 0, 0)$. This vector is $(r + 1, 2, 0, 0, 0, 0, 0, 0, 0)$ for $r(X) = 4$, and it is $(r + 1, 0, 2, 0, 0, 0, 0, 0, 0)$ for $r(X) = 6$. The mark vector is $(r + 1, 2, 0, 0, 0, 0, 0, 0, 0)$ for $r(X) = 10$, $(r + 1, 2, 2, 0, 0, 0, 0, 0, 0)$ for $r(X) = 12$, and $(r + 1, 2, 0, 0, 2, 0, 0, 0, 0)$ for $r(X) = 20$. It is $(r + 1, 0, 2, 0, 2, 0, 0, 0, 0)$ for $r(X) = 30$, and $(r + 1, 2, 2, 0, 2, 0, 0, 0, 0)$ for $r(X) = 60$. We obtain the following polynomials (in r):

$$\mu_{\chi,1}^2 = (r+1)/60, \quad \mu_{\chi,\bar{X}_i}^2 = 0 \text{ for } i > 1,$$

$$\mu_{\chi,1}^4 = (r-29)/60, \quad \mu_{\chi,\bar{X}_2}^4 = 1, \quad \mu_{\chi,\bar{X}_i}^4 = 0 \text{ for } i > 2,$$

$$\mu_{\chi,1}^6 = (r-19)/60, \quad \mu_{\chi,\bar{X}_3}^6 = 1, \quad \mu_{\chi,\bar{X}_i}^6 = 0 \text{ otherwise},$$

$$\mu_{\chi,1}^{10} = (r-11)/60, \quad \mu_{\chi,\bar{X}_5}^{10} = 1, \quad \mu_{\chi,\bar{X}_i}^{10} = 0 \text{ otherwise},$$

$$\mu_{\chi,1}^{12} = (r-49)/60, \quad \mu_{\chi,\bar{X}_2}^{12} = 1 = \mu_{\chi,\bar{X}_3}^{12}, \quad \mu_{\chi,\bar{X}_i}^{12} = 0 \text{ otherwise},$$

$$\mu_{\chi,1}^{20} = (r-41)/60, \quad \mu_{\chi,\bar{X}_2}^{20} = 1 = \mu_{\chi,\bar{X}_5}^{20}, \quad \mu_{\chi,\bar{X}_i}^{20} = 0 \text{ otherwise},$$

$$\mu_{\chi,1}^{30} = (r-31)/60, \quad \mu_{\chi,\bar{X}_3}^{30} = 1 = \mu_{\chi,\bar{X}_5}^{30}, \quad \mu_{\chi,\bar{X}_i}^{30} = 0 \text{ otherwise},$$

$$\mu_{\chi,1}^{60} = (r-61)/60, \quad \mu_{\chi,\bar{X}_i}^{60} = 1 \text{ for } i = 2,3,5 \text{ and } 0 \text{ otherwise}.$$

Hence there is a regular G-orbit on $V^\sharp$ if and only if r is different from $11, 19, 29, 31, 41, 49, 61$. One also can read off the minimal stabilizers listed in Theorem 7.2a. There is no real vector in V for G if $r \in \{11, 19, 31\}$, because then there are only stabilizers of order 3 or 5. For $r = 49$ observe that $\gcd(r-1, 60) = 12$, in which case we have four G-orbits on $V^\sharp$ with point stabilizers of order 2 (and $\frac{r-1}{6} \cdot \mu_{\chi,1}^6(49) = 4$ regular orbits for $X \circ Z_6$).

7.4. Sporadic Groups

Let L be a sporadic simple group. Then the (abbreviated) character table of L and all possible groups $n.L.a$ can be found in the [Atlas]. Using the information given in the Atlas one can bound $c(L)$. It turns out that $c(L) \le 6$ except possibly for the Fischer groups ($c(Fi_{22}) \le 8, c(Fi_{23}) \le 7, c(Fi'_{24}) \le 12$). In view of Lemma 7.3b this reduces the discussion to some few groups L admitting faithful projective representations of small degrees.

Let $L = M_{11}$ be the smallest Mathieu group. Then $L = E$ and $G = G_0 = E \times Z$. Consider the character $\chi = \chi_2$ of degree 10 [Atlas, p. 18]. Let x be one of the $|2A| = 165$ (conjugate) involutions in L. From $\chi(x) = 2$ we infer that $x_V = [1^{(6)}, -1^{(4)}]$. Let y be one of the $|3A| = 2 \cdot 220$ (conjugate) elements in L of order 3. From $\chi(y) = 1$ we infer that $y_V = [1^{(4)}, z_1^{(3)}, z_2^{(3)}]$ where the z_i are the primitive 3rd roots of unity (provided $3 \mid r-1$). Let u be one of the $|5A| = 4 \cdot 396$ (conjugate) elements in L of order 5. From $\chi(u) = 0$ we infer that each 5th root of unity occurs with multiplicity 2 as eigenvalue of t on V (provided $5 \mid r-1$). Each primitive 11th root of unity occurs with multiplicity 1 as eigenvalue of the elements of order 11 on V (if $11 \mid r-1$). Assume that there is no regular G-orbit on V. Then

$$r^{10} \le 165(r^6 + r^4) + 220(r^4 + 2r^3) + 396(5r^2) + 1.440(10r).$$

But this implies that $r < 5$, a contradiction. All other 9 irreducible characters χ_i of M_{11} are treated similarly. (Using that $c(M_{11}) \leq 4$ only the characters of degree $10, 11, 16$ need to be considered.)

Let $L = M_{12}$, where $\mathrm{M}(L)$ and $\mathrm{Out}(L)$ have order 2 [Atlas, p. 33]. One checks that $c(M_{12}) \leq 5$ by considering the maximal subgroups. So as before only the characters $\chi = \chi_i$ of small degrees have to be examined, namely χ_2, χ_3 of degree 11 and χ_4, χ_5 of degree 16 of M_{12} (which fuse in $\mathrm{Aut}(M_{12})$), and the characters of degree $10, 10, 12, 32$ of $2.M_{12}$ (extendible to $2.M_{12}.2$). In each case there is a regular orbit.

The other sporadic groups are treated in the same manner, except when $L = J_2$ is the second Janko group or $L = Suz$ is the Suzuki group. Let us treat these groups (briefly).

Let $L = J_2$. One checks that $c(J_2) \leq 5$. Using Lemma 7.3b this reduces the discussion to the case where $E = 2.J_2$ and $V = V_\chi$ with $\chi = \chi_{22}$ or χ_{23} in the Atlas notation [Atlas, p. 43], both characters of degree 6 fusing in $2.J_2.2$. Hence $X = E$ is the Schur cover of L and $G = G_0 = E \circ Z$. The usual counting technique yields the existence of a regular vector when $r \geq 26$. Since $p > 7$ by coprimeness, and since $F = \mathbb{F}_r$ contains a square root of 5 by the values of χ on 5-elements, we must have $r = p$ equal to 11 or to 19. On the basis of the table of marks for $L = J_2$ [GAP] we compute (observing that $\exp{(2.J_2)_3} = 3$)

$$\mu_{\chi,1}^2 = (r-11)(r+1)(r+11)(r^2 - 193)/|L|,$$
$$\mu_{\chi,1}^6 = \mu_{\chi,1}^2 - (r^2 - 14r + 265)/1.080,$$
$$\mu_{\chi,1}^{10} = \mu_{\chi,1}^2 - 7(r-11)/300.$$

This shows that there is a regular orbit for $r = 19$ but no regular orbit for $r = 11$. In this latter case the smallest stabilizer H is isomorphic to S_3 (as is seen by computing the corresponding polynomials). Since the characters χ_{22}, χ_{23} involved are conjugate under a group automorphism of $2.J_2$, we have a unique isomorphism type.

Let $L = Suz$. One checks that $c(Suz) \leq 6$. As before this reduces the discussion to the case where $E = X = 6.Suz$ and $\chi = \chi_{115}$ is of degree 12 [Atlas, p. 130]. Here $G = G_0 = X \circ Z$ and $V = V_\chi$ comes from the action of X on the complex Leech lattice. The usual counting argument yields that there is a regular orbit when $r \geq 23$. Since $p > 13$ and $6 \mid r - 1$, it remains the case where $r = p = 19$. (Since an involution x of $6.Suz$ belonging to the class $2A_0$ has the spectral pattern $x_V = [-1^8, 1^4]$ and $|2A_0| = 135.135$, the counting method fails for $r = 19$.) The table of marks for Suz is not available in [GAP], but [Köhler–Pahlings, 2001] indicate, on the basis of a different orbit algorithm, that a regular orbit exists also in the $r = 19$ case.

Remark. As for Theorem 7.1 it suffices to find a (strongly) real vector v such that $C_G(v)$ has a regular orbit on $V^\sharp$ ($r = 19$). The group $X = 6.Suz$ has a subgroup $Y \cong Z_2 \times 3^6 : M_{11}$, which in turn has a subgroup $H \cong M_{11}$ of pure permutations acting transitively on the 12 coordinates of the Leech lattice, a point stabilizer being a maximal $L_2(11)$ subgroup [Atlas, pp. 131, 18]. It follows that $\mathrm{Res}_H^G(\chi) = 1_H + \chi_5$ where χ_5 is the irreducible character of $H \cong M_{11}$ of degree 11 (which is rational-valued). Let $C_V(H) = Fv$ and $N = N_X(Fv)$. Either H is contained in a $2.M_{12}$ subgroup of X, on which χ is irreducible however, or N is in (a conjugate of) Y, with H acting irreducibly on $Y/Z(X) \cong 3^5$. We conclude that $C_X(v) = H$ and that $N = H \times Z(X)$. Application of Lemma 5.1b yields that $C_G(v) = H$. One also easily checks that H has a regular orbit on V.

7.5. Alternating Groups

Let $L = A_n$ for some $n \geq 5$, acting on $\Omega = \{1, \cdots, n\}$. Then $\mathrm{Aut}(L) = S_n$ unles $n = 6$ ($\mathrm{Out}(A_6) \cong V_4$) and $|\mathrm{M}(L)| = 2$ unless $n = 6, 7$ ($\mathrm{M}(A_6) = \mathrm{M}(A_7) = Z_6$). It is well known that $c(x) \leq n - 1$ if x is a transposition (as $(12), (13), \cdots, (1n)$ generate S_n). For $1 \neq x \in S_n$ we generally have

$$(7.5a) \qquad c(x) \leq \left\lceil \frac{n-2}{n - |\mathrm{orb}(\langle x \rangle \text{ on } \Omega)|} \right\rceil + 2,$$

except when $n = 6$ and x is an involution of cycle shape (2^3), where $c(x) = 5$ [Hall *et al.*, 1992]. In addition $c(x) \leq n/2$ for $n \geq 7$, unless x is a transposition [Guralnick–Saxl, 2003].

As usual we exclude the deleted permutation module, which is the Specht module $S^{(n-1,1)}$ for S_n and which for $n \geq 7$ is the unique faithful module of minimal degree. By [Rasala, 1977] for $n \geq 10$ the Specht module $S^{(n-2,2)}$ to the partition $(n - 2, 2)$ is the unique faithful module of the second minimal degree, $n(n - 3)/2$. This module is known to be irreducible for S_n [James–Kerber, 1981, 7.3.23]. Using the branching rule for Specht modules one shows, by induction, that $S^{(n-2,2)}$ is irreducible also for A_n. From [Schur, 1911, Sec. 50] one knows that for $n \geq 7$ the minimal faithful character degrees for $2.A_n$ and $2.S_n$ are $2^{\lfloor (n-2)/2 \rfloor}$ and $2^{\lfloor (n-1)/2 \rfloor}$, respectively.

Proposition 7.5b. *For $n \geq 8$ there exists a regular G-orbit on V.*

Proof. The character table for A_n and its related groups is in the [Atlas] for $n \leq 13$. Assume first that $n \geq 14$. Then by the above

(7.5c) $$d \geq n(n-3)/2$$

for $n \geq 16$, because then $2^{\lfloor \frac{n-2}{2} \rfloor} > n(n-3)/2$, and $d \geq 64$ for $n = 14, 15$.

Suppose $x \in S_n$ is nontrivial with $n - |\mathrm{orb}(\langle x \rangle \text{ on } \Omega)| > 3$. Then by (7.5a) $c(x) \leq \lceil \frac{n-2}{4} \rceil + 2 \leq (n+9)/4$ unless $n = 14$, x has cycle shape (2^7) and $c(x) \leq 6$. If $n - |\mathrm{orb}(\langle x \rangle \text{ on } \Omega)| \leq 3$, then x has cycle shape $(2), (3), (2^2), (4), (3, 2)$ or (2^3). Then $c(x) \leq n - 1, n/2, n/2$ and, in view of (7.5a), three times $c(x) \leq \lceil (n-2)/3 \rceil + 2 \leq (n+6)/3$. There are $\binom{n}{2}, 2 \cdot \binom{n}{3}$, $3 \cdot \binom{n}{4}$, $6 \cdot \binom{n}{4}$, $20 \cdot \binom{n}{5}$ and $15 \cdot \binom{n}{6}$ elements in S_n of these cycle shapes, respectively. Assume there is no regular G-orbit on V. Let first $n \geq 16$. Then by Lemma 7.3b

$$r^d/2 \leq |S_n| r^{d(1-\frac{4}{n+9})} + \binom{n}{2} r^{d(1-\frac{1}{n-1})} + \left[\binom{n}{3} + 3\binom{n}{4}\right] r^{d(1-\frac{2}{n})}$$

$$+ \left[6\binom{n}{4} + 20\binom{n}{5} + 15\binom{n}{6}\right] r^{d(1-\frac{3}{n+6})}.$$

Since $r > n \geq 16$ and $d \geq n(n-3)/2$ by (7.5c), it follows that

$$2n! > r^{4d/(n+9)} - r^{3d/(n+9)+2} - r^{2d/(n+9)+4} - r^{d/(n+9)+6}.$$

The second, third and fourth terms on the right are all less than $\frac{1}{6} r^{4d/(n+9)}$. Consequently $n^n > 4n! > r^{4d/(n+9)} > n^{4d/(n+9)}$ and so $n > 4d/(n+9) \geq 2n(n-3)/(n+9)$. But this implies that $n < 12$.

Let $n = 14$ or 15. We have the same estimate, replacing the term $|S_n| r^{d(1-\frac{4}{n+9})}$ by $|S_n| r^{d(1-\frac{1}{6})}$. Since then the difference of the terms on the left hand side and the right hand side is an increasing function of d, we may pick $d = 64$. We get that

$$r^{64}/2 \leq 15! r^{53} + 105 r^{59} + 5.005 r^{56} + 2.761.670 r^{54}.$$

But this implies that $r < 14$, which is a contradiction ($p > 14$).

Let finally $8 \leq n \leq 13$. Recall that $r \geq p > n$. By Lemma 7.3b we may assume that $r^d \leq 2\binom{n}{2} r^{d(1-\frac{1}{n-1})} + 2n! r^{d(1-\frac{2}{n})}$. So for $n = 8$ only the characters $\chi = \chi_3, \chi_{15}$ of degrees $d = 20, 8$, respectively, have to be examined [Atlas, p. 22]. Here, as in all further cases, the counting method works. $\square$

It remains to investigate the cases $n = 5, 6, 7$.

n = 5 : The Weil characters of $2.A_5 = \mathrm{Sp}_2(5)$ of degree 2 have been already treated in Example 7.3d. Let χ be one of the Weil characters of degree 3 of $\mathrm{PSp}_2(5) \cong A_5$. So $\chi = \chi_2$ or χ_3 [Atlas]. Further $E = A_5 = X$ and $G = G_0 = E \times Z$, and $\exp(X) = 30$. Let $V = V_\chi$ over $F = \mathbb{F}_r$ (so r is any power of a prime $p \geq 7$, and F contains $\sqrt{5}$). As above let $r(X) = \gcd(r-1, \exp(X))$. If $r(X) = 2$, then $r - 1 = 2u$ for an odd u divisible only by primes larger than 5, and the mark vector on $P_1(V)$ is the row vector given by $(r^2 + r + 1, r + 2, 1, 3, 1, 1, 1, 0, 0)$. This row vector is $(r^2 + r + 1, r + 2, 3, 3, 1, 1, 1, 0, 0)$ for $r(X) = 6$, is $(r^2 + r + 1, r + 2, 1, 3, 3, 1, 1, 0, 0)$ for $r(X) = 10$, and is $(r^2 + r + 1, r + 2, 3, 3, 3, 1, 1, 0, 0)$ for $r(X) = 30$. Using the table of marks for A_5 given in Example 7.3d we get:

$$\mu^2_{\chi,1} = (r-9)(r-5)/60, \ \mu^2_{\chi,\bar{X}_2} = (r-5)/2, \ \mu^2_{\chi,\bar{X}_i} = 1 \text{ for } i = 4,6,7,$$

$$\mu^6_{\chi,1} = (r^2 - 14r + 25)/60, \ \mu^6_{\chi,\bar{X}_2} = (r-5)/2, \ \mu^6_{\chi,\bar{X}_i} = 1 \text{ for } i = 3,4,6,7,$$

$$\mu^{10}_{\chi,1} = (r-11)(r-3)/60, \ \mu^{10}_{\chi,\bar{X}_2} = (r-5)/2, \ \mu^{10}_{\chi,\bar{X}_i} = 1 \text{ for } i = 4,5,6,7,$$

$$\mu^{30}_{\chi,1} = (r-13)(r-1)/60, \ \mu^{30}_{\chi,\bar{X}_2} = (r-5)/2, \ \mu^{30}_{\chi,\bar{X}_i} = 1 \text{ for } 3 \leq i \leq 7.$$

In all other cases $\mu^t_{\chi,\bar{X}_i} = 0$. Using that $\mathbb{F}_{13}$ does not contain a square root of 5, one concludes that there is a regular G-orbit on $V^\sharp$ if and only if $r \neq 11$. For $r = 11$ there are point stabilizers of order 2.

The character $\chi = \chi_4$ of A_5 [Atlas, p. 2] is afforded by the deleted permutation module, which case we have excluded. (There is no regular G_0-orbit if and only if $G_0 = S_5 \times Z$ with $r = 7$, in accordance with 5.1a.)

The character $\chi = \chi_5$ of A_5 of degree 5, which splits on S_5, is treated by the usual counting argument. For the characters $\chi = \chi_8$ and χ_9 of $2.A_5$ of degrees 4 and 6, which extend to any $2.A_5.2$, the counting argument applies as well.

n = 6 : The counting method rules out all cases except the following ones.

(i) $E = A_6$, $\chi = \chi_2$ resp. χ_3 of degree 5, which extend to $A_6.2_1 = S_6$ but fuse in $A_6.2_2 = \mathrm{PGL}_2(9)$ and $A_6.2_3 = M_{10}$. Here χ_2 is the character of A_6 afforded by the deleted permutation module. So we have to investigate χ_3 only. But χ_3 is obtained from χ_2 via an (exceptional) outer automorphism of A_6, and the same for the extensions to S_6. So the corresponding reduced pairs are isomorphic. (Therefore they do not appear in the list of Theorem 7.2a !) Of course the polynomials for χ_2, χ_3 agree. We know from Example 5.1a that there is no regular G_0-orbit on V_χ, for $\chi = \chi_2$ or χ_3, if and only if $r = 7$. From $\mu^6_{\chi,\langle(12)(34)(56)\rangle} = (r-1)(r-3)(r-5)/24$ we obtain two orbits with point stabilizers of order 2.

(ii) $E = 2.A_6$, $\chi = \chi_8$ resp. χ_9 of degree 4, which extend to both isoclinic variants $X = 2.A_6.2_1$ and fuse in the $2.A_6.2_2$. This has been already treated in Example 5.7b, but let us argue also on the basis of Proposition 7.3c. For $\chi = \chi_9$ and $X = 2^+S_6$ the polynomials have 7 as the only positive root coprime to $|E|$, and

$$\mu^6_{\chi,1} = (r-3)(r-7)(r+11)/720.$$

Hence there is no regular orbit for $r = 7$. Considering the subgroups $\bar{X}_3 = \langle (123)(345) \rangle$, $\bar{X}_4 = \langle (123)(45) \rangle$ and $\bar{X}_4 = \langle (123)(456), (123)(45) \rangle$ of $\bar{X} = S_6$ we have

$$\mu^6_{\chi,\bar{X}_3} = (r-1)/6, \quad \mu^6_{\chi,\bar{X}_4} = \mu^6_{\chi,\bar{X}_5} = 1.$$

Hence there are just three G-orbits on $V^\sharp$ for $r = 7$, and the point stabilizers are isomorphic to Z_3, Z_6, $Z_3 \times S_3$. This is as stated in Theorem 7.2a, and known from Example 5.7b. We have also seen that there exists no real vector in this case.

(iii) There are four algebraically conjugate characters, χ, of $E = 3.A_6$ of degree 3 which pairwise fuse in $3.A_6.2_1$, $3.A_6.2_2$ or $3.A_6.2_3$. So we have to consider $X = E$ and $G = G_0 = X \circ Z$. The splitting fields $F = \mathbb{F}_r$ for χ require the 3rd roots of unity and the square roots of 5. The counting argument yields that there are regular orbits when $r \geq 53$. Hence if there are no regular vectors, then $r = 19, 31$ or 49. It suffices to compute:

$$\mu^6_{\chi,1} = (r-19)(r-25)/360, \quad \mu^6_{\chi,\langle(12)(34)\rangle} = (r-15)/4;$$
$$\mu^{30}_{\chi,1} = (r-13)(r-31)/360, \quad \mu^{30}_{\chi,\langle(12)(34)\rangle} = (r-27)/4.$$

It follows that there is no regular orbit for $r = 19$ and $r = 31$, in which cases there are orbits with stabilizer of order 2. For $r = 49$ observe that there are $\frac{r-1}{6} \cdot \mu^6_{\chi,1}(47) = 16$ regular $E \circ Z_6$-orbits on $V^\sharp$ by Proposition 7.3c. Using that $\exp(E) = 2^2 \cdot 3 \cdot 5$ and that each involution of G is in $E \circ Z_6$ we see that these orbits fuse to 2 regular orbits for G. So the polynomial $\mu^{48}_{\chi,1} = \mu^6_{\chi,1}$ depends only on $t = 6$.

(iv) For $E = 3.A_6$ there is a pair of algebraically conjugate irreducible characters $\chi = \chi_{15}$ of degree 6 [Atlas, p. 5], which extend to $X = 3.A_6.2_3$ (requiring a square root of -2). So $G = G_0 = X \circ Z$. Assuming that there is no regular orbit,

$$r^6 \leq |2A_0|(r^4 + r^2) + |3A_0B_0|(3r^2) + \frac{1}{2}|5A_0B_0|(r^2 + 4r)$$
$$= 45(r^4 + r^2) + 120r^2 + 36(r^2 + 4r).$$

This implies that $r < 8$, hence $r = 7$. This is ruled out by the method of Proposition 7.3c; in fact, the polynomial $\mu^6_{\chi,1}$ has only the root 1.

n = 7 : The counting argument rules out all cases except when $E = 2.A_7$, $\chi = \chi_{10}$ resp. χ_{11} of degree 4, which fuse in $2.S_7$. We get regular orbits for $r \geq 23$. Since $p > 7$ and a splitting field $F = \mathbb{F}_r$ for χ requires a square root of -7, and since $SL_4(17)$ and $SL_4(19)$ are $7'$-groups, we must have $r = 11$. In this case the multiplicities $\mu^{10}_{\chi,1} = (r - 11)(r^2 + 12r - 7)/2.520$ and $\mu^{10}_{\chi,\langle(123)(456)\rangle} = (r - 5)/6$ give the result as stated in Theorem 7.2a.

7.6. Linear Groups

Let $L = L_m(q) = \mathrm{PSL}_m(q)$ for some integer $m \geq 2$ and some prime power q. Since $L_2(3)$ is solvable and $L_2(4) = L_2(5) \cong A_5$, $L_2(9) \cong A_6$ and $L_4(2) \cong A_8$, we may and do exclude the pairs $(m, q) = (2, 3), (2, 4), (2, 5), (2, 9)$ and $(4, 2)$. [Guralnick and Saxl, 2003] have given upper bounds for $c(L)$. Ignoring the groups excluded one has $c(L_m(q)) \leq 4, 4, 6$ for $m = 2, 3, 4$, respectively, and

(7.6a) $$c(L_m(q)) \leq m$$

in all other cases. From Appendix (C1) we know that

(7.6b) $$R_0(L_m(q)) = \begin{cases} (q - 1)/\gcd(q - 1, 2) & \text{if } m = 2 \\ (q^m - 1)/(q - 1) & \text{if } m > 2 \end{cases}$$

except when $(m, q) = (3, 2), (3, 4), (4, 3)$ (again ignoring the groups already ruled out). One has $R_0(L_3(2)) = 3$, $R_0(L_3(4)) = 6$ and $R_0(L_4(3)) = 26$, and otherwise $R_0(L_m(q))$ is just attained by the degree of Weil characters.

Recall that $\bar{G} \subseteq \mathrm{Aut}(L)$. The Schur multiplier and the automorphism group of L (and of any group of Lie type) has been determined in [Steinberg, 1967]. Each automorphism is a product of an inner, a diagonal, a field and a graph automorphism. In the present case the graph automorphism (of the Dynkin diagram) is just the inverse transpose automorphism, and the diagonal automorphisms are induced by conjugation with the diagonal matrices in $\mathrm{GL}_m(q)$ with determinant $\neq 1$. So $\mathrm{Out}(L)$ and $\mathrm{M}(L)$ are known.

Proposition 7.6c. *If there is no regular G-orbit on V, then $m \leq 3$ or $L \cong L_4(3)$ and $d \leq 52$, $L \cong L_5(2)$ and $d = 30$, or $L \cong L_6(2)$ and $d = 62$.*

This readily follows from the above information by applying Lemma 7.3b. Note that $L_6(2)$ is not an Atlas group. We know from Theorem 4.5b that $L_6(2)$ has a Weil character of degree $d = 2^6 - 2 = 62$. This is the unique irreducible character of this degree; the second lowest degree is 217.

m = 2 : Suppose that there is no regular G-orbit on V. Let first q be even. Then Lemma 7.3b implies that $q \leq 16$, hence $q = 16$ or $q = 8$. From the character table one reads off that, for each character $\chi \neq 1_L$, of $L = L_2(16)$ or a possible extension to $\mathrm{Aut}(L) = L.4$, no element of prime order in $\mathrm{Aut}(L)$ has eigenspaces of dimension greater than $\chi(1) - 6$. Thus (5.6d) yields that $r^6 \leq 2|\mathrm{Aut}(L)|$ in this case, giving the contradiction $r < 6$. It remains to exclude the case $L = L_2(8)$. Consider for instance the character $\chi = \chi_2$ of degree $d = 7$ of $E = L_2(8)$ extendible to $X = \mathrm{Aut}(L) = L.2$ [Atlas, p. 6], letting $G = X \times Z$ and $V = V_\chi$ over $F = \mathbb{F}_r$. Determining spectral patterns of elements of prime order we get

$$r^7 \leq |2A|(r^4 + r^3) + \frac{1}{2}|3A|(2r^3 + r) + \frac{1}{2}|7ABC|(7r) + \frac{1}{2}|3B|(r^3 + 2r^2).$$

Here $|2A| = 63, |3A| = 56, |7A| = |7B| = |7C| = 72, |3B| = 84$, which gives $r^7 \leq 63r^4 + 161r^3 + 84r^2 + 784r$. We obtain that $r < 5$, which is a contradiction since $|L| = 2^3 \cdot 3^2 \cdot 7$. The other characters are treated similarly.

So let q be odd, $q = s^f$ for some (odd) prime s. Assume $q \geq 37$. Consider first the case where $f = 1$ ($q = s$). We assert that then $d \geq q - 1$. For otherwise χ is one of the Weil characters $\xi_i \neq \bar\xi_i$ of $\mathrm{Sp}_2(q)$ for $i = 1, 2$, of degree $(q \pm 1)/2$, where the pairs of algebraically conjugate characters must fuse in $\mathrm{GL}_2(q)$ (see Theorem 4.5a and Appendix C2; the character table of $\mathrm{Sp}_2(q)$ can be found in [Schur, 1911]). Hence $\bar G \cong L = L_2(q)$. From Theorems 4.4 and 4.5a we infer that χ takes the value $(-1 \pm \sqrt{\mp q})/2$ on the noncentral elements of order q. So these elements have $(q \pm 1)/2$ distinct eigenvalues on $V = V_\chi$, whence no eigenspace has dimension greater than 1. For a noncentral element $g \in \mathrm{Sp}_2(q)$ of prime order $\neq q$ we have $\chi(g) = 0$ or ± 1, which implies that no eigenspace has dimension greater than $(q+5)/4$. Since $q = s$ and $|L| = q(q^2 - 1)/2$, it follows that

$$r^{(q-1)/2} \leq q(q^2 - 1)/2 \cdot q^{(q+5)/4}.$$

From $q \geq 37$ we infer that $r < 5$, a contradiction. So we indeed have $d \geq q - 1$. Using $c(L) \leq 4$, Lemma 7.3b yields that $r \leq \left(q(q^2 - 1)\right)^{\frac{4}{q-1}}$, which again leads to a contradiction. Hence we must have $f > 1$. Using that $|\mathrm{Aut}(L)| = fq(q^2 - 1)$ from Lemma 7.3b we now get

$$r \leq \left(2fq(q^2 - 1)\right)^{8/(q-1)},$$

and this gives a contradiction as before. Consequently $q < 37$, that is, $q \leq 31$. Now $L = L_2(q)$ is an Atlas group, and with the usual counting arguments one can rule out all cases except for $q = 7$.

So let $L = L_2(7)$. Let χ be one of the Weil characters $\chi_2, \chi_3 = \bar{\chi}_2$ of L of degree 3, which fuse in $\mathrm{PGL}_2(7)$ [Atlas, p. 3]. Hence $G = G_0 = L \times Z$ in this case. If there is no regular G-orbit on $V = V_\chi$, then

$$r^3 \leq |2A|(r^2 + r) + \frac{1}{2}|3A|(3r) + \frac{1}{3}|7AB|(3r) = 21r^2 + 129r.$$

Thus $r < 26$. Since χ, and hence $F = \mathbb{F}_r$, requires a square root of -7, this implies that $r = 11, 23$ or 25. Let H be generated by an involution of L (in the unique class $2A$). One finds:

$$\mu^2_{\chi,1} = (r - 9)(r - 11)/168, \quad \mu^2_{\chi,H} = (r - 7)/4,$$
$$\mu^6_{\chi,1} = (r^2 - 20r + 43)/168, \quad \mu^6_{\chi,H} = (r - 7)/4.$$

Hence there is no regular orbit when $r = 11$, in which case there are $\frac{r-1}{2} \cdot (r - 7)/4 = 6$ orbits with point stabilizers of order 2 (conjugate to H). Now $\gcd(r - 1, \exp(L))$ is 2 for $r = 23$, and is 12 for $r = 25$. It follows that there are regular orbits in these cases. For $r = 25$ observe that the $\frac{r-1}{6} \cdot \mu^6_{\chi,1}(r) = 4$ regular orbits for $L \times Z_6$ must fuse in $G = L \times Z_{24}$ since each involution in G is contained in $L \times Z_6$.

Let next $\chi = \chi_6$ be of degree 6, which extends to $X = \mathrm{PGL}_2(7)$ (affording a square root of 2). We have to consider $G = X \times Z$, and the counting method yields that there is a regular orbit when $r \geq 11$. Hence the only questionable case is $r = 5$, which is ruled out by showing that $\mu^4_{\chi,1} > 0$ at $r = 5$.

For all other characters the counting method applies and yields that there are regular orbits.

m = 3 : Using that $c(L) \leq 4$ and $d \geq R_0(L) = (q^3 - 1)/(q - 1)$, application of Lemma 7.3b shows that a regular orbit exists unless $q \leq 5$. Thus we only have to examine the cases $L = L_3(3)$ and $L = L_3(4)$.

Let $L = L_3(3)$. If $d \neq 12$ or 13, no prime order element of $\mathrm{Aut}(L)$ has an eigenspace of dimension greater than $d - 8$ [Atlas, p. 13]. Then $r^8 \leq 2|\mathrm{Aut}(L)|$ forces that $r < 4$. Hence there is a regular orbit. For the characters $\chi = \chi_2$ and χ_3 of degrees 12 and 13 the usual counting method works.

Let $L = L_3(4)$. Recall that $R_0(L) = 6$, and this minimal degree just happens for the faithful irreducible characters $\chi = \chi_{41}$ of $X = 6.L.2_1$ [Atlas, p. 25]. This χ is rational-valued on $E = 6.L$ and requires $\sqrt{2}$ or $\sqrt{-2}$ on X (depending on the isoclinism type of X). Of course also $6 \mid r - 1$. Let us describe the situation carefully. We have $\exp(X) = 2^3 \cdot 3 \cdot 5 \cdot 7$. The conjugacy class $2A$ of L lifts to two conjugacy classes $2A_0$, $2A_3$ of involutions in E ($\chi(2A_0) = -2, \chi(2A_3) = 2$). Since $Z(E) \subseteq Z$ and $|2A| = 315$, this gives the contribution $315(r^4 + r^2)$. From $\chi(2B_0) = 0$ and $|2B| = 280$ we get the contribution $280(2r^3)$. From $\chi(3A_0) = 0$ and $|3A| = 2.240$ we get the contribution $1.120(3r^2)$. The two conjugacy classes $5AB$ of L ($|5A| = |5B| = 4.032$) lift to two conjugacy classes $5A_0B_0$ of elements of order 5 in E, with $\chi(5A_0) = \chi(5B_0) = 1$. Hence the contribution $1.120(r^2 + 4r)$ if $5 \mid r - 1$ and $1.120(r^2)$ otherwise. The two conjugacy classes $7AB$ of L lift to two conjugacy classes $7A_0B_0$ of order 7, with $\chi(7A_0) = \chi(7B_0) = -1$. From $|7A| = |7B| = 2.280$ we obtain the contribution $960(6r)$ if $7 \mid r - 1$ and $960r$ otherwise. Thus if there is no regular orbit, then

$$r^6 \leq 315r^4 + 560r^3 + 4.795r^2 + 10.240r.$$

This forces that $r < 19$. Since $3 \mid r - 1$ and $r > 7$, we must have $r = 13$. But ± 2 are not squares mod 13. Therefore $G = E \circ Z$ or $G = G_0 = (E \circ Z).2$ (and $r = 13$). The table of marks for $L = L_3(4)$ is contained in the [GAP] library, and one gets that the polynomial $\mu_{\chi,1}^{12}$ does not vanish at $r = 13$. So there is a regular vector $v \in V$ for $E \circ Z$, and this is regular also for G_0 since otherwise $G_0 = (C_{G_0}(v)E) \circ Z$.

All other irreducible characters for groups $n.L_3(4).a$ can be treated by the usual counting method.

$\mathbf{m \geq 4}$: By the proposition only the following cases have to be treated.

(i) $L = L_4(3)$ and $d \leq 52$: In all cases where $26 < d \leq 52$, no eigenspace of a noncentral prime order element of G has dimension greater than $d - 13$ [Atlas]. Hence $r^{13} \leq 2 \cdot |L| \cdot 8$ forces that $r < 5$. For the characters $\chi = \chi_2$ resp. χ_3 of L, which fuse in $L.2_1$ but split on $X = L.2_2$, we similarly conclude from

$$r^{26} \leq |2D|r^{20} + |L.2_2|2r^{16}$$

that necessarily $r < 7$. Thus there is a regular orbit for $G = X \circ Z$.

(ii) $L = L_5(2)$ and $d = 30$: Here $\chi = \chi_2$ is the unique minimal (Weil) character (of degree $2^5 - 2$) of L discussed in Theorem 4.5b. It is rational and extends to $X = L.2$ requiring a square root of 2 [Atlas, p. 70]. So $G = L \times Z$ or $G = X \times Z$. The largest dimension of an eigenspace on V of a prime order element in X is 22 [Atlas]. From $r^8 \leq 2 \cdot |X|$ we obtain that $r < 9$. There are regular orbits.

(iii) $L = L_6(2)$ and $d = 62$: Then again $\chi = \xi$ is the Weil character of L studied in Theorem 4.5b (see also Appendix C1). Since $L_6(2)$ is not an Atlas group, we must go into some details. By uniqueness ξ extends to $X = L.2 = \mathrm{Aut}(L)$, and we consider $G = G_0 = X \times Z$. We assert that $\dim_F C_V(g) \leq 48$ for each noncentral element $g \in G$ of prime order. In order to prove this we may enlarge F, if necessary, by adjoining the 8th roots of unity. (If F_0 is an extension field of F and $V_0 = F_0 \otimes_F V$, then $\dim_{F_0} C_{V_0}(g) = \dim_F C_V(g)$.) Embed L into a standard holomorph T to $Q = 2_+^{1+12}$. By (4.5b) there is a subgroup $\langle L, \tau \rangle$ of T mapping onto X and satisfying $\tau^2 \in Z(Q)$. There is $z \in Z$ such that $(z\tau)^2 = 1$, so that X and G appear as subgroups of TZ. Since ± 2 are squares in F, by Theorem 4.3e there is a faithful irreducible $F[TZ]$-module W (of F-dimension 2^6), and V is a constituent of its restriction to G since W affords the character $\xi + 2 \cdot 1_L$ on L by Theorem 4.5b. Applying Theorem 6.2c we get

$$\dim_F C_V(g) \leq \dim_F C_W(g) \leq \frac{3}{4}\dim_F W = 48,$$

as asserted. There is a regular since $r^{62} > 2 \cdot |L.2|r^{48}$ for $r \geq 11$.

7.7. Symplectic Groups

Let $L = S_{2m}(q) = \mathrm{PSp}_{2m}(q)$ for some integer m and some prime power q. Since $S_2(q) = L_2(q)$ and $S_4(2) = A_6$ are already treated, we assume that $m \geq 2$ and exclude the pair $(m, q) = (2, 2)$. From [Guralnick–Saxl, 2003] one knows that then $c(L) \leq 2m + 1$ (and $c(L) \leq 2m$ if q is odd and $m \geq 3$). Up to few exceptions $|\mathrm{M}(L)| = (2, q - 1)$, and this is also the order of the group of outer diagonal automorphisms, and for $m \geq 3$ there is no proper graph automorphism [Steinberg, 1967]. As stated in Appendix (C2), one has $d \geq (q^m - 1)/2$ if q is odd and $d \geq (q^m - 1)(q^m - q)/(2(q+1))$ otherwise.

Proposition 7.7a. *There is a regular G-orbit on V except possibly when $m \leq 6$ and one of the following holds:*

• $(m, q) = (2, 3)$ *and* $d \leq 30$, $(m, q) = (2, 4)$ *and* $d \leq 34$, $(m, q) = (2, 5)$ *and* $d \leq 40$, $(m, q) = (3, 2)$ *and* $d \leq 35$, *or* $(m, q) = (4, 2)$ *and* $d \leq 85$.

• $(m, q) = (2, 7), (2, 9), (2, 11), (3, 3), (3, 5), (4, 3), (5, 3), (6, 3)$ *and* χ *is one of the irreducible Weil characters of the symplectic group.*

This follows from the above information by applying Lemma 7.3b.

The groups $S_4(7), S_4(9), S_4(11)$ and $S_6(5), S_8(3), S_{10}(3), S_{12}(3)$ are not Atlas groups. From Theorem 4.5a we know a great deal about Weil characters of symplectic groups (see also Appendix C). We need some more details on fixed point ratios.

Let $E = S_{2m}(q)$ or $E = \mathrm{Sp}_{2m}(q)$ for some $m \geq 2$ and some odd $q = q_0^f$ (q_0 prime), assuming that χ is on E one of the irreducible Weil characters ξ_1, ξ_2 of degree $(q^m - 1)/2$ and $(q^m + 1)/2$, respectively. Recall that just one of these characters is faithful for $\mathrm{Sp}_{2m}(q)$, and that ξ_1 is faithful if and only if $q^m \equiv 1 \pmod 4$. The characters are not invariant in $E.2$ and require a square root of $\pm q$ if q is not a square Theorem 4.5a. In particular $G = G_0 = E \circ Z$. Let U be the standard module for $S = \mathrm{Sp}_{2m}(q)$.

Lemma 7.7b. *Suppose the character χ of G afforded by V is a Weil character on the (symplectic) core E. Let $g \in G$ be a noncentral element of prime order, say s.*

(i) *If s is odd, then* $\dim {}_F C_V(g) \leq \frac{1}{2s}(q^m + (s-1)q^{m-1})$ *if $s \neq q$, and otherwise* $\dim {}_F C_V(g) \leq q^{m-1} - 1$ *or* q^{m-1} *depending on whether $\chi = \xi_1$ or $\chi = \xi_2$ on E.*

(ii) *Suppose $E = L = S_{2m}(q)$ and that g is an involution in L lifting to an element of order 4 in $S = \mathrm{Sp}_{2m}(q)$. Then $\chi(g) = \pm 1$.*

(iii) *Suppose either that $g = g'$ is a (noncentral) involution in $E = S = \mathrm{Sp}_{2m}(q)$ or that $E = S_{2m}(q)$ and g is the image of such an involution g' in S. Let $|C_U(g')| = q^{2m'}$, so that m' is an integer with $1 \leq m' < m$, and let $m'' = m - m'$. Then $C_S(g') \cong \mathrm{Sp}_{2m'}(q) \times \mathrm{Sp}_{2m''}(q)$, and*

$$\xi_1(g') = \pm\frac{1}{2}(q^{m'} - q^{m''}), \quad \xi_2(g') = \pm\frac{1}{2}(q^{m'} + q^{m''}),$$

the positive signs holding precisely when $q^{m''} \equiv 1 \pmod 4$.

Proof. Pick ξ_1, ξ_2 such that $\xi = \xi_1 + \xi_2$ is a Weil character of S (and $\chi = \xi_1$ or ξ_2 on E). Let j be the central involution of S, as usual. Let T be the standard holomorph to $Q = (q_0)_+^{1+2mf}$ (Theorem 4.3c). Recall from Theorem 4.5a that $Q : S$ is a subgroup of T, that $U = Q/Z(Q)$ is the standard module for S, and that ξ is the restriction to S of some faithful irreducible character of T, which will be also written ξ. Note that q is a divisor of $|E|$ and so $q_0 \neq p$ by coprimeness. Enlarging F if necessary (so that $q_0 \mid r - 1$) there is an FT-module W affording ξ. Write $\mathrm{Res}_S^T(W) = V_1 \oplus V_2$ with V_i affording ξ_i. Embed $Z(Q) = Z(T)$ into Z such that ξ and χ lie over the same linear character of $Z(Q)$.

(i) Since s is odd, we may identify $g = xz$ where $x \in S$ and $z \in Z$ is an sth root of unity. Let z act trivially on T, and extend ξ to a faithful character of $T_z = T\langle z \rangle$ such that $\xi(z)/\xi(1) = \chi(z)/\chi(1)$. (We have $T_z = T$ if $s = q_0$ and $T_z = T \times \langle z \rangle$ otherwise.) From Theorem 6.2c it follows that $\dim {}_F C_W(g) \leq \lfloor \frac{1}{s}(q^m + (s-1)q^{m-1}) \rfloor$ if $s \neq q$ and $\dim {}_F C_W(g) \leq 2q^{m-1}$ otherwise. Clearly

$$C_W(g) = C_{V_1}(g) \oplus C_{V_2}(g).$$

By Theorem 4.5a, $\xi_2 = \xi_1 + 1_S$ on $2'$-elements of S. The assertion follows by using that $\dim {}_F C_W(g) = \langle \xi, 1 \rangle_{\langle g \rangle} = \langle \xi, \nu \rangle_{\langle x \rangle}$ where ν is the linear character of $\langle x \rangle$ defined by $\bar{\nu}(x) = \chi(z)/\chi(1)$, and similar statement for the V_i, ξ_i. Note also that $\lfloor (2q^{m-1} - 1)/2 \rfloor = q^{m-1} - 1$ and $\lfloor (2q^{m-1} + 1)/2 \rfloor = q^{m-1}$.

(ii) Let $\tilde{g} \in S$ have image g, so that $\tilde{g}^2 = j$. Let $V = V_a$ afford the Weil character ξ_a which is not faithful for S, and let V_b afford the the other irreducible Weil character ξ_b (with $\xi = \xi_a + \xi_b$). Then j acts as -1 on V_b, so that $C_{V_b}(\tilde{g}) = 0$ and $C_V(g) = C_W(\tilde{g})$. By Theorem 4.5a, $\xi(\tilde{g})$ is an integer, and $\xi_a(g) = \xi_a(\tilde{g})$ is an integer as g is an involution. Hence $\xi_b(\tilde{g})$ is an integer. Since $\tilde{g}^2 = j$, the eigenvalues of $\tilde{g}$ on V_b are $\pm i$. We infer that $\xi_b(\tilde{g}) = 0$. Further $|C_U(\tilde{g})| = 1$ and so $\xi(\tilde{g}) = \pm 1$ by Theorem 4.4, because $\tilde{g}$ is good for U. Hence $\chi(g) = \xi_a(g) = \xi(\tilde{g}) = \pm 1$.

(iii) Notice that g' and $g'j$ belong to different conjugacy classes of S (but both mapping onto g if $g \neq g'$). So the following depends on the choice of the inverse image. We have a proper (orthogonal) decomposition $U = [U, g'] \oplus C_U(g') = U^- \oplus U^+$, where U^-, U^+ are the eigenspaces of g' to the eigenvalues -1 and $+1$, respectively. This corresponds to a central decomposition $Q = Q' \circ Q''$, where $Q' = (q_0)_+^{1+2m'f}$ and $Q'' = (q_0)_+^{1+2m''f}$, and yields embeddings of the standard holomorphs, and of the "Young group" $Y = \mathrm{Sp}_{2m'}(q) \times \mathrm{Sp}_{2m''}(q)$ into S. Since ξ is (absolutely) irreducible on $(Q' : \mathrm{Sp}_{2m'}(q)) \circ (Q'' : \mathrm{Sp}_{2m''}(q))$, we obtain a tensor decomposition

$$\mathrm{Res}_Y(\xi) = \xi' \otimes \xi''$$

of Weil characters. Just by considering character degrees we get that $\mathrm{Res}_Y(\xi_1) = \xi_1' \otimes \xi_2'' + \xi_2' \otimes \xi_1''$ and $\mathrm{Res}_Y(\xi_2) = \xi_1' \otimes \xi_1'' + \xi_2' \otimes \xi_2''$. Here the irreducible Weil characters have to be chosen properly.

One knows that there are precisely $m - 1$ conjugacy classes of noncentral involutions in $S = \mathrm{Sp}_{2m}(q)$, the conjugacy class of an involution being determined by the order of its fixed point group on U [Dieudonné, 1971, pp. 25, 26]. This readily gives the statement for $C_S(g')$. Moreover, the conjugacy class of g' is determined by the integer m', and g' is conjugate in S to the element $1' \times j''$ in $\mathrm{Sp}_{2m'}(q) \times \mathrm{Sp}_{2m''}(q)$, j'' being the central involution in $\mathrm{Sp}_{2m''}(q)$. The character values are obtained by using that ξ_1'' is faithful on $\mathrm{Sp}_{2m''}(q)$ by Theorem 5.4a, satisfying $\xi_1''(j'') = -(q^{m''} - 1)/2$, if and only if $q^{m''} \equiv 1 \pmod 4$. ☐

We now discuss the cases remaining by virtue of Proposition 7.7a.

m = 2 : We have to examine the following.

(i) $L = S_4(3)$ and $d \le 30$: Let us consider first the Weil characters. So let χ be faithful for $E = \mathrm{Sp}_4(3)$ of degree 4 (characters χ_{21}, χ_{22} in [Atlas, p. 27]). We have $G = G_0 = E \circ Z$, and $F = \mathbb{F}_r$ requires a square root of -3, that is, $r - 1$ is divisible by 3. Of course r is coprime to $|L|$ and so not divisible by $2, 3, 5$. The usual counting argument (on the basis of the character table) yields that there is a regular G-orbit on V if $r \ge 61$. Hence $r \in \{7, 13, 19, 31, 37, 43, 49\}$. We proceed by computing the relevant polynomials $\mu_{\chi,1}^t$. The table of marks for $L \cong U_4(2)$ is in the [GAP] library. We obtain:

$$\mu_{\chi,1}^6 = (r - 7)(r - 13)(r - 19)/|L|,$$
$$\mu_{\chi,1}^{12} = (r + 11)(r - 13)(r - 37)/|L|,$$
$$\mu_{\chi,1}^{30} = (r - 31)(r^2 - 8r + 223)/|L|.$$

We conclude that there is no regular orbit if and only if $r \in \{7, 13, 19, 31, 37\}$. For $r = 43$ note that $\gcd(r - 1, \exp(E)) = 6$, and for $r = 49$ by Proposition 7.3c we have $4 \cdot \mu_{\chi,1}^{12}(49) = 4$ regular orbits for $E \circ Z_{12}$ which fuse in $G = E \circ Z_{48}$ as each involution of G is in $E \circ Z_{12}$.

Computing the polynomials for nontrivial subgroups of $\bar{X} = L$ leads to the results stated in Theorem 7.2a. We know from Corollary 7.2b that for $r = 31$ the pair (G, V) admits a real (but not strongly real) vector. The information given in Sec. 7.1 for $r = 7, 13, 19$, basically obtained by a study of the [Atlas], manifest that we have six nonreal reduced pairs here.

Let next $\chi = \chi_2$ or χ_3 for $E = L = S_4(3)$, a Weil character of degree $d = 5$. Here $G = G_0 = L \times Z$ and $3 \mid r - 1$. By the counting principle we get regular orbits unless $r < 56$. So r is as before. Computation with [GAP] yields

$$\mu^2_{\chi,1} = (r - 7)(r - 9)(r - 13)(r - 15)/|L|,$$
$$\mu^6_{\chi,1} = (r - 5)(r - 7)(r - 13)(r - 19)/|L|,$$
$$\mu^{10}_{\chi,1} = \mu^2_{\chi,1} - \tfrac{1}{5} \text{ and } \mu^{30}_{\chi,1} = \mu^6_{\chi,1} - \tfrac{1}{5}.$$

We conclude there are no regular orbits if and only if $r \in \{7, 13, 19\}$. We similarly get the minimal stabilizers as stated in Theorem 7.2a.

Let $\chi = \chi_4$ in the Atlas notation, a rational (Weil) character of $L \cong U_4(2)$ of degree 6 which extends rationally to $X = L.2$. So $G = G_0 = X \times Z$. The counting argument yields the existence of a regular orbit unless $r < 47$. The method of Proposition 7.3c shows that there are no regular orbits if and only if $r \in \{7, 11, 13\}$, and one gets the minimal stabilizers listed in Theorem 7.2a. For the remaining projective representations of $S_4(3)$ the counting principle works.

(ii) $L = S_4(4)$ and $d \leq 34$: Consider first the minimal character $\chi = \chi_2$ of degree 18, which is rational and extends to $X = \text{Aut}(L) = L.4$ requiring $\sqrt{-1}$ [Atlas, p. 45]. Assuming that there is no regular orbit we have

$$r^{18} \leq |2AB|(r^{12} + r^6) + |2C|(r^{10} + r^8) + |2D|(2r^9) + |L|(3r^6)$$

$$= 510(r^{12} + r^6) + 3.825(r^{10} + r^8) + 2.720r^9 + 2^8 \cdot 3^3 \cdot 5^2 \cdot 7r^6.$$

This implies that $r < 9$, which is impossible (by coprimeness). For the representations of $L.2$ of dimension 34, no prime order element has an eigenspace of dimension greater than 22, and the result follows.

(iii) $L = S_4(5)$ and $d \leq 40$: For the minimal Weil characters (of degree 12) of $2.L = \text{Sp}_4(5)$ the counting argument (based on the character table [Atlas, pp. 62–63]) gives a regular orbit (unless $r < 6$). For the other Weil characters (of degree 13) of L the result is similar. For the character $\chi = \chi_4$ of L of degree 40, which extends to $L.2$, the largest dimension of an eigenspace of a prime order element is 26, and the result follows.

(iv) $L = S_4(7)$ and χ is a Weil character: Here L is not an Atlas group. We argue on the basis of Lemma 7.7b. By Theorem 4.5a, F requires a square root of -7. Either $\chi = \xi_1$ is faithful for $E = \mathrm{Sp}_4(7)$ (of degree $d = (7^2 - 1)/2 = 24$) and $G = G_0 = E \circ Z$, or $\chi = \xi_2$, $d = 25$ and $G = G_0 = L \times Z$. Let $g \in G$ be a noncentral element of prime order s.

If g is not an involution, then $\dim {}_F C_V(g) \leq \frac{1}{2s}(7^2 + (s - 1)7)$ if $s \neq 7$ and $\dim {}_F C_V(g) \leq 7$ otherwise. So the fixed point ratio $f(g, V) < \frac{1}{2}$, at any rate. Suppose next that $\chi = \xi_2$ and x is an involution in $E = L$ that lifts to an element of order 4 in $\mathrm{Sp}_4(7)$. Then $\chi(x) = \pm 1$ by the lemma, whence $f(g, V) \leq \frac{13}{25}$ when $g = xz$ for some $z \in Z$. Let finally $x \in E$ be a noncentral involution which is not of this type. By the lemma

$$\chi(x) = -\frac{1}{2}(7 + 7) = -7$$

if $\chi = \xi_2$, and $\chi(x) = 0$ otherwise. So $f(g, V) \leq \frac{1}{2}$ when $\chi = \xi_1$ (for any g), in which case there is a regular orbit since $r^{12} > 2|\mathrm{Sp}_4(7)|$ for $r \geq 6$. So let $\chi = \xi_2$. Then $f(g, V) \leq \frac{16}{25}$ for each g, so that we have to inspect when the inequality

$$r^9 > 2|S_4(7)| = 2^9 \cdot 3^2 \cdot 5^2 \cdot 7^4$$

does hold. This indeed holds for $r \geq 9$, implying that there is a regular orbit also in this case.

(v), (vi) $L = S_4(9)$ and $L = S_4(11)$, with χ being a Weil character: These are treated in exactly the same manner as (iv).

m = 3 : By Proposition 7.7a the following cases have to be examined.

(i) $L = S_6(2)$ and $d \leq 35$. Consider first the character $\chi = \chi_2$ of L of degree 7 [Atlas, p. 47]. Here we have $G = G_0 = L \times Z$. The counting argument shows that there is a regular orbit when $r \geq 80$. So $r = p$ can be any prime between 11 and 79 at this stage (by coprimeness). We use projective marks. The table of marks for $L = \mathrm{Sp}_6(2)$ is in [GAP]. We get

$$\mu_{\chi,1}^2 = (r - 5)(r - 7)(r - 9)(r - 11)(r - 13)(r - 17)/|L|,$$
$$\mu_{\chi,1}^6 = (r - 7)(r - 13)(r - 19)(r^3 - 23r^2 + 187r - 325)/|L|,$$
$$\mu_{\chi,1}^{14} = \mu_{\chi,1}^2 - \frac{1}{7} \text{ and } \mu_{\chi,1}^{42} = \mu_{\chi,1}^6 - \frac{1}{7}.$$

We deduce that there are no regular orbits if and only if $r \in \{11, 13, 17, 19\}$. The minimal stabilizers listed in Theorem 7.2a are obtained from the corresponding polynomials to the appropriate nontrivial subgroups of L.

Consider next the character $\chi = \chi_{31}$ of degree 8 of $E = 2.\mathrm{Sp}_6(2)$. The counting method yields that there are regular orbits for $G = E \circ Z$ except possibly when $r = 11$. This is handled by showing that the polynomial $\mu_{\chi,1}^{10}$ is positive at $r = 11$.

(ii) $L = S_6(3)$ and χ is a Weil character: Here L is an Atlas group [Atlas, pp. 110–113), and the usual counting arguments work. There is a regular orbit.

(iii) $L = S_6(5)$ and χ is a Weil character: Here L is not an Atlas group. By Theorem 4.5a the field F requires a square root of 5. Either $\chi = \xi_1$ is faithful (of degree $(5^3 - 1)/2 = 62$) for $E = \mathrm{Sp}_6(5)$ and $G = G_0 = E \circ Z$, or $\chi = \xi_2$ and $G = G_0 = L \times Z$. If g is a noncentral element in G of odd prime order, then $\dim {}_F C_V(g) \leq \frac{1}{6}(5^3 + 2 \cdot 5^2)$ by Lemma 7.7b (taking the worst case $o(g) = 3$), whence $f(g, V) < \frac{1}{2}$. If $G = L \times Z$ and $g \in L$ is an involution which lifts to an element of order 4 in $\mathrm{Sp}_6(5)$, we have $\chi(g) = \xi_2(g) = \pm 1$ by the lemma and so $f(gz, V) \leq \frac{32}{63}$ for any $z \in Z$. If $g \in L$ is an involution which is not of this type, then $\chi(x) = \xi_2(x) = 15$. If $\chi = \xi_1$ and $x \in E$ is a noncentral involution, then $\chi(g) \in \{10, -10\}$ by Lemma 7.7b.

We conclude that $f(g, V) \leq \frac{36}{62}$ for any noncentral element $g \in G$ if $G = E \circ Z$, and $f(g, V) \leq \frac{39}{63}$ if $G = L \times Z$. In the second (worse) case we verify that

$$r^{24} > 2|S_6(5)| = 2^{10} \cdot 3^4 \cdot 5^9 \cdot 7 \cdot 13 \cdot 31$$

for $r \geq 5$. Hence there are regular orbits.

m = 4 : By Proposition 7.7a the following has to be treated.

(i) $L = S_8(2)$ and $d \leq 85$: Then $d = 35, 51$ or 85 [Atlas, p. 124]. If $d = 35$ or 85 (characters χ_3, χ_4 in the [Atlas]), then no prime order element has an eigenspace on V of dimension greater than $d - 15$. From $r^{15} \leq 2 \cdot |S_8(2)|$ we obtain that $r < 6$. Hence there is a regular orbit. For $d = 35$ (character χ_2) no prime order element, except the involutions in class $2A$, have eigenspaces of dimension greater than 23, and the involutions in $2A$ have the spectral pattern $[-1^{(28)}, 1^{(7)}]$. From

$$r^{35} \leq |2A|r^{28} + 2|L|r^{23} = 255r^{28} + 2^{17} \cdot 3^5 \cdot 5^2 \cdot 7 \cdot 17$$

we obtain that $r < 9$. Hence there is again a regular orbit.

(ii) $L = S_8(3)$ and χ is a Weil character: The field F requires a square root of -3 by Theorem 4.5a, and ξ_1 is faithful for $E = \mathrm{Sp}_8(3)$ (of degree

40). So $G = G_0 = E \circ Z$ if $\chi = \xi_1$, and $G = G_0 = L \times Z$ if $\chi = \xi_2$. Let g be a noncentral element of G of prime order s. If s is odd, by Lemma 7.7b we have $\dim {}_F C_V(g) \le 3^3 - 1 = 26$ when $\chi = \xi_1$ and $\dim {}_F C_V(g) \le 27$ otherwise, considering the worst case $s = 3$. If $\chi = \xi_2$ and $g \in L$ is an involution lifting to an element of order 4 in $\mathrm{Sp}_8(3)$, then $\chi(g) \pm 1$ by the lemma. If $g \in L$ is an involution which is not of this kind, then $\chi(g) = \xi_2(g) \in \{9, -15\}$. If $\chi = \xi_1$ on E and $g \in E$ is a (noncentral) involution, then $\chi(g) \in \{0, 12, -12\}$ by Lemma 7.7b.

We conclude that, for all g, $f(g, V) \le \frac{26}{40}$ when $\chi = \xi_1$, and $f(g, V) \le \frac{28}{41}$ when $\chi = \xi_2$. Assuming that there is no regular G-orbit on V, in the first case we have

$$r^{15} \le 2|\mathrm{Sp}_8(3)| = 2^{16} \cdot 3^{16} \cdot 5^2 \cdot 7 \cdot 13 \cdot 41,$$

which forces that $r < 16$. Since r is coprime to 13 (and to $2 \cdot 3 \cdot 5 \cdot 7$) and since -3 is not a square modulo 11, we obtain the desired contradiction. In the second case the inequality $r^{13} \le 2|S_8(3)|$ only holds if $r < 21$, and it remains to rule out the case $r = 19$. We have to improve the estimate. By (5.6d) we may replace the factor 2 by $(1 + \frac{1}{r^{15}})$, which is not sufficient, however.

Consider the conjugacy class $2B$ of involutions in $S_8(3)$ where $\chi = \xi_2$ takes the value -15. By Lemma 7.7b, $|2B| = |\mathrm{Sp}_8(3)|/(|\mathrm{Sp}_6(3)| \cdot |\mathrm{Sp}_2(3)|) = 2 \cdot 3^6 \cdot 5 \cdot 41$. We may replace the above inequality by

$$r^{41} \le |2B|(r^{28} + r^{13}) + |S_8(3)|(r^{27} + r^{14}).$$

This implies that $r < 19$, as desired.

$\mathbf{m = 5, 6}$: By Proposition 7.7a we have merely to treat the Weil characters for $\mathrm{Sp}_{2m}(3)$. One argues as for $\mathrm{Sp}_8(3)$.

7.8. Unitary Groups

Let $L = \mathrm{U}_m(q) = \mathrm{PSU}_m(q)$ for some integer $m \ge 3$ and some prime power q. The case $(m, q) = (3, 2)$ cannnot happen since L is simple, and we exclude $(m, q) = (4, 2)$ since $\mathrm{U}_4(2) \cong S_4(3)$ has been already treated in Sec. 7.7. The automorphism group of L, and its Schur multiplier, is known [Steinberg, 1967]. Recall that the centre $Z(\mathrm{SU}_m(q))$ is (cyclic) of order

$\gcd(q+1, m)$. By [Guralnick–Saxl, 2003] $c(L) \le m+1$ except when $m = 4$, in which case $c(L) \le 6$, and $c(L) \le m$ for $m \ge 5$. By [Tiep–Zalesskii, 1996]

$$R_0(U_m(q))) = \begin{cases} (q^m - q)/(q+1) & \text{if } m \text{ is odd} \\ (q^m - 1)/q + 1 & \text{otherwise} \end{cases},$$

except when $(m, q) = (4, 3)$ (see Appendix C3). Then if $d = \chi(1) \le R_0(L) + 1$, χ is a Weil character of $\mathrm{SU}_m(q)$. Using that $d \ge R_0(L)$ one gets the following, on the basis of Lemma 7.3b.

Proposition 7.8a. *There is a regular G-orbit on V except possibly in the following cases:*

- *$(m, q) = (3, 3)$ and $d \le 21$; $(m, q) = (4, 3)$ and $d \le 45$; $(m, q) = (5, 2)$ and $d \le 55$; $(m, q) = (6, 2)$ and $d \le 56$.*
- *$(m, q) = (3, 4), (3, 5), (3, 7), (4, 4), (5, 3), (7, 2), (8, 2), (9, 2)$ and χ is one of the Weil characters of $\mathrm{SU}_m(q)$.*

The groups $U_4(4), U_5(3), U_7(2), U_8(2)$ and $U_9(2)$ are not Atlas groups. So we need some general information on their Weil characters. Some crucial facts have been described in Appendix (C3). Let U be the standard module for $G_u = \mathrm{GU}_m(q)$, and let λ_u be a fixed linear character of $Z_u = Z(G_u)$ of order $q + 1$ (noting that Z_u is cyclic of order $q + 1$). There is a *generic* Weil character ξ_u of G_u, which is rational-valued and satisfies $\xi_u^2 = \pi_U$. There are precisely $q + 1$ irreducible constituents $\xi_0, \xi_1, \cdots, \xi_q$ of ξ_u, all remaining irreducible and distinct when restricted to $S_u = \mathrm{SU}_m(q)$. The Weil character ξ_j is the (irreducible) constituent of ξ_u lying above the linear character λ_u^j, so that $\xi_u = \sum_{j=0}^q \xi_j$. We have $\xi_0(1) = \frac{q^m + (-1)^m q}{q+1}$ and $\xi_j(1) = \xi_0(1) - (-1)^m$ for $j > 0$. Let $\varepsilon = e^{2\pi i/(q+1)}$, and let z_u be the generator of Z_u satisfying $\lambda_u(z_u) = \varepsilon$.

Lemma 7.8b. *Suppose the character χ of G afforded by V agrees on the (unitary) core E with an (irreducible) Weil character. Let $g \in G$ be a noncentral element of prime order s.*

(i) *Suppose g induces on E an inner or diagonal automorphism, and assume that $s \nmid q + 1$. Then $\dim {}_F C_V(g) \le \frac{1}{s(q+1)}(q^m + (s-1)q^{m-1})$ when $s \nmid q$ or $s = 2 \mid q$, and $\dim {}_F C_V(g) \le \frac{2}{q+1} q^m$ otherwise.*

(ii) *Suppose g is an involution inducing an outer field automorphism on E. Then either $\chi(g) = 0$ or $\chi(g) = \pm q^a$ for some integer $a \le \frac{m}{2}$.*

Proof. Let $q = q_0^f$, q_0 prime. Since q is a divisor of $|E|$, $q_0 \neq p$ by coprimeness. Enlarging F, if necessary, we may assume that F contains the q_0th roots of unity when q is odd, and the 8th otherwise. G_u acts faithfully on a q_0-group Q such that $Q/Z(Q) = U$ is the standard module and $Z(Q)$ is centralized by G_u, where $Q \cong (q_0)_+^{1+2fm}$ for odd q and $Q \cong 2_0^{1+2fm}$ otherwise. Let T be the standard holomorph to Q, and let ξ be a faithful irreducible character of T of degree q^m. Then $Q : G_u$ is a subgroup of T, and $\mathrm{Res}_{G_u}^T(\xi) = \xi_u \cdot \mu$ where μ is the linear character of G_u of order $\gcd(q+1, 2)$ (Appendix C3). Embed $Z(Q) = Z(T)$ into Z such that ξ and χ lie above the same linear character.

(i) By assumption $g = xz$ for some sth root of unity $z \in Z$ and some $x \in E$, because G_u induces all diagonal automorphisms on E and $|G_u/S_u| = q + 1$. Since $Z(S_u) \subseteq Z_u$ and $s \nmid |Z_u| = q + 1$, we may let $\langle z \rangle$ act trivially on T and regard ξ as a faithful character of $T_0 = T\langle z \rangle$ with $\xi(z)/\xi(1) = \chi(z)/\chi(1)$. (Either $s = q_0$ and $T = T_0$ or $T_0 = T \times \langle z \rangle$.) Identify x with the (unique) element of S_u of order s mapping onto x (so that $g = xz$ gets an element of T_0). Let ν be the linear character of $\langle x \rangle$ with $\nu(x) = \frac{\chi(z)}{\chi(1)}$. From Theorem 6.2c it follows that

$$\langle \xi_u, \nu \rangle_{\langle x \rangle} = \langle \xi, 1 \rangle_{\langle g \rangle} \leq \frac{1}{s}(q^m + (s-1)q^{m-1})$$

if $s \neq q$ or $s = 2 \mid q$, and $\langle \xi_u, \nu \rangle_{\langle x \rangle} \leq 2q^{m-1}$ otherwise. By hypothesis no eigenvalue $\neq 1$ of x on U has order dividing $q + 1$. Hence letting d_0 be the dimension over $\mathbb{F}_{q^2}$ of $C_U(x)$, and letting $d_k = 0$ otherwise, by Appendix (C3)

$$\xi_j(x) = \frac{(-1)^m}{q+1} \sum_{k=0}^{q} (-q)^{d_k} \varepsilon^{kj} = \frac{(-1)^m}{q+1} \cdot \begin{cases} \left((-q)^{d_0} + q \right) & \text{if } j = 0 \\ \left((-q)^{d_0} - 1 \right) & \text{if } j > 0 \end{cases}.$$

We just use that $\varepsilon^j \neq 1$ for $j = 1, \cdots, q$ and that $\sum_{k=0}^{q} \varepsilon^{jk} = 0$. We infer that the $\xi_j(x)$ for $j > 0$ agree, and that $\xi_0(x) - \xi_j(x) = (-1)^m$. There is a corresponding statement replacing x by any power $x^i \neq 1$. The assertion follows noting that $\xi_u = \sum_{j=0}^{q} \xi_j$.

(ii) Exclude (also) the case $(m, q) = (6, 2)$ (where $\chi(g) = 0$ or $\chi(g) = \pm 2^3$ by [Atlas, p. 117]). Then some distinguished group Y appears in T, where $Y = G_u.2$ is a field extension group in the odd case and $Y/Z(T) \cong G_u.2$ otherwise (Appendix C3) Since $\mathrm{Out}(L)$ is the semidirect product of the cyclic groups of outer diagonal automorphisms (induced by G_u) with that of field automorphisms, by assumption $\langle L, Z(E)g \rangle$ is isomorphic to a subgroup of $G_u.2/Z_u$. So $\langle E, g \rangle$ is a section of Y. Observe that $G_u/S_u \cong Z_u$ as modules for the group of field automorphisms. The outer field automorphism of order 2 inverts the elements of $Z_u = \langle z_u \rangle$, because it sends z_u to $z_u^q = z_u^{-1}$.

Consider first the case that q is even (hence $q + 1$ odd). By the construction of the irreducible Weil characters the outer field automorphism of G_u of order 2 leaves ξ_0 invariant and fuses the other characters pairwise. Hence by assumption $\chi = \xi_0$ on E and $E = L$ (as χ is faithful and Z_u is the kernel of ξ_0). Let $g' \in Y$ map onto g. Then $\chi(g) = \xi_0(g')$, the character ξ_0 suitably extended to Y. We have $\xi(g') = \pm\xi_0(g')$ as the ξ_j for $j > 0$ are fused pairwise by g'. Hence $\chi(g) = \pm\xi(g')$. By Theorem 4.4 either $\xi(g') = 0$ or g' is good for U and $\xi(g')^2 = |C_U(g)|$. Suppose $\chi(g) \neq 0$. Then

$$\chi(g)^2 = \xi(g')^2 = |C_U(g')| = |C_U(g'z)|$$

for all $z \in Z_u$. Certainly $C_U(g') \cap C_U(g'z_u) = 0$. Hence $\dim_{\mathbb{F}_q} C_U(g') \leq \frac{1}{2}\dim_{\mathbb{F}_q} U = \frac{1}{2}2m = m$ and $\chi(g) = \pm q^a$ for some integer $a \leq m/2$.

Let q be odd. Then the outer field automorphism of G_u of order 2 leaves ξ_0 and $\xi_{(q+1)/2}$ invariant and fuses the other irreducible Weil characters pairwise. Hence $\chi = \xi_0$ or $\chi = \xi_{(q+1)/2}$ on E. Let ξ_a, ξ_b denote the irreducible constituents of $\mathrm{Res}_Y^T(\xi)$ having the kernels $Z_u = \langle z_u \rangle$ and $\langle z_u^2 \rangle$, respectively. So $\xi_a = \xi_0$ and $\xi_b = \xi_{(q+1)/2}$ on G_u if m is even $(\mathrm{Ker}(\mu) \supseteq Z_u)$, and vice versa otherwise. Also, $E = L$ unless m is even, $\chi = \xi_b$ on E and $q \equiv 3 \,(\mathrm{mod}\, 4)$, in which case $\langle E, g \rangle$ is isoclinic to a subgroup of $Y/\langle z_u^2 \rangle$. There is a 2-element g' of Y mapping onto $Z_u g$ (identified with $Z(E)g$). We have $\chi(g) = \pm\xi_a(g')$ or $\pm\xi_b(g')$, possibly $\chi(g) = \pm i\xi_b(g')$ when $E \neq L$ and the isoclinism is proper. By Theorem 4.4, g' is good for U (as q is odd). From Theorem 4.5a we infer that $\xi(g'z)$ is rational for all $z \in Z_u$, because $g'z$ is a q_0'-element.

We have $\xi(g') = \xi(g'z) = \xi_a(g') + \xi_b(g')$ when z is a square in Z_u, and $\xi(g'z) = \xi_a(g') - \xi_b(g')$ otherwise. As before $C_U(g') \cap C_U(g'z_u) = 0$. The map $u \mapsto u(1 + z_u)$ is an injection from $C_U(g')$ into $C_U(g'z_u)$, because

$$u(1 + z_u)g' = u(1 + z_u^{-1}) = \big(u(1 + z_u)\big)z_u^{-1}$$

for $u \in C_U(g')$. Similarly, the map $v \mapsto v(1 + z_u^{-1})$ is an injection from $C_U(g'z_u)$ into $C_U(g')$. Hence $\dim_{\mathbb{F}_q} C_U(g') = \dim_{\mathbb{F}_q} C_U(g'z_u) \leq m$. Moreover $\xi(g')^2 = |C_U(g')| = |C_U(g'z_u)| = \xi(g'z_u)^2$ and so $\xi(g') = \pm\xi(g'z_u)$ as both are integers. It follows that

$$\xi_a(g') + \xi_b(g') = \pm\big(\xi_a(g') - \xi_b(g')\big).$$

Consequently $\xi_a(g') = 0$ or $\xi_b(g') = 0$. Thus if $\chi(g) \neq 0$, then $\chi(g) = \pm\xi(g')$ as both are integers, and

$$\chi(g)^2 = |C_U(g')| = q^{2a}$$

for some integer $a \leq m/2$, as desired.   □

Because of Proposition 7.8a we have to examine only some few cases. As before we proceed starting with the lowest rank of the groups involved.

m = 3 : We have to consider the following.

(i) $L = U_3(3)$ and $d \leq 21$: Suppose first $\chi = \chi_2$ is the Weil character of (minimal) degree $d = 6$ [Atlas, p. 14]. This χ is rational-valued but extends to $X = L.2 \cong G_2(2)$ requiring the 3rd roots of unity. If there is no regular orbit for $X \times Z$, then

$$r^6 \leq |2A|(r^4 + r^2) + \frac{1}{2}|3A|(2r^3) + \frac{1}{2}|3B|(3r^2) + \frac{1}{2}|7AB|(6r) + |2B|(2r^3)$$

$$= 63(r^4 + r^2) + 56r^3 + 1.008r^2 + 2.592r + 504r^2.$$

This implies that $r < 11$. Hence $r = 5$ (by coprimeness), and these case cannot happen for $X \times Z$. So consider $G = L \times Z$ and $r = 5$. One checks that $\mu_{\chi,1}^4$ vanishes at $r = 5$, so there is no regular orbit for G. Also, the minimal point stabilizers in G are cyclic groups of order 4 which however get larger in $G_0 = N_{GL(V)}(L) = G.2$. (Consider elements in the classes $4C$ and $8C$ of $L.2$). We show that there is $v \in V$ such that $C_{G_0}(v) \cong S_3$, as stated in Theorem 7.2a.

Let x be an element of L in the class $3B$. Then $N_L(\langle x \rangle) \cong S_3$ has no fixed points on $V^\sharp$, but x does. Let v be a nonzero vector fixed by x. Then $C_L(v) = \langle x \rangle$, and from Lemma 5.1b it follows that $H = C_G(v) \cong S_3$. We assert that $C_{G_0}(v) = H$. Otherwise H is of index 2 in this group, which forces that $C_{G_0}(v) \cong H \times Z_2$ since $Z(S_3) = 1$ and $\mathrm{Aut}(S_3) = S_3$. Here the generator of Z_2 belongs to the conjugacy class $2B$ of $L.2$, whose elements do not centralize elements in the class $3A$ [Atlas]. Hence the assertion.

Let next $\chi = \chi_3$ be the Weil character of degree 7 which extends (rationally) to $X = L.2 = G_2(2)$. So $G = G_0 = X \times Z$. The counting argument yields regular vectors unless $r < 8$. For $r = 5$ we compute

$$\mu_{\chi,1}^4 = (r - 5)(r^5 + 6r^4 + 31r^3 - 159r^2 - 1.760r + 2.457)/|X|,$$

and we get some point stabilizer of order 2. For the algebraically conjugate (Weil) characters $\chi = \chi_4, \chi_5$ of L, which fuse in $G_2(2)$, the counting argument again gives regular vectors unless $r = 5$. As before there is no regular orbit for $G = G_0 = L \times Z_4$ when $r = 5$, and a point stabilizer of order 2.

For all other irreducible characters of $L = U_3(3)$ no prime order element of $\mathrm{Aut}(L) = G_2(2)$ has an eigenspace of dimension greater than $d - 8$, and from $r^8 \leq 2|G_2(2)|$ it follows that $r < 4$.

(ii) $L = U_3(4)$ and χ is a Weil character: The character $\chi = \chi_2$ of degree 12 extends to $X = L.4$, and for $G = X \times Z$ the counting method works. Similar statement for the other four Weil characters of L which fuse pairwise in $L.2$ and together in $L.4$. There are always regular orbits.

(iii) $L = U_3(5)$ and χ is a Weil character: In each case no noncentral element of $\mathrm{Aut}(L) = L.3$ or of $3.L.3$ of prime order has an eigenspace of dimension greater that $d - 8$. From $r^8 \leq 2|3.L.3|$ it follows that $r < 7$.

(iv) $L = U_3(7)$ and χ is a Weil character: In the representations of degree at most 43 no prime element of $\mathrm{Aut}(L) = L.2$ has an eigenspace of dimension greater than $n - 18$. From $r^{18} \leq 2|L.2|$ it follows that $r < 3$.

$\mathbf{m = 4}$: By Proposition 7.8a we have to examine the following.

(i) $L = U_4(3)$ and $d \leq 45$: Assume there is no regular orbit. For the 21-dimensional representations of $\mathrm{Aut}(L) = L.D_8$ (character $\chi = \chi_2$ in the [Atlas, p. 54]) we then have

$$r^{21} \leq |2D|r^{15} + |2B|r^{14} + |2A|r^{13} + 2|\mathrm{Aut}(L)| \cdot r^{11}$$

$$= 126r^{15} + 540r^{14} + 2.835r^{13} + 52.254.720r^{11},$$

which implies that $r < 7$, a contradiction. For the 35-dimensional representations of $L.2_1$ and $L.2_2$ no prime order element has an eigenspace of dimension greater than $d - 10$, and from $r^{10} \leq 2|\mathrm{Aut}(L)|$ it follows that $r < 6$. The 20-dimensional representations of $2.L$ and $4.L$ are treated similarly. For the 15-dimensional representations of $3_1.L.2_2$ (characters $\chi = \chi_{56}$) the counting method yields that $r < 7$, a contradiction. For the 36-dimensional representations of $3_2.L$ (or $3_2.L.2_3$) and $12_2.L$, and for the 45-dimensional representations of the former group(s), the same method applies.

It remains to investigate the (faithful) 6-dimensional representations of $X = 6_1.L.2_2$ (as a complex reflection group). The isoclinism type of X does not matter, and the character χ of X is an extension of one of the two algebraically conjugate characters χ_{72} (requiring $\sqrt{-3}$) [Atlas]. The counting argument gives the existence of regular orbits for $G = G_0 = X \circ Z$ when $r \geq 157$. Now the table of marks for $L = U_4(3)$ (and for $L.2_1$) is available in [GAP], and from that it is not difficult to compute the table of marks for $\bar{X} = L.2_2$ (extension by a diagonal automorphism). One obtains that

$$\mu_{\chi,1}^6 = (r - 13)(r - 19)(r - 25)(r - 31)(r - 37)/|L.2_2|,$$

and that $\mu_{\chi,1}^t = \mu_{\chi,1}^6$ for all multiples t of 6 dividing $\exp(X) = 2^3 \cdot 3^2 \cdot 5 \cdot 7$. So there are no regular orbits if and only if $r \in \{13, 19, 31, 37\}$. Similar computation leads to the minimal stabilizers as given in Theorem 7.2a.

The two isoclinic variants of $X = 6_1.U_4(3).2_2$ get isomorphic in $G = X \circ Z$ when $r = 13$ or $r = 37$, because then $(r - 1)/6$ is even.

(ii) $L = U_4(4)$ and $\chi = \xi_j$ is a Weil character: Here $L = E = SU_4(4)$ is not an Atlas group, and $G_u = GU_4(4) = L \times Z_u$ where $Z_u = \langle z_u \rangle$ is cyclic of order 5. $\mathrm{Aut}(L) = L.4$ permutes cyclically the nontrivial elements of $Z_u = Z(G_u)$, hence the Weil characters $\xi_1, \xi_2, \xi_3, \xi_4$ (of degree 51). Let $g \in G$ be a noncentral element of prime order s. If $s \neq 5$, by Lemma 7.8b (picking $s = 2$)

$$\dim {}_F C_V(g) \leq \frac{1}{5s}(4^4 + (s-1)4^3) \leq 32.$$

Let $s = 5$. Write $g = xz$ where $x \in L$ and $z \in Z$. Let $\varepsilon = e^{2\pi i/5}$, and let $d_k = d_k(x) = \dim_{\mathbb{F}_{16}}(z_u^{-k}x)$ for $k = 0, \cdots, 4$, where U is the standard module for G_u. By Appendix (C3) $\xi_j(x) = \frac{1}{5}\sum_{k=0}^4 (-4)^{d_k}\varepsilon^{kj}$. Since x is faithful on U, $d_k \leq 3$ for each k. We conclude that the eigenspaces of x (and of g) on V are at most 13-dimensional when $\chi = \xi_0$ on L, and at most 12-dimensional otherwise.

Let $g \in G$ be an involution inducing an outer field automorphism on $E = L$. Then $\chi = \xi_0$ on L (of degree 52), and we consider the worst case $G = G_0 = L.4 \times Z$. By Lemma 7.8b either $\chi(g) = 0$ or $\chi(g) = \pm 4^a$ for some integer $a \leq \frac{m}{2} = 2$. Hence $\dim {}_F C_V(g) \leq \frac{1}{2}(52 + 4^2) = 34$. From

$$r^{18} \leq 2 \cdot |L.4| = 2^{15} \cdot 3^2 \cdot 5^3 \cdot 13 \cdot 17$$

we obtain that $r < 4$. Hence there is a regular orbit, in all cases.

m = 5 : We only have to examine the following.

(i) $L = U_5(2)$ and $d \leq 55$: We have to deal with the three Weil characters $\xi_0 = \chi_2$ (degree 10) and $\xi_1 = \chi_3$, $\xi_2 = \chi_4$ [Atlas, p. 72]. For the other characters (χ_5 of degree 44 and χ_6 of degree 55), no prime order element of $L.2$ has an eigenspace of dimension larger than $d - 16$, and the result follows.

Let us consider first $\chi = \xi_0$. Suppose $X = L.2$ and $G = G_0 = X \times Z$. The usual counting argument yields that there are regular vectors unless $r < 15$. (The involutions in the class $2A$ have weak spectral pattern $[8, 2]$, so one cannot improve the estimate without additional arguments. On the other hand, the counting argument yields the existence of strongly real vectors.) The only possibilities are $r = 7$ and $r = 13$ (by coprimeness). But the character ξ_0 requires $2\sqrt{-2}$ on $L.2$ outside L, and -2 is not a square mod 7 or 13. Hence we are left with the cases $r = 7, 13$, but we have to consider $G = L \times Z$ (and $G_0 = (L \times Z).2$). The table of marks for $L = U_5(2)$ is in [GAP] (but not that for $U_5(2).2$). For $U_5(2)$ we obtain that the only roots of the polynomials $\mu_{\chi,1}^6$ and $\mu_{\chi,1}^{12}$ are 1, 5 and 7. So there is no regular orbit just when $r = 7$. In this case all polynomials $\mu_{\chi,H}^6$ for subgroups H of order 2, 3 vanish at $r = 7$ but not for a Klein 4-group $H \cong V_4$. So there is $v \in V$ such that $C_G(v) = H$. We assert that $C_{G_0}(v) = H$ as well.

From the [Atlas] we read off that two involutions in H must belong to the class $2B$ and one to $2A$. If $C_{G_0}(v) \neq H$, it is a dihedral group of order 8 containing an involution in the class $2C$, which centralizes the involution in H in class $2A$ and interchanges the other ones. Notice that the elements in class $4D$ square to elements in class $2B$ and cannot exchange two involutions in different classes. But $N(2A)$ is enlarged when passing from L to $L.2$ by replacing a split extension with $2.A_4$ on the top by a $2.S_4$ [Atlas], and for this one needs an element of order 8. Hence the assertion.

For $\chi = \xi_j$, $j > 0$, we have to consider $G = L \times Z$. The counting argument yields regular vectors unless $r = 7$. Here the polynomial $\mu_{\chi,1}^6$ is irreducible over $\mathbb{F}_7$. Hence there is a regular orbit.

(ii) $L = U_5(3)$ and χ is a Weil character: Here L is not an Atlas group. Let $G_u = GU_5(3) = L \times Z_u$ where $Z_u = Z(G_u) = \langle z_u \rangle$ has order 4. Recall that $\xi_0(1) = 60$ and $\xi_j(1) = 61$ for $j > 0$. Either $G_0 = L \times Z$ and $\chi = \xi_1$ or ξ_3 on L, or $G_0 = L.2 \times Z$ and $\chi = \xi_0$ or ξ_2. Let $G = G_0$. Suppose g is a noncentral element of prime order s in G. If $s \neq 2$, then by Lemma 7.8b (picking $s = 3$)

$$\dim {}_FC_V(g) \leq \lfloor \frac{1}{4s}(3^5 - (s-1)3^4) \rfloor \leq \lfloor \frac{1}{12}(3^5 + 2 \cdot 3^4) \rfloor = 33.$$

If $g = xz$ is an involution, with $x \in L$ and $z \in Z$, then $\xi_j(x) = \frac{-1}{4}((-3)^{d_0} + (-3)^{d_1}i^j + (-3)^{d_2}i^{2j} + (-3)^{d_3}i^{3j})$ by the formula given in Appendix (C3). Here $d_k = d_k(x) = \dim_{\mathbb{F}_9}C_U(z_u^{-k}x)$ (and $i^2 = -1$). This yields that $|\chi(x)| \leq \frac{1}{4}(3^4 + 3^2 + 2) = 23$ and so $\dim {}_FC_V(g) \leq 42$.

Suppose that g is an involution inducing an outer field automorphism on L. Then by Lemma 7.8b either $\chi(g) = 0$ or $\chi(g) = \pm 3^a$ for some integer $a \leq \frac{m}{2} = \frac{5}{2}$. So the eigenspaces of g on V are at most of dimension 30. Since for any g the fixed point ratio $f(g, V) \leq \frac{42}{60}$, and since $r^{17} > 2|L.2| = 2^{13} \cdot 3^{10} \cdot 5 \cdot 7 \cdot 61$ for $r > 10$, there exist regular orbits in each case.

m = 6 : By the proposition we only have to examine $L = U_6(2)$, with $d \leq 56$. For the 56-dimensional representations of $2.L.2$, the largest dimension of an eigenspace of a noncentral element of prime order is 40, and we obtain a regular orbit. So it remains to consider the Weil characters ($\xi_0 = \chi_2$ of degree 22, the others corresponding to χ_{78} in the Atlas notation [Atlas, p. 116–121]. The counting method applies.

m = 7 : By Proposition 7.8a we just have to examine the Weil characters of $L = SU_7(2)$. Recall that $G_u = GU_7(2) = L \times Z_3$ and that $\xi_0(1) = 42$, $\xi_j(1) = 43$ for $j > 0$. Consider $G = G_0$, so either $G = L \times Z$ or $G = L.2 \times Z$ and $\chi = \xi_0$ on L. Let $g \in G$ be a noncentral element of prime order s. If $s \geq 5$, then for any possibility for $\chi = \xi_j$ by Lemma 7.8b

$$\dim {}_FC_V(g) \leq \lfloor \frac{1}{3s}(2^7 + (s-1)2^6) \rfloor \leq 25.$$

Let $s = 3$, and let $g = xz$ for $x \in L$ and $z \in Z$. Let $\varepsilon = e^{2\pi i/3}$ and $d_k = d_k(x) = \dim_{\mathbb{F}_4}C_U(z_u^{-k}x)$ for $k = 0, 1, 2$, where U is the standard module for G_u. By Appendix (C3), $\chi_j(x) = \frac{-1}{3}((-2)^{d_0} + (-2)^{d_1}\varepsilon^j + (-2)^{d_2}\varepsilon^{2j})$. Using that $\varepsilon + \varepsilon^2 = -1$ and that $\xi_j(x)$ is an algebraic integer one obtains that $\xi_0(x) \in \{0, -3, 6, 9, 12, -21\}$ and that $\xi_j(x) \in \{-2, 4, 10, 1 + 12\varepsilon, 1 + 12\varepsilon^2, -5 + 3\varepsilon, -5 + 3\varepsilon^2, 1 - 21\varepsilon, 1 - 21\varepsilon^2\}$ for $j = 1, 2$. We deduce that no eigenspace of x (or g) on V has dimension greater than 21 when $\chi = \xi_0$, and greater than 22 when $\chi = \xi_1$ or ξ_2.

Let $s = 2$. Consider first the case that g induces an outer field auto-morphism on L. Then $\chi = \xi_0$ on L, and by Lemma 7.8b either $\chi(g) = 0$ or $\chi(g) = \pm 2^a$ for some integer $a \leq \frac{m}{2} = \frac{7}{2}$. It follows that

$$\dim {}_F C_V(g) \leq \frac{1}{2}(42 + 2^3) = 25$$

in this case. Let finally g be an involution inducing an inner or diagonal automorphism on L. We consider the worst case that $\chi = \xi_0$ on L, as before. From Lemma 7.8b it follows that $\dim {}_F C_V(g) \leq 32$. There is a unique conjugacy class $2A$ of unitary transvections t in G_u (which generates $SU_m(q)$; see [Aschbacher, 1986, (22.3) and (22.4)]). For this we have $d_0(t) = \dim {}_{\mathbb{F}_{q^2}} C_U(t) = m - 1 = 6$ and $\xi_0(t) = -22$ (C3). Hence if $g = (-1)t$, then indeed $\dim {}_F C_V(g) = 32$. For the other noncentral involutions t_1, t_2 in G_u, up to conjugacy, we have $d_0(t_1) = 5$, $\chi(t_1) = 10$, and $d_0(t_2) = 4$, $\chi(t_2) = -6$, so that the F-dimensions of their eigenspaces on V are at most 26. Hence, combining the previous estimates, there is a regular G-orbit on V provided

$$r^{42} > 2|L.2|r^{26} + |2A|(r^{32} + r^{10})$$

for $r \geq 13$. This is true since $2|L.2| = 2^{23} \cdot 3^8 \cdot 5 \cdot 7 \cdot 11 \cdot 43$ and $|2A| = 2.709$. Observing that the centre $[U, t]$ of the transvection t consists of isotropic vectors, the size of the conjugacy class $2A$ is nothing but $(|I(U)| - 1)/(q^2 - 1)$ where $I(U)$ is the set of isotropic vectors in U. As in Lemma 4.6b one computes $|I(U)| = q^{2m-1} + (-1)^m(q^m - q^{m-1})$.

The remaining cases (Weil characters for $SU_8(2)$ and $SU_9(2)$) are treated in the same manner.

7.9. Orthogonal Groups

Let L be a simple orthogonal group with associated module of dimension m. Then the automorphism group and the Schur multiplier of L is known [Steinberg, 1967]. By [Guralnick–Saxl, 2003] the covering number $c(L) \leq m + 1$, and $c(L) \leq m$ for $m \geq 5$. Furthermore $R_0(L)$ has been computed by [Tiep–Zalesskii, 1996]. On the basis of Lemma 7.3b this readily gives the following.

Proposition 7.9. *There exists a regular orbit except possibly when* $L \cong \Omega_7(3)$ *and* $d \leq 78$, *or* $L \cong \Omega_8^-(2)$ *and* $d \leq 52$, *or* $L \cong \Omega_8^+(2)$ *and* $d \leq 84$.

We have to examine these three cases and show that regular orbits exist except when $L \cong \Omega_8^+(2), d = 8$.

Case 1: $L = \Omega_7(3)$ and $d \leq 78$: We use [Atlas, pp. 106–108]. In the 78-dimensional representations of $L.2$, the largest eigenspace of an element of prime order is 56. Since $r^{22} > 2|L.2|$ for $r > 3$ (even), we are done. For the 27-dimensional representations of $3.L$ the usual counting argument works as well.

Case 2: $L = \Omega_8^-(2)$ and $d \leq 52$: For the representations of $L.2$ of degree 34 and 51, the largest eigenspace of elements in the class $2D$ ([Atlas, p. 88]; $|2D| = 136$) has dimension at most $d - 7$, and no other element of prime order has an eigenspace of dimension greater than $d - 12$. From $r^{12} \leq 2|2D| + 2|L.2|$ it follows that $r < 7$. Hence the result.

Case 3: $L = \Omega_8^+(2)$ and $d \leq 84$: For the characters of L [Atlas, p. 86] χ_2 (degree $d = 28$, extendible to $\text{Aut}(L) = L.S_3$), χ_3 ($d = 35$, extendible to $L.2$), χ_6 ($d = 50$, extendible to $L.S_3$), and χ_7 ($d = 84$, extendible to $L.2$) observe that the elements in the class $2F$ have no eigenspaces of dimension greater than $d - 7$. All other elements of prime order have no eigenspaces of dimension greater than $d - 12$, which holds also for the characters of the above degrees which do not split in a proper extension of L. Since

$$r^{28} > |2F|(r^{21} + r^7) + 2|L.S_3|r^{16}$$

for $r > 7$, there is a regular orbit. Similarly, for the 56-dimensional representations of $2.L.2$ none of the noncentral elements of prime order has an eigenspace of dimension greater than 35, and the result follows.

It remains to consider some faithful character $\chi = \chi_{54}$ of degree 8 of $E = 2.L$, which is extendible to any $X = 2.L.2$. The multiplier $\text{M}(L)$ is elementary of order 4 but $\text{Aut}(L)$ permutes its involutions. The isoclinic type being irrelevant it suffices to consider $X = W(E_8)$, the Weyl group of the root lattice E_8. The spectral pattern of an involution g in the class $2F_0$ is $g_V = [1^{(7)}, -1^{(1)}]$, and $|2F| = 120$. So it is clear that the counting method cannot work. (We only get in this manner regular vectors when $r > 275$.) The table of marks for $\bar{X} = X/Z(X)$ is available in [GAP]. One gets:

$$\mu_{\chi,1}^2 = (r - 7)(r - 11)(r - 13)(r - 17)(r - 19)(r - 23)(r + 91)/|L|,$$
$$\mu_{\chi,1}^4 = \mu_{\chi,1}^2 - (r - 13)(r - 17)(r - 29)/11.520,$$
$$\mu_{\chi,1}^6 = \mu_{\chi,1}^2 - (r - 7)(r - 13)(r - 19)/38.880,$$

$\mu_{\chi,1}^{10} = \mu_{\chi,1}^2 - (r-11)/150,$

$\mu_{\chi,1}^{12} = \mu_{\chi,1}^6 - (r-9)(r-13)(r-37)/11.520,$

$\mu_{\chi,1}^{20} = \mu_{\chi,1}^4 - (r-41)/150,$

$\mu_{\chi,1}^{30} = \mu_{\chi,1}^6 - (r-31)/150,$ and

$\mu_{\chi,1}^{7t} = \mu_{\chi,1}^t - \frac{1}{7}$ for all positive even integers t.

One concludes that $G = G_0 = X \circ Z$ has no regular orbit on V if and only if $r \in \{11, 13, 17, 19, 23\}$. Further inspection gives the minimal stabilizers listed in Theorem 7.2a.

The existence of a (strongly) real vector can be seen also as follows. The representation of X of degree 8 comes from the action of X as automorphism group on the 8-dimensional root lattice E_8. X is transitive on the 240 minimal vectors, and a point stabilizer is isomorphic to the real group $S_6(2) \times Z_2$.

7.10. Exceptional Groups

Let L be a simple exceptional group of Lie type (including the twisted Suzuki and Ree groups). Then $\mathrm{Aut}(L)$ and $\mathrm{M}(L)$ are known. By [Guralnick and Saxl, 2003] the covering number $c(L) \leq \ell + 3$, where ℓ is the untwisted rank of L, except possibly when $L = F_4(q)$ with $c(L) \leq 8$. From [Seitz–Zalesskii, 1993] one has lower bounds for $R_0(L)$. On the basis of Lemma 7.3b this yields the following.

Proposition 7.10. *There are regular orbits except possibly when* $L = G_2(3)$ *and* $d \leq 27$, *or* $L = G_2(4)$ *and* $d = 12$, *or* $L = \mathrm{Sz}(8)$ *and* $d \leq 56$, *or* $L = {}^3D_4(2)$ *and* $d \leq 52$, *or* $L = {}^2F_4(2)'$ *and* $d \leq 52$, *or* $L = F_4(2)$ *and* $d = 52$.

The remaining groups to be examined are all Atlas groups, and the counting method applies in each case.

This completes the proof of Theorem 7.2a.

At this stage, the $k(GV)$ problem is settled in all characteristics p different from $3, 5, 7, 11, 13, 19$ and 31. This is a consequence of the classification of the nonreal reduced pairs in Theorems 6.1, 7.1, in terms of the Robinson–Thompson criterion (Theorem 5.2b) and Clifford reduction (Theorem 5.4). The challenge now is to describe effectively the pairs (G, V) admitting no real vectors, which must "involve" nonreal reduced pairs in some Clifford-theoretic sense.

Chapter 8

Modules without Real Vectors

In solving the $k(GV)$ problem we may assume that G is irreducible on V (Proposition 3.1a) and that no vector in V is real for G (Theorem 5.2b). We show that then (G, V) must be a "nonreal induced" pair, that is, obtained by module induction from a nonreal reduced pair. This is an important step towards the solution of the problem.

8.1. Some Fixed Point Ratios

As usual $F = \mathbb{F}_r$ is a finite field of characteristic p not dividing the order of the finite group G, and V is a FG-module. Let $Z \cong F^\star$ be the group of scalar multiplications on V.

Lemma 8.1a. *Suppose $V = W_1 \otimes_F \cdots \otimes_F W_n$ is a tensor decomposition into FG-modules. Let $g_i \in \mathrm{GL}(W_i)$ and $g = g_1 \otimes \cdots \otimes g_n$ in $\mathrm{GL}(V)$.*

(i) *If $f(zg_i, W_i) \leq m$ for all $z \in F^\star$ and all i, then $f(g, V) \leq m$.*

(ii) *Let $v = w_1 \otimes \cdots \otimes w_n$ for nonzero vectors $w_i \in W_i$. If G induces all scalar transformations on the W_i, then $C_G(v)/C_G(V)$ is isomorphic to a subgroup of $C_G(w_1)/C_G(W) \times \cdots \times C_G(w_n)/C_G(W)$.*

Proof. (i) If some g_i is a scalar transformation on W_i, say with z_i, then $f(z_i^{-1} g_i, W_i) = 1$ and so the result is obvious ($m \geq 1$). So exclude this. We argue by induction on n, the statement being obvious for $n = 1$. So let $n > 1$, and let the $\varepsilon_j \in Z$ be the distinct eigenvalues of g_1 on W_1. Then

$$f(g, V) = \sum_j f(\varepsilon_j^{-1} g_1, W_1) \cdot f(\varepsilon_j g_2 \otimes g_3 \otimes \cdots \otimes g_n, W_2 \otimes \cdots \otimes W_n) \leq m$$

by induction since $\sum_j f(\varepsilon_j^{-1} g_1, W_1) \leq 1$.

(ii) Let $g \in C_G(v)$, and let $g_i \in \mathrm{GL}(W_i)$ be induced by g. Then $w_i g = w_i g_i = z_i w_i$ for some $z_i \in Z$, and $z_1 \cdots z_n = 1$. By hypothesis $z_i^{-1} g_i$ is induced on W_i by some element in G (centralizing w_i). Hence the assignment $g \mapsto (z_1^{-1} g_1, \cdots, z_n^{-1} g_n)$ is a homomorphism from $C_G(v)$ into

148

$C_G(w_1)/C_G(W_1) \times \cdots \times C_G(w_n)/C_G(W_n)$. Its kernel is $C_G(V)$, because if $g_i = z_i$ on W_i for each i, then g is the identity on V. $\qquad\square$

Lemma 8.1b. *Let $X = H\mathrm{wr}\,S_n$ for some finite group H and some integer $n \geq 2$. Suppose $V = W^{\otimes n}$ for some faithful FH-module W of dimension $d = \dim_F W \geq 2$, with X acting on V as usual (1.2). Let $N = H^{(n)}$ be the base group of the wreath product. Then $f(x, V) \leq 1 - \frac{1}{d}$ for $x \in N \smallsetminus C_X(V)$, and $f(x, V) \leq 1 - \frac{1}{2d}$ for $x \in X \smallsetminus N$. Moreover, if $x \in X \smallsetminus N$ is a p'-element, then $f(x, V) \leq \frac{d+1}{2d}$.*

Proof. Since $f(h, W) \leq (d-1)/d = 1 - \frac{1}{d}$ for each $h \in H^\sharp$, the first statement follows from Lemma 8.1a. Let $x \in X \smallsetminus N$, and observe that $C_X(V) \subseteq N$. There exists $y \in N$ such that $[x, y] \notin C_X(V)$. We have $\dim_F C_V(x) = \dim_F C_V(x^{-1}) = \dim_F C_V(y^{-1}xy)$ and $C_V(x^{-1}) \cap C_V(y^{-1}xy) \subseteq C_V([x, y])$. Hence

$$2 \dim_F C_V(x) \leq \dim_F V + \dim_F C_V([x, y]) \leq \dim_F V(1 + (1 - \frac{1}{d}))$$

as $[x, y] \in N \smallsetminus C_X(V)$. Hence $f(x, V) \leq 1 - \frac{1}{2d}$.

Suppose finally that $x \in X \smallsetminus N$ is a p'-element and $Nx \in S_n$ is a product of r disjoint cycles. Let χ be the Brauer character of X afforded by V, and let θ be that of H afforded by W. Then $\chi = \mathrm{Ten}_Y^X(\theta)$ where Y is the stabilizer in X of some tensor factor W of V ($Y/N \cong S_{n-1}$). By the character formula (1.2e), $\chi(x)$ is the product of $r \leq n-1$ values of θ. Hence $|\chi(x)| \leq \theta(1)^r \leq \chi(1)/d$. Now $\dim_F C_V(x) \leq |\chi(x)| + \sum_i \dim_F C_V(z_i^{-1}x)$, the sum taken over all eigenvalues $z_i \neq 1$ of x on V, and $\dim_F C_V(x) + \sum_i \dim_F C_V(z_i^{-1}x) \leq \dim_F V = \chi(1)$. This gives the asserted estimate $2f(x, V) \leq \frac{1}{d} + 1$. $\qquad\square$

8.2. Tensor Induction of Reduced Pairs

Recall the Frobenius embedding of a group into a wreath product (Theorem 1.2a). We need a slight improvement.

Lemma 8.2a. *Let Y be a subgroup of the group X with index n. Assume that there is a normal subgroup C of Y such that $\mathrm{Core}_X(C) = 1$. Then there is an embedding of X into $\bar{Y}\mathrm{wr}\,S_n$ where $\bar{Y} = Y/C$.*

In the discussion of Theorem 1.2a replace the elements $x_i \in Y$ by their cosets Cx_i.

Proposition 8.2b. *Suppose* $V = \text{Ten}_H^G(W)$ *where* $(H/C_H(W), W)$ *is a nonreal reduced pair of quasisimple type. If H is a proper subgroup of G, then there is a strongly real vector in V for G.*

Proof. As usual V is a coprime FG-module. We know from Theorems 7.1 and 7.2 that $F = \mathbb{F}_p$ with $p \in \{7, 11, 13, 19, 31\}$ and that $d = \dim_F W = 2$ or 4. We may assume that $H/C_H(W) = E \circ Z$ is large, that is, $Z \cong F^\star$. Here E is one of $2.A_5$, $2^{\pm}S_6$ or $\text{Sp}_4(3)$. There exist pairs which are not large only when $E = \text{Sp}_4(3)$ and $p = 13$ or 19. But we may certainly assume that G induces all scalar multiplications on $V = W^{\otimes n}$ (see the comment after 5.3b). If $g \in G$ induces the scalar c on V, it preserves each tensor factor and induces certain scalars c_i, with $c = c_1 \cdots c_n$. Using that $F^\star$ is cyclic we see that H induces all scalar transfomations on W.

Let $n = |G : H|$. By hypothesis $n \geq 2$. By Lemma 8.2a there is a (Frobenius) embedding of $G/C_G(V)$ into $T = (EZ))\text{wr}\,\text{S}_n$. We regard $G/C_G(V)$ as a subgroup of $G_0 = T/C_T(V)$ (the wreath product acting in the obvious way). Then

$$G_0 = X \circ Z$$

where X is the image in G_0 of $E\text{wr}\,\text{S}_n$. Let N and N_0 be the images in G_0 of the base groups of $E\text{wr}\,\text{S}_n$ and T, respectively. Then $N \subseteq X$ and $Z(N) = Z(X) \cong Z(E)$ has order 2, and $N_0 = N \circ Z$. Of course $G_0/N_0 \cong X/N \cong S_n$. The elements of N will be written as n-tuples $(h_1, \cdots, h_n)$ with $h_i \in E$ (and the obvious central amalgamation).

Let Y be the stabilizer in X of some tensor factor W of V, so that $Y/N \cong S_{n-1}$. Let χ be the Brauer character of X afforded by V, and let θ be that of E afforded by W. Then $\chi = \text{Ten}_Y^X(\theta)$ as in (1.2e). We sometimes view θ as a (faithful) character of $N_0 = N \circ Z$ and χ as a character of $G_0 = X \circ Z$.

We have $|W| = p^d$ and $|V| = p^{d^n}$. Let f be the maximum of the $f(x) = f(x, V)$ taken over all noncentral $x \in X$. Since $|X/N| = |S_n| = n!$ and $|N/Z(N)| = |E/Z(E)|^n$, by (5.6d) there is a regular G_0-orbit on V, hence a regular G-orbit, provided

$$n! |E/Z(E)|^n (p^{\lfloor f d^n \rfloor} + p^{d^n - \lfloor f d^n \rfloor}) < p^{d^n} = |V|.$$

Now we examine each possible case.

Type (2.A₅) : Here $E \cong 2.A_5 \cong \mathrm{Sp}_2(5)$, θ is a (faithful) Weil character of E of degree $d = 2$, and $p = 11, 19$ or 31 (Sec. 7.1). Also $|E/Z(E)| = 60$, and $f \leq \frac{3}{4}$ by Lemma 8.1b. There is a regular G-orbit on V if $n!60^n < \frac{p^{2^{n-2}}}{1+1/p^{2^{n-1}}}$. This holds true for $n \geq 6$, and all p. It remains to investigate the cases $n \leq 5$. Then G_0 is a p'-group, and we may take $G = G_0$. We use that E, $E\mathrm{wr}\,\mathrm{S}_n$ and, therefore, X are real groups.

Let $p = 31$. Then $G = X \times Z_{15}$. The above estimate also holds for $n = 5$. By Lemma 8.1b we have $f(g) \leq \frac{1}{2}$ if $g \in N$ is noncentral. Let $n = 4$. Then all 5-elements of G are in N_0, and if $x \in X \smallsetminus N$ has order 3, then $Nx \in S_4$ has two cycles (orbits) and so $|\chi(x)| \leq \theta(1)^2 = d^2 = 4$. It follows that x has the eigenvalue 1 on V with multiplicity at most 8, because the primitive 3rd roots of unity have real part $-\frac{1}{2}$ (and $9 - 7/2 > 4$). Thus $f(x) \leq \frac{1}{2}$, as before. Now $|G| = 24 \cdot 60^4 \cdot 30 < 31^8 = |V|^{\frac{1}{2}}$, whence $|G| \cdot |V|^{\frac{1}{2}} < |V|$. We infer that there is $v \in V$ such that $C_G(v)$ contains no elements of order 3 or 5, that is, $C_G(v)$ is a 2-group. Thus $C_G(v) \subseteq X$ and v is strongly real for G.

Let $n = 3$ (and $p = 31$). For $x \in X \smallsetminus N$ of order 3 we have $|\chi(x)| = \theta(1) = 2$ (as Nx is a 3-cycle). Hence $x_V = [4, 2, 2]$. Fix an element $h \in E$ of order 5, and let $g = (h, h, h)$ in N. Then $C_X(g)$ has order $2 \cdot 5^3 \cdot 6$, and $C_X(g)$ contains x. We infer that there are at most $2 \cdot 5^3 = 250$ subgroups in X of order 3 outside N. There are $20^3 + 3 \cdot 20^2 + 3 \cdot 20 = 2 \cdot 4.630$ elements $y \in N$ of order 3, all satisfying $\chi(y) = \pm 1$ (as $\theta(3A_0) = -1$). We get $y_V = [3, 3, 2]$. There are $24^3 + 3 \cdot 24^2 + 3 \cdot 24 = 4 \cdot 3.906$ elements of order 5 in N, all having eigenvalues on V with multiplicity at most 4 by Lemma 8.1b. Since

$$250(p^4 + 2p^2) + 4.630(2p^3 + p^2) + 3.906(2p^4) < p^8 = |V|,$$

there exists $v \in V$ such that $C_G(v)$ is a 2-group. We are done as before. The $n = 2$ case is treated similarly.

Let $p = 19$. Then every $3'$-subgroup of G is a real group. Let $g \in G$ be of order 3. If $g \in N_0$, then $f(g) \leq \frac{1}{2}$ by Lemma 8.1b. If $g \notin N_0$ then its image in $S_n \cong G/N_0$ is a product of at most $n - 2$ disjoint cycles. Consequently the real part

$$\mathrm{Re}(\chi(g)) \leq |\chi(g)| \leq \theta(1)^{n-2} = 2^{2^{n-2}}.$$

Let $g_V = [1^{(a)}, z_1^{(b)}, z_2^{(c)}]$ where $z_1 \neq z_2$ are the primitive 3rd roots of unity in F. Then $f(g) = a/2^n$ and $\mathrm{Re}(z_i) = -\frac{1}{2}$. For $n = 5$ we have $\mathrm{Re}(\chi(g)) \leq 8$, which implies that $a \leq 16$ and $f(g) \leq \frac{1}{2}$. Similarly $a \leq 8$ for $n = 4$, hence $f(g) \leq \frac{1}{2}$ as before. For $n = 4, 5$ we have $|G| = n!60^n \cdot 18 < 19^{2^n/2} = |V|^{\frac{1}{2}}$ and so $|G| \cdot |V|^{\frac{1}{2}} < |V|$. We deduce that there is $v \in V$ such that $C_G(v)$ is a $3'$-group, as desired.

The case $p = 19$, $n = 3$ is handled by showing that $N_X(Fv)$ is a $3'$-group for some $v \in V$. By Lemma 5.1b then v is strongly real for G. There are $2 \cdot 10^3$ elements of order 3 in N. Since $\theta(h) = -1$ for each $h \in E$ of order 3 [Atlas], for any $y \in N$ of order 3 we have $\chi(y) = \pm 1$ and necessarily $y_V = [3, 3, 2]$ (so $\chi(y) = -1$). As before we see that there are at most $4 \cdot 30^3$ subgroups of order 3 in X outside N (where we know the spectral pattern already). Since

$$250(p^4 + 2p^2) + 4.630(2p^3 + p^2) < p^8 = |V|,$$

there is v as required. In the $n = 2$ case we pick an element $x = (h, h)$ in X of order 5 ($h \in E$). If T is a subgroup of X containing x, then $N_X(T)/T$ is a $3'$-group. The character table of $E \cong 2.A_5$ gives that $\dim {}_F C_V(\mathrm{x}) = 2$. Now let v be any nonzero vector in $C_V(x)$, and let $T = C_X(v)$. By Lemma 5.1b this v is strongly real for G_0.

Let $p = 11$. Then $G = X \times Z_5$. Each $5'$-subgroup of G is real. For $n = 5, 4$ we claim that there is $v \in V$ such that $C_G(v)$ is a $5'$-group. Let $g \in G$ be an element of order 5. If $g \in N_0$ then $f(g) \leq \frac{1}{2}$ by Lemma 8.1b. Otherwise $n = 5$ and $N_0 g$ is a 5-cycle in S_5. Then $|\chi(g)| = \theta(1) = 2$ and $\dim {}_F C_V(\mathrm{g}) = 8$ as each primitive 5th root of unity in F appears as eigenvalue of g on V with multiplicity 6. Hence $f(g) = \frac{1}{4}$ in this case. Now use that $|G| \cdot |V|^{\frac{1}{2}} < |V|$, which gives the claim. Let $n = 3, 2$, and let $y = (h, h, h)$ resp. $y = (h, h)$ for an element $h \in Y$ of order 3. Then $\chi(y) = \theta(h)^n = (-1)^n$, and we deduce that $\dim {}_F C_V(\mathrm{y}) = 2$ in both cases. Let v be a nonzero vector in $C_V(y)$, and let $C = C_X(v)$. Then $N_X(C)/C$ is a $5'$-group. Now use that χ takes only real values on X, and use Lemma 5.1b. This shows that v is a strongly real vector for G.

Type $(2.A_6)$:   Here $E \cong 2^{\pm} S_6$, $d = 4$ and $p = 7$. By Lemma 8.1b, $f \leq \frac{7}{8}$. We obtain that there is a regular G_0-orbit on V provided $n \geq 6$. So let $n \leq 5$. Then G_0 is a p'-group, and we take $G = G_0$. By Lemma 8.1b now $f \leq \frac{3}{4}$. This yields the existence of a regular G-orbit on V for $n \geq 3$. So let $n = 2$. Inspection of the character table [Atlas, p. 5], noting that $\theta = \chi_9$ for $E = 2^+ S_6$ and $\theta = \chi_8$ for $E = 2^- S_6$, yields that $f(g) \leq \frac{1}{2}$ for each noncentral element $g \in G$ of order 3. (If $g \in N$ either

$\chi(g) = 1$ and $g_V = [6,5,5]$, or $\chi(g) = -2$ and $g_V = [4,6,6]$, or $\chi(g) = 4$ and $g_V = [8,4,4]$.) The number of noncentral elements of order 3 in G is at most $3 \cdot (80^2 + 2 \cdot 80) < 7^8 = |V|^{\frac{1}{2}}$. It follows that there is $v \in V$ such that $C_G(v)$ is a $3'$-group (contained in N). The restriction to $C_G(v)$ of χ is rational-valued [Atlas].

Type $(\mathbf{Sp_4(3)})$: Here $E \cong Sp_4(3)$, $d = 4$ and $p \in \{7,13,19\}$ (Sec. 7.1). Recall also that $\theta = \chi_{21}$ or χ_{22} in the notation of the Atlas [Atlas, p. 27]. From the general estimate we get the existence of a regular G-orbit on V when $p = 13$ or 19 and $n \geq 3$, and when $p = 7$ and $n \geq 4$.

Let $p = 19$ and $n = 2$. Then $G_0 = X \times Z_9$ is a p'-group. Each element of order 3 in G_0 lies in N_0, and the number of such elements is at most $|G_0|/2 < 19^8 = |V|^{\frac{1}{2}}$. Hence there is $v \in V$ such that $C_{G_0}(v)$ is a $3'$-group. Then the restriction to $C_{G_0}(v)$ of χ is rational-valued [Atlas].

Let $p = 13$ and $n = 2$. Then $G_0 = X \times Z_6$ is a p'-group. Let $G = G_0$. Each element in X of order 3 lies in N. If $x = (h,1)$ or $(1,h)$ with $h \in E$ belonging to one of the conjugacy classes $3A_0B_0$, then $x_V = [12,4]$. There are $2 \cdot 80$ such elements. For the remaining $80^2 + 240^2 + 2 \cdot 240 + 480^2 + 2 \cdot 480 = 2 \cdot 147.920$ elements y of order 3 in X we have $\sum_{z \in Z} |C_V(zy)| \leq p^9 + p^6 + p$. It follows that

$$\sum_{\gamma \in \beta_3(G)} |C_V(\gamma)| \leq 80(p^{12} + p^4) + 147.920(p^9 + p^6 + p).$$

Let $g \in N$ be a noncentral involution. Then $g = (h_1, h_2)$ where either both $h_i \in E$ belong to the class $2A_0$ or one h_i lies in $2A_0$ and the other is 1. Then $\chi(g) = 0$ [Atlas] and so $g_V = [8,8]$. There are $45^2 + 2 \cdot 45 = 2.115$ such involutions in N. It follows that $\sum_{\gamma \in \beta_2(G)} |C_V(\gamma)| \leq 2.115(2p^8)$. We have $\sum_{\gamma \in \beta_{2,3}(G)} |C_V(\gamma)| < |V| = 13^{16}$. We conclude that there is $v \in V$ such that $C_G(v) = C_X(v)$ is a $\{2,3\}'$-group. The restriction to $C_X(v)$ of χ is rational-valued [Atlas].

Let $p = 7$ and $n = 2, 3$. As before we prove that $\sum_{\gamma \in \beta_3(G_0)} |C_V(\gamma)| < |V| = 7^{2^n}$. There is $v \in V$ such that $C_{G_0}(v) = C_X(v)$ is a $3'$-group, and the restriction to $C_X(v)$ of χ is rational-valued [Atlas]. We are done. $\square$

Proposition 8.2c. *Suppose* $V = \mathrm{Ten}_H^G(W)$ *is a coprime FG-module where* $(H/C_H(W), W)$ *is a reduced pair of quasisimple type. There is $v \in V$ such that $C_G(v)/C_G(V)$ has a regular orbit on V.*

Proof. We may and do assume that H induces all scalar transformations on W. Let $d = \dim_F W$ and let $n = |G : H|$. In view of Corollary 7.2c we may assume that $n \geq 2$. Pick $w \in W$ such that $C = C_H(w)/C_H(W)$ is a point stabilizer of minimal order. That is, either $C = 1$ or C is as listed in Theorem 7.2a, or $(H/C_H(W), W)$ is a permutation pair (Example 5.1a). In this latter case either $C = 1$ or $r = p = d + 2$ or $d + 3$, in which case we may choose w such that C is cyclic of order dividing $p - 1$. In each case let

$$v = w \otimes \cdots \otimes w \in W^{\otimes n} = V.$$

Let $N = \mathrm{Core}_G(H)$. So $S = G/N$ is a transitive permutation group of degree n, which is a p'-group (like G).

By Lemma 8.1a, $C_N(v)/C_N(V)$ is isomorphic to a subgroup of $C^{(n)}$. It follows that $C_G(v)/C_G(V)$ is isomorphic to a subgroup of $C \,\mathrm{wr}\, S$. In particular, $|C_G(v)/C_G(V)|$ is a divisor of $|C|^n \cdot n!$. Let $F = \mathbb{F}_r$, so that $|W| = r^d$ and $|V| = r^{d^n}$.

Suppose first that $d \geq 3$. Then $f(g) = f(g, V) \leq 1 - \frac{1}{d}$ for each $g \in G \smallsetminus C_G(V)$ by Lemma 8.1b. Hence the number of vectors in V centralized by some nontrivial element in $C_G(v)/C_G(V)$ is at most equal to $n!|C|^n|V|^{1-\frac{1}{d}}$. There is a regular vector in V for $C_G(v)$ provided this number is less than $|V|$, that is, if

$$n!|C|^n < r^{d^{n-1}}.$$

This is fulfilled for all $n \geq 2$ when $C = 1$ or when $|C| \leq p - 1$ (handling the cases where $(H/C_H(W), W)$ is a permutation pair). Otherwise $r = p$ and $|C| \leq 72$ by Theorem 7.2a. Then the above inequality is fulfilled for all n if $d \geq 5$. For $d = 3$ we even have $p \geq 11$ and $|C| \leq 5$, which gives the result. Let $d = 4$. Then the above inequality holds for all $n \geq 3$. So let $n = 2$. If $|C| \leq 18$, the result holds. Otherwise by Theorem 7.2a, $H/C_H(W) \cong \mathrm{Sp}_4(3) \times Z_3$, $d = 4$, $r = 7$, and by Corollary 7.2c there exists a regular vector $\widetilde{w}_1 \in W^\sharp$ for $C \cong \mathrm{SL}_2(3) \times Z_3$ such that $\langle \widetilde{w}_1 \rangle \cap \widetilde{w}_1 C = \{\widetilde{w}_1\}$. Let $\widetilde{w}_2 \in W$ be any vector linearly independent from $\widetilde{w}_1$. Then $\widetilde{w}_1 \otimes \widetilde{w}_2$ is a regular vector for $C \,\mathrm{wr}\, S_2$ and hence for $C_G(v)/C_G(V)$.

Let $d = 2$. Then $f(g, V) \leq \frac{3}{4}$ for each nontrivial element in $G/C_G(V)$ by Lemma 8.1b. There is a regular $C_G(v)/C_G(V)$-orbit on V provided

$$n!|C|^n < r^{2^{n-2}}.$$

Since $r \geq 3$ is odd, this inequality holds when $C = 1$. So let $C \neq 1$, in which case $H/C_H(W) \cong 2.A_5 \circ Z$ by Theorem 7.2a. If $r \neq 11$, then $|C| \leq 3$, and the inequality holds for all $n \geq 2$. So let $F = \mathbb{F}_{11}$ and $|C| = 5$. Then the estimate holds for $n \geq 5$. Let $n = 4$. We have $|P_1(W)| = 12$, and C has on $P_1(W)$ two fixed points $Fw = Fw_1$ and Fw_2, say, and two orbits of length 5 represented by Fw_3 and Fw_4 (say). The vector $w_1 \otimes w_2 \otimes w_3 \otimes w_4$ is regular for $C \operatorname{wr} S_4$ and hence for $C_G(v)/C_G(V)$. The cases $n = 3, 2$ are treated similarly. $\qquad \square$

8.3. Tensor Products of Reduced Pairs

Proposition 8.3a. *Suppose* $V = U \otimes_F W$ *where both* $(G/C_G(U), U)$ *and* $(G/C_G(W), W)$ *are reduced pairs. Then there is a real vector in V for G.*

Proof. Without loss we may assume that G induces on U and on W all scalar transformations. By Proposition 5.3b we may also assume that one of the reduced pairs is nonreal, say the first one. Then $d = \dim_F U = 2, 3$ or 4 and $r = p \in \{3, 5, 7, 11, 13, 19, 31\}$ by the classification of the nonreal reduced pairs ($F = \mathbb{F}_r$). By Theorems 6.1 and 7.1, and by Corollary 7.2c, there is $w \in W$ such that $C_G(w)/C_G(W)$ has a regular orbit on W. Hence the assertion follows from part (iii) of Proposition 5.3b provided $\dim_F W \geq$ d. But when $\dim_F W < d$, we may exchange the roles of U and W. $\qquad \square$

Proposition 8.3b. *Let* $W = \operatorname{Ten}_H^G(\widetilde{W})$ *where* $(H/C_H(\widetilde{W}), \widetilde{W})$ *is a reduced pair of quasisimple type. Suppose* $V = U \otimes_F W$ *where* $(G/C_G(U), U)$ *is a reduced pair. Then there is a real vector in V for G.*

Proof. Without loss of generality we may assume that G induces all scalar transformations on U and on W. If $H = G$ the result follows from Proposition 8.3a. So let $n = |G : H| \geq 2$. Combining Propositions 8.2b and 5.3c we see that there is a real vector in W for G, at any rate. If, in addition, there is a real vector in U for G, the result follows from part (i) of Proposition 5.3b.

Hence we may assume that $(G/C_G(U), U)$ is a nonreal reduced pair. By the results in Chapters 6, 7 then $\dim_F U = 2, 3$ or 4. On the other hand, $\dim_F \widetilde{W} \geq 2$ and so $\dim_F W = (\dim_F \widetilde{W})^n \geq 4$. By Proposition 8.2c there is $w \in W$ such that $C_G(w)/C_G(W)$ has a regular orbit on W. Therefore part (iii) of Proposition 5.3b applies again. $\qquad \square$

Proposition 8.3c. *Let* $V = U_1 \otimes_F U_2 \otimes_F U_3$ *where the* $(G/C_G(U_i), U_i)$ *are nonreal reduced pairs. Then there exists a real vector in V for G.*

Proof. We may assume that G induces all scalar transformations on the U_i. Suppose first that there are abelian vectors $u_i \in U_i$ for G at least two times, say for the indices $i = 2, 3$. Let $W = U_2 \otimes_F U_3$ and let $w = u_2 \otimes u_3$. By part (ii) of Lemma 8.1a, w then is an abelian vector for $G/C_G(W)$. Since $\dim_F W \geq 4 \geq \dim_F U_1$ by Theorems 6.1, 7.1, the assertion follows from part (iii) of Proposition 5.3b.

It remains to examine the situations where there are no such abelian vectors. By Theorems 6.1, 7.1 then at least two of the pairs are of type $(\mathrm{Sp}_4(3))$, say the latter ones, and $p = 7$ or $p = 13$. Then $U = U_2 = U_3$ and G acts on $W = U_2 \otimes_F U_3 = U^{\otimes 2}$ like the base group of $(G/C_G(U))\mathrm{wr}\,S_2$. Hence the result follows from Proposition 8.3b. $\square$

8.4. The Riese–Schmid Theorem

We call a pair (G, V) *nonreal induced* provided V is a faithful coprime FG-module which admits no real vector for G and which is induced from some nonreal reduced pair (H, W), that is, $V = \mathrm{Ind}_{G_0}^G(W)$ for some subgroup G_0 of G satisfying $H \cong G_0/C_{G_0}(W)$. From the classification of the nonreal reduced pairs it follows that then V is an absolutely irreducible FG-module and $F = \mathbb{F}_p$ for one of the primes $p = 3, 5, 7, 11, 13, 19$ or 31. Of course this also forces that $|G : G_0|$ is not divisible by p.

Theorem 8.4 (Riese–Schmid). *Let V be a faithful, irreducible, coprime FG-module admitting no real vector for G. Then (G, V) is nonreal induced.*

Proof. Assume V is a counterexample with $d = \dim_F V$ minimal. So there is no real vector in V for G, hence G is nonabelian and $d \geq 2$. Furthermore, whenever V_0 is an irreducible coprime F_0G_0-module for which $\mathrm{char}(F_0) = p = \mathrm{char}(F)$ and $\dim_{F_0} V_0 < d$, then there is a real vector in V_0 for G_0, that is, for $G_0/C_{G_0}(V_0)$, or (G_0, V_0) is nonreal induced. We argue, at first, like for Theorem 5.4.

(1) *V is an absolutely irreducible FG-module:*

For otherwise embed F (properly) into the field $F_0 = \mathrm{End}_{FG}(V)$. Then $F_0 \otimes_F V$ is the direct sum of the distinct Galois conjugates over F of some absolutely irreducible F_0G-module V_0. This V_0 is a faithful module and $\dim_{F_0} V_0 < d$. Thus either V_0 contains a real vector v_0 for G, or $V_0 = \mathrm{Ind}_H^G(W)$ where $(H/C_H(W), W)$ is a nonreal reduced pair (over F_0). In the former case the sum of the distinct Galois conjugates over F of v_0 is a real vector in V for G. The second case cannot occur by the classification of the nonreal reduced pairs, because F_0 is not a prime field.

(**2**) V *is a primitive FG-module:*

Otherwise $V = \operatorname{Ind}_H^G(U)$ for some proper subgroup H of G and some FH-module U. Then $\dim_F U < d$. Hence either there is a real vector in U for H, which implies that there is a real vector in V for G by part (i) of Proposition 5.3a, or $U = \operatorname{Ind}_Y^H(W)$ where $(Y/C_Y(W), W)$ is a nonreal reduced pair. By transitivity of module induction $V = \operatorname{Ind}_Y^G(W)$ in the latter case.

(**3**) *The irreducible constituents of* $\operatorname{Res}_N^G(V)$ *are absolutely irreducible for all normal subgroups N of G:*

Otherwise choose N maximal such that (3) is false. Then $N \neq G$ by (1). As in the proof for Theorem 5.4 one gets that $\operatorname{Res}_N^G(V) = W$ is an irreducible FN-module with $F_0 = \operatorname{End}_{FN}(W)$ being a proper extension field of F. Moreover $G/N \cong \operatorname{Gal}(F_0|F)$ and

$$F_0 \otimes_F V \cong \operatorname{Ind}_N^G(U),$$

where U is an (absolutely) irreducible constituent of $F_0 \otimes_F W$. If there is a real vector in U for N, then there is a real vector in $F_0 \otimes_F V$ for G by Proposition 5.3a, and we get a real vector in V. Hence by the choice of V we have $U = \operatorname{Ind}_Y^N(U_0)$ for some subgroup Y of N and some F_0Y-module U_0, with $(Y/C_Y(U_0), U_0)$ being a nonreal reduced pair. But such a pair does not exist. Hence the assertion.

From (1), (2), (3) it follows that every abelian normal subgroup of G is cyclic and central in G (acting by scalar multiplications). The generalized Fitting subgroup of G is nonabelian for otherwise G were cyclic and so had a regular orbit on V.

(**4**) *There is a nonabelian normal subgroup N of G such that* $\operatorname{Res}_N^G(V)$ *is not (absolutely) irreducible:*

Assume the contrary. Let E be a minimal nonabelian normal subgroup of G. By assumption and (2), (3), $\operatorname{Res}_E^G(V) = W$ is absolutely irreducible. Hence $C_G(E) = Z = Z(G)$ acts via scalar multiplications on V and EZ is the generalized Fitting subgroup of G. Suppose first that E is solvable. Then it is a q-group of "symplectic type" for some prime $q \neq p$. Either q is odd and $E = \Omega_1(EZ)$ is of exponent q, or E is an extraspecial 2-group or the central product of such one with a cyclic group of order 4 (when F contains the 4th roots of unity). Hence (G, V) is a nonreal reduced pair with core E, a contradiction.

Suppose next that E is not solvable. If E is quasisimple, we get a contradiction as before. Hence E is the central product of $n \geq 2$ distinct G-conjugates of some quasisimple group E_0. Let $G_0 = N_G(E_0)$, so that $n = |G : G_0|$. Then W is the tensor product of n distinct G-conjugates of the unique *absolutely* irreducible constituent W_0 of $\operatorname{Res}_{E_0}^{E}(W)$. Let θ and θ_0 be the Brauer characters of E, E_0 afforded by W, W_0, respectively. By Theorem 1.9c there is an $FG_0(\theta_0)$-module $\widehat{W_0}$ extending W_0 (in the usual sense). The pair $\big(G_0(\theta_0)/C_{G_0(\theta_0)}(\widehat{W_0}), \widehat{W_0}\big)$ is reduced with quasisimple core isomorphic to E_0. By Theorem 1.9d there is a finite extension $\widehat{G}$ of $G = G(\theta)$ by an abelian p'-group, containing a subgroup $\widehat{G}_0$ mapping onto $G_0(\theta_0)$, such that

$$V = \operatorname{Ten}_{\widehat{G}_0}^{\widehat{G}}(\widehat{W_0}).$$

If there is a real vector in $\widehat{W_0}$ for $\widehat{G}_0$, there is a real vector for G in V by Proposition 5.3c. If there is no real vector in $\widehat{W_0}$ for $\widehat{G}_0$, by Proposition 8.2b there is a real vector in V for G as well. This proves (4).

(5) *Over some central extension $\widetilde{G}$ of G we have $V = U \otimes_F W$ where the pair $(\widetilde{G}/C_{\widetilde{G}}(U), U)$ is nonreal reduced and where W is an absolutely irreducible and primitive $F\widetilde{G}$-module with* $\dim_F W \geq 2$:

By (4), (3) there is a nonabelian normal subgroup N of G such that the restriction of V to N is a proper multiple eU of some absolutely irreducible FN-module U ($e \geq 2$). This U is faithful. Now argue as in the proof for Theorem 5.4, on the basis of stable Clifford theory. This gives the tensor decomposition $V = U \otimes_F W$ as asserted where $\dim_F U \geq 2$ and $\dim_F W \geq 2$. By (1), (2) both U and W are absolutely irreducible and primitive. For if $W = \operatorname{Ind}_H^{\widetilde{G}}(\widetilde{W})$ were imprimitive, say, then $V = \operatorname{Ind}_H^{\widetilde{G}}\big(\operatorname{Res}_H^{\widetilde{G}}(\widetilde{U}) \otimes_F \widetilde{W}\big)$ and (2) applies. Since both $\dim_F U$ and $\dim_F W$ are less than d, the theorem holds for U and for W. At least one of these modules does not contain a real vector by Proposition 5.3b, hence defines a nonreal reduced pair by the choice of V. The situation being symmetric we assume that $(\widetilde{G}/C_{\widetilde{G}}(U), \widetilde{U})$ is a nonreal reduced pair.

(6) *We have $F = \mathbb{F}_p$ for a certain odd prime $p \leq 31$. There is a real vector in W for $\widetilde{G}$, and $(\widetilde{G}/C_{\widetilde{G}}(W), W)$ is neither reduced nor is obtained by (proper) tensor induction from a reduced pair with quasisimple core:*

The first statement follows from (5) by the classification of the nonreal reduced pairs. Apply furthermore Propositions 8.3a and 8.3b.

Now we proceed by showing that statements (3), (4) hold for W in place of V.

(7) *The irreducible constituents of* $\mathrm{Res}^{\widetilde{G}}_{\widetilde{N}}(W)$ *are absolutely irreducible for all normal subgroups* $\widetilde{N}$ *of* $\widetilde{G}$:

Otherwise there exists a proper normal subgroup $\widetilde{N}$ of $\widetilde{G}$ which is maximal subject to the property that $\overline{W} = \mathrm{Res}^{\widetilde{G}}_{\widetilde{N}}(W)$ has an irreducible summand which is not absolutely irreducible. Then $\overline{W}$ is irreducible, $F_0 = \mathrm{End}_{F\widetilde{N}}(\overline{W})$ is a proper extension field of F and $\widetilde{G}/\widetilde{N} \cong \mathrm{Gal}(F_0|F) = \Gamma$ (as before). Furthermore, denoting by $\overline{W}_0$ an (absolutely) irreducible constituent of $F_0 \otimes_F \overline{W}$, we have $F_0 \otimes_F W \cong \mathrm{Ind}^{\widetilde{G}}_{\widetilde{N}}(\overline{W}_0)$. Let $\widetilde{Z} = C_{\widetilde{G}}(V)$. Then $G = \widetilde{G}/\widetilde{Z}$, as G is faithful on V, and $\widetilde{Z}$ induces scalars from F^* on both U and W. By the (maximal) choice of $\widetilde{N}$ therefore $\widetilde{Z} \subseteq \widetilde{N}$. Let $N = \widetilde{N}/\widetilde{Z}$. Since $\widetilde{G}/\widetilde{N} \cong G/N$ is cyclic, $\widetilde{G}/C_{\widetilde{G}}(U)$ contains a normal subgroup mapping onto the core of $\widetilde{G}/C_{\widetilde{G}}(U)$. It follows that $\widetilde{U} = \mathrm{Res}^{\widetilde{G}}_{\widetilde{N}}(U)$ is absolutely irreducible. Let $\widetilde{U}_0 = F_0 \otimes_F \widetilde{U}$. We conclude that

$$F_0 \otimes_F V = (F_0 \otimes_F U) \otimes_{F_0} (F_0 \otimes_F W) = \mathrm{Ind}^{\widetilde{G}}_{\widetilde{N}}(\widetilde{U}_0 \otimes_{F_0} \overline{W}_0).$$

Hence, in view of (1), $\widetilde{V}_0 = \widetilde{U}_0 \otimes_{F_0} \overline{W}_0$ is an absolutely irreducible $F_0\widetilde{N}$-module, which is centralized by $\widetilde{Z}$. By Mackey decomposition, identifying $\Gamma = G/N$,

$$F_0 \otimes_F \mathrm{Res}^G_N(V) = \mathrm{Res}^G_N(F_0 \otimes_F V) \cong \bigoplus_{\tau \in \Gamma} \widetilde{V}_0^{\tau}.$$

By virtue of (3) and (2) this implies that $\mathrm{Res}^G_N(V) \cong e\widetilde{V}$ for some absolutely irreducible FN-module $\widetilde{V}$ satisfying $F_0 \otimes_F \widetilde{V} \cong \widetilde{V}_0$, where the ramification index $e = |\Gamma| = |G/N|$. However, $e = 1$ since Γ is cyclic. This may be seen by observing that if θ is the Brauer character of N afforded by $\widetilde{V}$, then the representation group $\Gamma(\theta)$ is a central extension of the cyclic group Γ. Hence $\Gamma(\theta)$ is abelian, and the Brauer character of G afforded by V is of the form $\chi = \hat{\theta} \otimes \zeta$ for some *linear* character ζ of $\Gamma(\theta)$, where $\hat{\theta}(1) = \theta(1)$. We have the contradiction $N = G$.

(8) *There is a nonabelian normal subgroup* $\widetilde{N}/C_{\widetilde{G}}(W)$ *of* $\widetilde{G}/C_{\widetilde{G}}(W)$ *such that* $\mathrm{Res}^{\widetilde{G}}_{\widetilde{N}}(W)$ *is not absolutely irreducible:*

Otherwise consider a normal subgroup $\widetilde{E}$ of $\widetilde{G}$ which is minimal subject to the condition that its image in $\widetilde{G}/C_{\widetilde{G}}(W)$ is nonabelian. As in step (4) either $(\widetilde{G}/C_{\widetilde{G}}(W), W)$ is a reduced pair, its core being the image of $\widetilde{E}$, or W is obtained by (proper) tensor induction from a reduced pair with quasisimple core. Both cases are excluded by (6).

(9) *Conclusion:*

By (8), (7) and stable Clifford theory we may write $W \cong U_2 \otimes_F U_3$ over some central extension $\widehat{G}$ of $\widetilde{G}$, where both modules U_i are absolutely irreducible and primitive with $\dim_F U_i < \dim_F W$. Viewing $U_1 = U$ as an $F\widehat{G}$-module we have

$$V \cong U_1 \otimes_F U_2 \otimes_F U_3.$$

Recall that the pair $(\widehat{G}/C_{\widehat{G}}(U_1), U_1)$ is nonreal reduced by (5). If the pairs $(\widehat{G}/C_{\widehat{G}}(U_i), U_i)$ are nonreal reduced also for $i = 2, 3$, then V contains a real vector for G by Proposition 8.3c. Hence, by the choice of V, one of these modules, say U_3, has a real vector.

Now $\widehat{U} = U_1 \otimes_F U_2$ is an absolutely irreducible and primitive $F\widehat{G}$-module with dimension less than $d = \dim_F V$. Thus either $\widehat{U}$ contains a real vector for $\widehat{G}$ or $(\widehat{G}/C_{\widehat{G}}(\widehat{U}), \widehat{U})$ is a nonreal reduced pair. In the first case by Proposition 5.3b there is a real vector for G in $V = \widehat{U} \otimes_F U_3$, a contradiction. By the classification of the nonreal reduced pairs the latter case can hold only when $F = \mathbb{F}_7$, $\dim_F \widehat{U} = 4$ and $(\widehat{G}/C_{\widehat{G}}(U_1), U_1)$ is nonreal of type (Q_8). However, then $|U_1| = |U_2| = 7^2$. Since $\mathrm{GL}_2(7)$ is a $5'$-group, $\widehat{G}/C_{\widehat{G}}(\widehat{U})$ is a $5'$-group too. But $(\widehat{G}/C_{\widehat{G}}(\widehat{U}), U)$ is of type (2^5_-), $(2.A_6)$ or $(\mathrm{Sp}_4(3))$, and $|\widehat{G}/C_{\widehat{G}}(\widehat{U})|$ is divisible by 5 in each case. $\square$

8.5. Nonreal Induced Pairs, Wreath Products

Let (G, V) be a nonreal induced pair, say induced from the nonreal reduced pair (H, W). Suppose

$$V = W_1 \oplus \cdots \oplus W_n$$

is an imprimitivity decomposition, where the G-conjugate reduced pairs $(N_G(W_i)/C_G(W_i), W_i)$ are isomorphic to (H, W). Assume that $n \geq 2$. Let E be the core of (H, W), and let $E_i \cong E$ be the core of $H_i = N_G(W_i)/C_G(W_i)$. Fix an isomorphism $(H_1, W_1) \cong (H, W)$ of reduced pairs. Inspection of Theorems 6.1, 7.1 yields that distinct normal subgroups of H

are not isomorphic (and in most cases even determined by their order). Let $N = \bigcap_{i=1}^{n} N_G(W_i)$, and suppose that $S = G/N$ acts (transitively) on the set $\Omega = \{1, \cdots, n\}$ like on the $\{W_i\}$. For each $i \in \Omega$ let $T_i = C_N(W_i)$ and $N_i = N/T_i$. Identifying each N_i with $NC_G(W_i)/C_G(W_i)$ we have normal subgroups of the H_i, which are G-conjugate, hence have a common image $N_0 \cong N_i$ in $H = H_0$.

Since G is faithful on V, by Lemma 8.2a we may identify G with a subgroup of $H \operatorname{wr} S$. Our objective is to show that G is not far from being the full wreath product. Define $R_i = \bigcap_{j \neq i} T_j$ for each $i \in \Omega$, which are G-conjugate normal subgroups of N (and which would agree with a direct factor H_i in the base group $B = H^{(n)} = H_1 \times \cdots \times H_n$ of the wreath product). We have $R_i \cap T_i = 1$, and

$$R = R_1 \times \cdots \times R_n$$

is a normal subgroup of G. Let $R_0 \cong R_i \cong RC_G(W_i)/C_G(W_i)$ be the common image of the R_i and of R in $H = H_0$. Observe that R_0, N_0 are normal subgroups of H.

Lemma 8.5a. N_0 *has no regular orbit on* W. *Hence* N_0 *is a nonabelian normal (characteristic) subgroup of* H *and so* $N_0 \supseteq E$.

Proof. Assume that, for some $i \in \Omega$, there is $w_i \in W_i$ such that $C_{N_i}(w_i) = 1$. Since G is transitive on the $\{W_i\}$, this is true then for all $i \in \Omega$. Then $v = \sum_i w_i$ is a regular vector for N as $C_N(v) = \bigcap_{i \in \Omega} C_N(w_i) = \bigcap_{i \in \Omega} C_N(W_i) = 1$. But now by Proposition 5.3b (ii) there is a real vector in V for G, which contradicts the (implicit) hypothesis in the lemma. The remainder follows from (3.4b) and the definition of a reduced pair. $\square$

For a subgroup Y of H we denote by $Y^{\mathfrak{S}}$ the "stabilizer residual" of Y, the intersection of all normal subgroups T of Y for which Y/T is isomorphic to a subgroup of $C_H(w)$ for some $w \in W^{\sharp}$. We have to examine these residuals only when (H, W) is of extraspecial type.

Lemma 8.5b. *Let* (H, W) *be of extraspecial type. Suppose* Y *is a subgroup of* H *which has no regular orbit on* W. *Then* $Y^{\mathfrak{S}} \not\subseteq Z(H)$ *when* $p \neq 3, 5$, *and* $Y \supset E$ *for* $p = 5$. *If* Y *is assumed to be subnormal in* H, *then* $Y \supseteq E$. *Further, then* Y *is irreducible on* $U = E/Z(E)$ *except when* $p = 3$ *or* 5, *or when* $p = 7$ *and* (H, W) *is of type* (3_+^3) *with* $Y = E\langle j \rangle$ *or* $E\langle j \rangle Z(H)$ *for some involution* j *inverting the elements of* U.

Proof. Clearly Y is nonabelian. Let $Z = Z(H)$. From Theorem 6.1 we know that $Z \cong \mathbb{F}_p^\star$ unless $p = 7$, $E \cong 3_+^{1+2}$ and $H = X = E : \mathrm{Sp}_2(3)$ is the standard holomorph of E. Otherwise we have $H = X \circ Z$ where X is the standard holomorph of E when $4 \nmid p - 1$ and $p \neq 3$, and of $E \circ Z_4$ otherwise, or we have $p = 7$, $E \cong 2_-^{1+2m}$ and $X/E \cong \Omega_{2m}^-(2)$ for $m = 1, 2$. Recall also that for $p = 3$ we have type (2_-^5), and $X \cong 2_-^{1+4} : (Z_5 : Z_4)$ is related to the largest Bucht group.

By Theorem 6.1 the point stabilizers either are abelian or (H, W) is of type (2_-^5) with $p = 7$, or of type (3_+^3) with $p = 13$, where $\mathrm{SL}_2(3) = Q_8 : Z_3$ can appear. Hence either $Y^{\mathfrak{S}} \supseteq Y'$ or we are in these latter situations, in which cases $Y^{\mathfrak{S}} \supseteq Y'''$ at least.

Let χ denote the Brauer character of H afforded by W. We work through the various types, making use of the knowledge of the H-orbit structure on $W^\sharp$ given in Sec. 6.1.

Type $(\mathbf{Q_8})$: Here $E = Q_8$ and $X \cong 2^- S_4$ or $\mathrm{SL}_2(3)$ for $p = 7$, and $X \cong (Q_8 \circ Z_4).S_3$ for $p = 5$ and $p = 13$. Let $T = N_H(P$ for some Sylow 3-subgroup P of H. Then $H = ET$, $E \cap T = Z(E)$ and $T/Z \cong S_3$ or A_3. If $y \in H$ is a noncentral element of order 3, then y is irreducible on U and so $|\chi(y)| = 1$ by Theorem 4.4. Hence $\langle Zy \rangle$ fixes 2 points in $P_1(W)$ if $3 \mid p - 1$, and at most 1 otherwise. Each noncentral involution in H fixes 2 points in $P_1(W)$. It follows that T has a regular orbit on W for $p = 13$. The same holds true for $p = 7$, because then $P = Z_3 \times C_H(w)$ for some $w \in W^\sharp$ (Sec. 6.1), and the involutions in T/Z leave Fw invariant. For $p = 5$ any such stabilizer $C_H(w)$ is cyclic of order 4, and either T is regular on $W^\sharp$ or has two orbits of size 12. But in the latter case $H/Z = (EZ/Z) : (T/Z)$ has an element of order 4 whose square is in T/Z, which is impossible.

Let next T be a subgroup of H such that $Z(E) \subset E \cap T \subset E$. Then T contains no element of X of order 3 that is irreducible on U. It follows that T/Z has order 4 and so has a regular orbit on $P_1(W)$. For $p = 5$ argue as above, in which case T can be dihedral of order 8 (having two regular orbits).

Thus $Y \supseteq E$. Of course $E \cong Q_8$ has a regular orbit on W, and EZ has a regular orbit on W if and only if $p \neq 5$. Hence $Y^{\mathfrak{S}} \supseteq Y' \not\subseteq Z$ for $p \neq 5$, as desired. Finally, if Y is subnormal in H and not irreducible on U, then $Y = EZ$ and $p = 5$.

Type $(\mathbf{2_-^5})$: Here $E \cong 2_-^{1+4}$ and $X = E : (Z_5 : Z_4)$ for $p = 3$, and $X \cong E.S_5$ or $E.A_5$ for $p = 7$. The core E has 10 noncentral involutions corresponding to the 5 singular points in the orthogonal space $U = E/Z(E)$. Each such involution t has two eigenspaces on W, both of dimension 2, and so fixes exactly $2 \cdot (p+1)$ points in $P_1(W)$. Since t and tz for $z \in Z(E)$ have the same eigenspaces, the number of points in $P_1(W)$ fixed by some noncentral involution in E is at most $5 \cdot 2 \cdot (p+1)$. This number is smaller than $|P_1(W)| = p^3 + p^2 + p + 1$ for $p = 7$ but not for $p = 3$.

Indeed for $p = 3$ there is no regular E-orbit on W. Here H is transitive on $W^\sharp$, a point stabilizer being cyclic of order 8 (Sec. 6.1). Since E is normal in H, the E-orbits on $W^\sharp$ have equal size, but $|W^\sharp| = 80$ is not divisible by $|E| = 2^5$. Each proper subgroup of E has a regular orbit on W, however.

Suppose Y is subnormal in H. Then $(Y \cap E)/Z(E)$ cannot be a proper subgroup of U, because the normal closure of Y in YE then were a proper subgroup of YE and so Y centralizes a nontrivial quotient group of U. But this implies that Y is a $5'$-group, and that $Y \subset EZ$ by the structure of H. This in turn forces that $p = 3$ and $Y = E$. For $p = 7$ we obtain that $Y = X$ or X' and so Y is irreducible on U.

Let $p = 7$ in what follows, and assume that Y is not subnormal in H. In the proof for Proposition 6.6b, part (ii), we have seen that there is, up to conjugacy, a unique subgroup $T \cong 2.S_5$ of X such that $X = TE$ and $T \cap E = Z(E)$, and $\operatorname{Res}_T^X(\chi)$ is the character obtained by fusing χ_6 and χ_7 in the Atlas notation (p. 2). Let $y \in T$ be an element of order 3. Then $\chi(y) = -2$ and $y_W = [z_1^{(2)}, z_2^{(2)}]$ where $z_1 \neq z_2$ are the primitive 3rd roots of unity. Hence y fixes $2 \cdot 8 = 16$ points in $P_1(W)$. The group $EZ\langle y \rangle$ has a regular orbit on W, because $EZ\langle y \rangle/Z$ contains $|E : C_E(y)| = |\chi(y)|^2 = 4$ (conjugate) subgroups of order 3 and $80 + 4 \cdot 16 < |P_1(W)|$.

Each noncentral involution $x \in T$ ($\chi(x) = 0$) fixes at most $2 \cdot 8 = 16$ points in $P_1(W)$. There are 16 subgroups isomorphic to $Z\langle x \rangle/Z$ in $EZ\langle x \rangle/Z$ complementary to EZ/Z. Hence $EZ\langle x \rangle$ has a regular orbit on W since $80 + 16 \cdot 16 < 400 = |P_1(W)|$. The subgroup $T_0 \cong 2.A_5$ of T has a regular orbit on W but $T \circ Z$ does not (see the remark to Theorem 7.2a). However this does not matter since $Y/Y^{\mathfrak{G}}$ is solvable.

It remains to examine the situation that $Y \cap E \supset Z(E)$. We know that YZE/ZE is not trivial and not of order 2 or 3. Suppose first that Y maps onto $SL_2(3)$ or onto Q_8. Then YZE/ZE is isomorphic to A_4 resp. to $Z_2 \times Z_2$. In both cases Y centralizes a singular point in U but no nonsingular one [Atlas, p. 2]. By definition

$$Y^{\mathfrak{G}} \subseteq Y \cap E \text{ and } [Y \cap E, Y] \subseteq Y^{\mathfrak{G}}.$$

Thus from $Y^{\mathfrak{S}} \subseteq Z$ it follows that $(Y \cap E)/Z(E)$ does not contain nonsingular points. In other words, $Y \cap E$ is elementary of order 4. But then a simple counting argument shows that Y has a regular orbit on W.

Hence we are left with the case that $Y^{\mathfrak{S}} \supseteq Y'$. If $|Y|$ is divisible by 5, then Y is irreducible on U. Then even $Y' \supseteq E$. Then $Y^{\mathfrak{S}} \subseteq Z$ implies that YZ/Z is abelian and that Y is a $\{2,3\}$-group. Moreover, YZE/ZE must be isomorphic either to $Z_2 \times Z_2$ or to $S_2 \times A_3$. The intersection of the centralizers in $S_5 \cong O_4^-(2)$ of a singular and a nonsingular point of U has order at most 2 [Atlas]. We conclude that in the former case $Y \cap E$ either is elementary abelian of order 4, which is handled as above, or $(Y \cap E)/Z(E)$ does not contain singular points. In the $S_2 \times A_3$ case Y also centralizes no nonsingular point. Thus $Y \cap E$ is cyclic (of order 4) or isomorphic to Q_8 in these cases. Again a simple counting argument shows that Y has a regular orbit on W, in contrast to our hypothesis.

Type (3_+^3) : Here $E \cong 3_+^{1+2}$, $X \cong E : \mathrm{Sp}_2(3)$ and $p = 7$ or 13. Each noncentral element of E has 3 eigenspaces on W to different eigenvalues. Hence every nontrivial subgroup of $E/Z(E)$ fixes just 3 points in $P_1(W)$. There are $3 + 1 = 4$ such subgroups. Thus the number of points in $P_1(W)$ which are fixed by some nontrivial element of $E/Z(E)$ is at most $4 \cdot 3 = 12$, which is less than $|P_1(W)| = (p^3 - 1)/(p - 1)$ for $p = 7$ and for $p = 13$. So EZ has a regular vector in W.

If x is an element of order 3 in X outside E, then by Theorem 4.4 either $C_E(x) = Z(E)$ and $\chi(x) = 0$, or $C_E(y)/Z(E) = C_U(x)$ has order $|\chi(x)|^2 = 3$. In the former case x fixes just three points in $P_1(W)$, otherwise at most $1 + (p + 1)$ points, among them the eigenspaces of the noncentral elements of E contained in $C_E(x)$. If j is an involution in X, then j inverts the elements of U and $|\chi(j)| = 1$ by Theorem 4.4. By Theorem 4.5a, or by inspection of the character table of $\mathrm{Sp}_2(3)$, $\chi(j) = -1$. Since $\mathrm{Sp}_2(3)$ has four elements of order 3 and a unique involution, and since $4 \cdot (p + 2) + 1 \cdot (p + 2) < |P_1(W)|$ (for $p = 7$ and $p = 13$), every complement to EZ/Z in H/Z has a regular vector in W.

If T is a subgroup of H with $Z(E) \subset E \cap T \subset E$, then T/Z cannot contain an element of order 4. So consider $T = ZE_0\langle x, j \rangle$ where x, j are as above and E_0 is a subgroup of E of order 3^2 normalized by x. Either x centralizes E_0, and then E_0Z/Z and $Z\langle x \rangle/Z$ together fix at most $p + 2$ points in $P_1(W)$, or these fix $2 \cdot 3 \le p + 2$ points. There are four subgroups in T/Z of order 3, and there are 3 subgroups in T/Z of order 2. Thus at most $3 \cdot (p+2) + 3 \cdot (p+2)$ points in $P_1(W)$ are fixed by some nontrivial element of T/Z. It follows that T has a regular orbit on W.

Consequently $Y \supset E$. Assume that Y is subnormal in H but not irreducible on U. Then Y does not contain an element of order 4, and no element of order 3 outside E, whence $|YZ/EZ| = 2$. There are $|U| = 9$ involutions j in Y, hence at most $9(p+2)$ elements in $P_1(W)$ are fixed by any of them. Since the number of points in $P_1(W)$ fixed by nontrivial subgroups of EZ/Z is at most 12, and since $12 + 9 \cdot (p+2) < |P_1(W)|$ for $p = 13$, we have $p = 7$. In the $p = 7$ case the result is as stated.

In general $Y \supset E$ contains an involution j inverting the elements of U, or an element $x \in X$ of order 3 outside E. In each case $Y/Y \cap Z$ is nonabelian, and $Y^{\mathfrak{S}} \not\subseteq Z$ as the nonabelian point stabilizers are isomorphic to $\mathrm{SL}_2(3)$. $\qquad\square$

Remark. Let (H, W) be the unique, up to isomorphism, nonreal reduced pair in characteristic $p = 5$. Then H is a 5-complement in $\mathrm{GL}_2(5)$; the pair (H, W) is of type (Q_8) and will turn out to be exceptional (Chapter 10). We have seen above that if Y is a subgroup of H having no regular orbit on W, then Y properly contains the core $E \cong Q_8$ of H. Indeed either Y is one (of three) normal subgroups of H containing $O_2(H) \cong Q_8 \circ Z_4$, or $Y = P \cong Z_4 \mathrm{wr} Z_2$ is a Sylow 2-subgroup of H. Here P' is cyclic of order 4 and subnormal but not normal in H (see Lemma 10.4a below).

Theorem 8.5c (Riese–Schmid). *Let (G, V) be nonreal induced. Keeping the notation introduced above we have $R_0 \supseteq E$, except possibly when $p = 3$ or 5. In the exceptional cases at least $R_0 \supseteq Z(E)$ (and $N_0 \supseteq E$).*

Proof. Since there is no real vector in V for G, N has no regular orbit on V by Proposition 5.3b. From Lemma 8.5a we know that N_0 has no regular orbit on W, and $N_0 \supseteq E$. Our argumentation will be different depending on whether the core E of (H, W) is quasisimple or of extraspecial type.

Case 1: E is quasisimple

Assume the assertion is false. Then $R_0 \not\supseteq E$. Since R_0 is normal in H, by definition of reduced pairs this implies that R_0 is abelian and $R_0 \subseteq Z(H)$. So for each $i \in \Omega$, using that G is transitive on Ω and identifying $N_i = N/T_i$ with a subgroup of H_i, we have $N_i \supseteq E_i$ and $\bar{R}_i \subseteq Z(N_i)$ where $\bar{R}_i = R_i T_i / T_i$. Also, for $i \neq j$ in Ω, $T_{ij} = T_i T_j$ is a normal subgroup of N

with $T_{ij}/T_i \cong T_{ij}/T_j$ and

$$T_{ij}/(T_i \cap T_j) \cong T_{ij}/T_i \times T_{ij}/T_j.$$

Either $T_{ij}/(T_i \cap T_j)$ is abelian and hence central in $N/(T_i \cap T_j)$, or $E_i \subseteq T_{ij}/T_i$ and $E_j \subseteq T_{ij}/T_j$.

Define the subgroup Z_i of N by $Z_i/T_i = Z(N/T_i)$ ($i \in \Omega$), and define a binary relation on Ω by letting $i \sim j$ if $Z_i = Z_j$. This is a G-invariant equivalence relation on Ω. We assert that each equivalence class J has at least two elements. In order to prove this we use that E is quasisimple. Note that $\bigcap_i T_i = 1$ and that $\bigcap_i Z_i = Z(N)$.

There exists a quasisimple subnormal subgroup Y of N, a component of N. Either $YT_i/T_i \cong E_i$ or $Y \subseteq T_i$. In the former case $Y \cong E_i$. (Use that if the multiplier $M(E) \neq 1$, then $E \cong 2.A_6$ and $M(E) = Z_3$, but Y maps into $N_i' = E_i$ and $Z(Y) \subseteq \bigcap_i \tilde{Z}_i$ where $\tilde{Z}_i/T_i$ corresponds to $Z(E_i)$.) Let $J = J_Y$ be the set of those $j \in \Omega$ for which Y is mapped isomorphically onto E_j. Then $J \neq \varnothing$, and if $J = \{j\}$ then $Y \subseteq R_j = \bigcap_{i \neq j} T_i$ were abelian (as $\bar{R}_j \subseteq Z(H_j)$). Hence $|J| > 1$. Let $i \neq j$ be in J. Assume $i \nsim j$. Then $Y = Y'$ maps isomorphically onto a subnormal subgroup of $T_{ij}/(T_i \cap T_j)$ which in turn maps onto the normal subgroup E_i of T_{ij}/T_i and on the normal subgroup E_j of T_{ij}/T_j. This is impossible since $T_{ij}/(T_i \cap T_j)$ has only 2 components [Aschbacher, 1986, (31.7)]. Alternately use that in a direct decomposition of perfect groups the direct factors are uniquely determined.

Thus $J = J_Y$ is an equivalence class with respect to $\sim$, hence a G-block on Ω, with $|J| \geq 2$. For if $g \in G$ is such that $Y^g = Y$, then g stabilizes J, and if $Y^g \neq Y$, then $[Y, Y^g] = 1$ and $J^g = J_{Y^g}$ intersects J trivially. Pick $i \neq j$ in J, and write $Z_J = Z_i = Z_j$. We assert that there are nonzero vectors $w_i \in W_i$ and $w_j \in W_j$ such that

$$C_N(w_i) \cap C_N(w_j) \subseteq Z_J.$$

By Theorem 7.1 we may choose w_i such that $C_N(w_i)/T_i$ is cyclic, except when (H, W) is of type $(\mathrm{Sp}_4(3))$ with $p = 7, 13$. In the cyclic case the group $C = C_N(w_i)Z_J/T_j$ is abelian, and we pick w_j in a regular C-orbit on W_j. In the exceptional cases we treat the (worst) situation that $N_i = N/T_i$ is isomorphic to $\mathrm{Sp}_4(3) \times Z_{(p-1)/2}$ so that $Z_J/T_i \cong \mathbb{F}_p^*$. By Theorem 7.1 we may pick $w_i \in W_i$ such that $C_N(w_i)/T_i$ is isomorphic to $\mathrm{SL}_2(3) \times Z_3$

$(p = 7)$ resp. to $Z_3 \mathrm{wr}\, S_2$ $(p = 13)$. By Corollary 7.2c, $C_N(w_i)Z_J/Z_J$ has a regular orbit on $P_1(W_i)$, and this depends only on the isomorphism type of the point stabilizer. It follows that $C_N(w_i)Z_J/T_j$ has a regular orbit on W_j likewise. Hence the assertion.

Thus there is $v_J = \sum_{j \in J} w_j$ consisting of nonzero vectors $w_j \in W_j$ such that $C_N(v_J) = \bigcap_{j \in J} C_N(w_j) \subseteq Z_J$. Let v be the sum of the distinct G-conjugates of v_J. Then

$$C_N(v) \subseteq \bigcap_{g \in G} (Z_J)^g = Z(N).$$

So $C_N(v)$ acts on each W_i as a group of scalar multiplications fixing a nonzero vector. This implies that $C_N(v) = 1$, a contradiction.

Case 2: E is extraspecial

Here we argue as follows. Let the base group $B = H_1 \times \cdots \times H_n$ of the wreath product $H \mathrm{wr}\, S$ act on $V = W_1 \oplus \cdots \oplus W_n$ diagonally (as usual). Suppose N is any subgroup of B which has no regular orbit on V and for which the image N_i of N in H_i is a *subnormal* subgroup of H_i, for all $i \in \Omega$. Let $T_i = C_N(W_i)$, $R_i = \bigcap_{j \neq i} T_j$ and $\bar{R}_i = R_i T_i / T_i$ (as usual). We assert that $N_i \supseteq E_i$ and $\bar{R}_i \supseteq Z(E_i)$ for *some* $i \in \Omega$, and even $\bar{R}_i \supseteq E_i$ for *some* i if $p \neq 3$ and $p \neq 5$. We argue by induction on $|N|$.

So we ignore, at first, the group G and its transitive action on Ω. Only in a concluding step this will be used (yielding then the assertion for *all* $i \in \Omega$, hence for the common images N_0 of the N_i and R_0 of the $R_i \cong \bar{R}_i$ in $H = H_0$).

Assume that this assertion is false (so that $\bar{R}_i \not\supseteq E_i$ for all $i \in \Omega$, and even $\bar{R}_i \not\supseteq Z(E_i)$ when $p = 3$ or 5). If for each i, N_i has a regular orbit on W_i, so does N on V. (Argue as for Lemma 8.5a.) So we may assume that N_1, say, does not have a regular vector in W_1. Then $N_1 \supseteq E_1$ by Lemma 8.5b. We have $R_1 \cap T_1 = 1$, and T_1 contains $R_2, \cdots, R_n$. Similarly T_1 acts faithfully on $\widetilde{V} = W_2 \oplus \cdots \oplus W_n$. The group $\widetilde{N} = R_1 \cdot T_1$ fulfills all requirements. Since N_1 contains E_1 but $\widetilde{N}/T_1 \cong R_1$ does not, we have $|\widetilde{N}| < |N|$. Since $C_{\widetilde{N}}(\bigoplus_{j \neq i} W_j) \subseteq R_i$ for all i, by induction there must be a regular vector $v \in V$ for $\widetilde{N}$. We may write uniquely $v = w_1 + \widetilde{v}$ with $w_1 \in W_1$ and $\widetilde{v} = \sum_{i=2}^{n} w_i$ for $w_i \in W_i$. Here without loss we may assume that all $w_i \neq 0$. Let

$$C_1 = C_N(\widetilde{v}) = \bigcap_{i=2}^{n} C_N(w_i).$$

Then C_1 contains R_1 as a normal subgroup. Clearly $C_1 \cap T_1 = C_N(v) \cap T_1 = 1$. So $\bar{C}_1 = C_1 T_1 / T_1$ is a subgroup H_1 which is isomorphic to C_1 and contains $\bar{R}_1$. If there exists w_1 in W_1 which is regular for C_1, then $w_1 + \tilde{v}$ is a regular vector in V for N, a contradiction. Consequently C_1 has no regular orbit on W_1.

Consider the action of C_1 on W_i for $i = 2, \cdots, n$. Let $\bar{C}_i = C_1 T_i / T_i$. Then $\bar{C}_i \subseteq C_{H_i}(w_i)$ for $i \geq 2$. These point stabilizers are well understood by Theorem 6.1. It follows that $C_1^{\mathfrak{S}} \subseteq R_1$.

Let first $p \neq 3, 5$. Suppose that N_1 is irreducible on $U_1 = E_1 / Z(E_1)$. Then $\bar{R}_1 \subseteq Z(N_1) \subseteq Z(H_1)$. But then $\bar{C}_1^{\mathfrak{S}} \subseteq Z(H_1)$, which contradicts Lemma 8.5b. Thus N_1 is not irreducible on U_1 (but still $N_1 \supseteq E_1$). From Lemma 8.5b it follows that then $(H, W) \cong (H_i, W_i)$ must be of type (3_+^3) and $p = 7$. Moreover, then $N_1 = E_1 \cdot \langle j \rangle$ or $E_1 \langle j \rangle Z(H_1)$ for some involution j of H_1 inverting the elements of U_1. Since C_1 has no regular orbit on W_1, $\bar{C}_1^{\mathfrak{S}} \not\subseteq Z(H_1)$ by Lemma 8.5b. Since $\bar{C}_1^{\mathfrak{S}}$ is contained in all (normal) subgroups of $\bar{C}_1$ with index 3 (as Z_6 is a point stabilizer) and E_1 is a 3-group, this implies that $\bar{C}_1 = E_1 \langle j \rangle$ or $E_1 \langle j \rangle Z(H_1)$. Since j inverts the elements of U_1 and $\bar{R}_1 \not\supseteq E_1$, C_1 / R_1 has an S_3 quotient group. However, by the structure of the point stabilizers $C_1 / C_1^{\mathfrak{S}}$ does not have an S_3 quotient group. This is the desired contradiction.

Let finally $p = 3$ ($E \cong 2_-^{1+4}$) or $p = 5$ ($E \cong Q_8$). Then all point stabilizers of nonzero vectors are cyclic 2-groups by Theorem 6.1. By assumption $\bar{R}_1 \not\supseteq Z(E_1)$. Since $\bar{R}_1$ is normal in $N_1 \supseteq E_1$, this forces that $R_1 = 1 = C_1^{\mathfrak{S}}$. Hence C_1 is an abelian 2-group. But then C_1 has a regular orbit on W_1, a final contradiction. This completes the proof. $\square$

In the nonreal induced (imprimitive) situation $k(GV) \leq k(NV) \cdot k(S)$ by (1.7b). Here $S = G/N$ is a transitive permutation group of degree $n \geq 2$. The class number $k(S)$ will be investigated in the next chapter. When $p \neq 3, 5$ by Theorem 8.5c the group N contains a normal subgroup of G which is the direct product of n copies of the core E of (H, W) permuted transitively by S. So G is indeed not far from being the full wreath product $H \operatorname{wr} S$. Observe that if $G = H \operatorname{wr} S$, then $GV = (HW) \operatorname{wr} S$.

Let us compute $k(X \operatorname{wr} S)$ for an arbitrary finite group X and an arbitrary transitive permutation group S of degree $n \geq 2$. Let $k = k(X)$, and let $\Sigma_{n,s}$ be the set of partitions of the n-set $\Omega = \{1, \cdots, n\}$ into s parts, so that $\sigma_{n,s} = |\Sigma_{n,s}|$ is the *Stirling number* of the second kind to n

and s. Counting the ordered n-sets of a k-set with repetition one gets the combinatorial identity

$$\sum_{s=1}^{n} k(k-1)\cdots(k-s+1)\sigma_{n,s} = k^n.$$

This identity is valid even when $k < n$ (where the summands for $s > k$ vanish). For a partition σ the stabilizer S_σ is the set of all elements of S belonging to the Young subgroup of S_n determined by σ.

Proposition 8.5d. *For $k = k(X)$ we have*

$$k(X \operatorname{wr} S) = \sum_{s=1}^{n} k(k-1)\cdots(k-s+1)\sum_{\sigma \in \Sigma_{n,s}} k(S_\sigma)/|S : S_\sigma|.$$

If $S = Z_n$ is generated by an n-cycle, then $k(X \operatorname{wr} S) = (k^n - k)/n + kn$ if n is a prime, and $k(X \operatorname{wr} S) \leq k^n - k + kn$ in general.

Proof. Every irreducible character χ of the base group $X^{(n)}$ of $G = X \operatorname{wr} S$ is of the form $\chi = \chi_1 \otimes \cdots \otimes \chi_n$ for unique irreducible characters $\chi_i \in \operatorname{Irr}(X)$. This determines a partition $\sigma(\chi)$ of Ω consisting of those subsets on which the components χ_i agree and which are maximal with this property. The inertia group $I_S(\chi) = S_{\sigma(\chi)}$. One knows that χ can be extended to the inertia group $X \operatorname{wr} S_{\sigma(\chi)}$ of χ in $X \operatorname{wr} S$ [James–Kerber, 1981, p. 154]. If $\sigma \in \Sigma_{n,s}$ is a partition of Ω into s parts, there are $k(k-1)\cdots(k-s+1)$ irreducible characters χ of $X^{(n)}$ with $\sigma(\chi) = \sigma$ if $k \geq s$, and none otherwise. Hence $k(G)$ is as stated.

Let $S = Z_n = \langle x \rangle$. If n is a prime, then $S_\sigma = 1$ unless $\sigma = \Omega$. Hence $k(G) = k|S| + (k^n - k)/|S|$ by virtue of the above combinatorial identity. In general the estimate follows, in a similar manner, once we have shown that

$$\sum_{\sigma \in \Sigma_{n,s}} k(S_s)/|S : S_\sigma| \leq \sigma_{n,s}$$

for each $s \geq 2$. Suppose $\sigma \in \Sigma_{n,s}$ is such that $S_\sigma \neq 1$. Let Δ be a set in σ of smallest cardinality. Then $|\Delta| = t > 1$ is a divisor of n and $S_\sigma = \langle x^{\frac{n}{t}} \rangle$ (and each set belonging to σ has cardinality divisible by t). Pick any $\alpha \in \Omega \setminus \Delta$, say $\alpha \in \Gamma \in \sigma$, and replace Δ by $\Delta \cup \{\alpha\}$, Γ by $\Gamma \setminus \{\alpha\}$ and leave the other sets in σ unchanged. We obtain $\sigma_\alpha \in \Sigma_{n,s}$ with $S_{\sigma_\alpha} = 1$. Noting that $t \leq \frac{n}{s}$ and that at most $\frac{1}{n-\frac{n}{s}}\sigma_{n,s}$ partitions in $\Sigma_{n,s}$ have nontrivial stabilizer in S, the result follows since obviously $\frac{n}{s^2}\frac{\sigma_{n,s}}{n-\frac{n}{s}} + \frac{\sigma_{n,s}}{n} \leq \sigma_{n,s}$. $\square$

Chapter 9

Class Numbers of Permutation Groups

Let $\Omega = \{1, \cdots, n\}$ for some integer $n \geq 2$. Permutation groups of degree n are regarded as acting on Ω. For each such group S one has the upper bound $k(S) \leq 2^{n-1}$, due to [Liebeck and Pyber, 1997]. We shall give a sketch of proof for this theorem, and present a slight improvement.

9.1. The Partition Function

Since two elements of the symmetric group S_n are conjugate if and only if they have the same cycle type, $k(S_n) = p(n)$ where $p(n)$ is the partition function on positive integers. There is no simple formula for $p(n)$. Hardy and Ramanujan showed that asymptotically $p(n) \sim \frac{e^{\pi\sqrt{2n/3}}}{4n\sqrt{3}}$, which readily yields the estimate

$$(9.1a) \qquad p(n) < \frac{\pi}{\sqrt{6(n-1)}} \cdot e^{\pi\sqrt{2n/3}}$$

for $n \geq 2$ (cf. [Erdös, 1942] for an elementary approach). We get the upper bound $p(n) < 0.2735 \cdot 2^{3.701\sqrt{n}}$, which is fairly good for $n \geq 23$. With elementary means one can establish the following.

Proposition 9.1b. *For $n \geq 5$ we have $k(A_n) < k(S_n) \leq 2^{n-2}$.*

Proof. The assertion is verified for $5 \leq n < 10$ by inspection [Atlas]. Let $n \geq 10$. Let k_d be the number of conjugacy classes of elements in S_n with shortest cycle length d, so conjugate to $(12\cdots d)$. Then either $d = n$ ($k_n = 1$) or $d \leq \lfloor \frac{n}{2} \rfloor$. Consider the elementwise stabilizer $S_{(d)} \cong S_{n-d}$ of $\{1, 2, \cdots, d\}$ for $d \neq n$ in S_n. Evidently $k_d \leq k(S_{(d)})$. By induction $k(S_{(d)}) < 2^{n-d-2}$ for $n \neq d$ since then $n - d \geq 5$. Thus

$$k(S_n) < 1 + 2^{n-3} + 2^{n-4} + \cdots + 2^{\lceil \frac{n}{2} \rceil - 2},$$

which is certainly less than 2^{n-2}.

Let $x \in A_n$. If $x^{S_n} = x^{A_n}$, then associate to x the partition σ defined by $x \in S_n$. If $x^{S_n} \neq x^{A_n}$, then σ is of the form $(a_1, \cdots, a_s)$ with $a_1 > a_2 >$

$\cdots > a_s$ and all a_j being odd. In this case x^{S_n} consists of two A_n-classes, and we associate to these the different partitions σ and $\sigma' = (a_1-1, a_2, \cdots, a_s, 1)$. Hence $k(A_n) \leq k(S_n)$. Since the partition $(n-3, 3)$ is not associated to a conjugacy class of A_n in this manner, we even have $k(A_n) < k(S_n)$. □

One knows that roughly $k(A_n)$ is half of $k(S_n)$ (Erdös).

9.2. Preparatory Results

We need some preparations. For a simple group L we denote by $P(L)$ the smallest degree of a faithful permutation representation of L, that is, the index of a proper subgroup of L of largest order. Thus $P(L) \geq R_0(L) + 1$ and $P(L) \geq R_p(L) + 1$ for each prime p.

Lemma 9.2a. *Suppose L is a finite simple group of Lie type of (untwisted) rank $\ell = \mathrm{rk}(L)$ over $\mathbb{F}_r$, where $r = q^t$ for some prime q.*

(i) *If $\ell = 1$, then $L = L_2(r) = \mathrm{PSL}_2(r)$ and $P(L) \geq r + 1$ unless $r \leq 11$.*

(ii) *If $L = {}^2B_2(r) = \mathrm{Sz}(r)$ is a Suzuki group, then $P(L) = r^2 + 1$, and if $L = {}^2G_2(r)$ is a Ree group, then $P(L) = r^3 + 1$.*

(iii) *In general $P(L) \geq (r^{\ell+1} - 1)/(r - 1)$ except when $L = L_2(7)$, $L_2(11)$, $G_2(4)$, ${}^3D_4(2)$, ${}^2F_4(2)'$, ${}^2B_2(r)$ or ${}^2G_2(r)$.*

Statement (i) is a classical result of Galois (and Dickson) [Huppert, 1967, II.8.28]. The result on the Suzuki groups in (ii), where $r = 2^t$ with $t \geq 3$ odd, is due to [Suzuki, 1962], and the result on the Ree groups, where $r = 3^t$ with $t \geq 3$ odd, can be found in [Ward, 1966]. (The Ree groups ${}^2F_4(2^t)$, $t \geq 3$ odd, are not exceptional here.) The general lower bounds for $P(L)$ can be found in Table 5.2A for L classical, and in Table 5.3A for L exceptional in [Kleidman–Liebeck, 1990]. In the cases $L = G_2(4)$, ${}^3D_4(2)$ and for the Tits group ${}^2F_4(2)'$ one can take $P(L)$ from the [Atlas], namely $P(L) = 416$, 819 and 1.600 respectively.

Lemma 9.2b. *Let L, $\ell = \mathrm{rk}(L)$ and $r = q^t$ be as before. Then $|\mathrm{Out}(L)| \leq 2(\ell + 1)t$ exept when $L \cong \mathrm{P}\Omega_8^{\pm}(r)$, in which case $|\mathrm{Out}(L)| \leq 24t$.*

This is immediate from Steinberg's description of the automorphisms of the groups of Lie type in Chapter 10 of [Steinberg, 1967]. For $L = L_2(r)$ we have $|\mathrm{Out}(L)| = 2t$ if r is odd and $|\mathrm{Out}(L)| = t$ otherwise.

Lemma 9.2c. *Let L, $\ell = \mathrm{rk}(L)$ and $r = q^t$ be as before.*

(i) $k(L_2(r)) \le r + 1$, $k(\mathrm{Sz}(r)) = r + 3$ *and* $k(^2G_2(r)) = r + 8$.

(ii) *In general* $k(L) \le (6r)^\ell$.

[Schur, 1911] has already computed the character table of $L_2(r)$, [Suzuki, 1962] for the Suzuki groups (see also [Huppert–Blackburn, 1982, XI.5.10]), and [Ward, 1966] for the Ree groups. Hence the statements in (i). Statement (ii) is due to [Liebeck and Pyber, 1997]. It is based on general results on conjugacy classes in algebraic groups [Springer–Steinberg, 1970].

For a proof of the following fundamental result we refer to [Dixon–Mortimer, 1996].

Theorem 9.2d (O'Nan–Scott). *Suppose S is a primitive permutation group of degree n. Then the socle of S has the form $L^{(m)}$ for some simple group L, and one of the following holds:*

• Affine type: $|L| = q$ *for some prime q, S is a subgroup of the affine group $\mathrm{AGL}_m(q)$ containing the translations, and a point stabilizer in S is an irreducible subgroup of $\mathrm{GL}_m(q)$.*

• Diagonal type: L *is nonabelian (simple), $n = |L|^{m-1}$ with $m \ge 2$ and S is a subgroup of a wreath product with the diagonal action. Also, $S/L^{(m)}$ is isomorphic to a subgroup of $\mathrm{Out}(L) \times S_m$, and a point stabilizer has a primitive action of degree m.*

• Product type: L *is nonabelian, $n = b^d$ for integers b, d greater than 1 and S is a subgroup of $T \mathrm{wr}\, S_d$, in product action, for some primitive subgroup T of S_b.*

• Almost simple type: S *is almost simple $(m = 1$ and $S \subseteq \mathrm{Aut}(L))$.*

9.3. The Liebeck–Pyber Theorem

Theorem 9.3 (Liebeck–Pyber). *If S is a permutation group of degree n, then $k(S) \le 2^{n-1}$.*

Proof. We argue by induction on n. Suppose S is not transitive, and let Ω_1 be an S-orbit and $\Omega_2 = \Omega \setminus \Omega_1$. Then $N = C_S(\Omega_1)$ is faithfully represented on Ω_2 as a permutation group of degree $n_2 = |\Omega_2| < n$, hence $k(N) \le 2^{n_2-1}$ by induction. S/N is isomorphic to a subgroup of S_{n_1} where $n_1 = |\Omega_1|$, whence $k(S/N) \le 2^{n_1-1}$. By (1.7b)

$$k(S) \le k(N) \cdot k(S/N) \le 2^{n_1-1+n_2-1} = 2^{n-2}.$$

Let $T = S_1$ be a point stabilizer, and assume T is not maximal in S. Choose a proper intermediate group $T \subset H \subset S$, and let $C = \mathrm{Core}_G(H)$. Let $|S : H| = b$, $|H : T| = d$, and let $b_0 = |S : CT|$, $d_0 = |CT : H|$. By induction $k(S/C) \leq 2^{b-1}$ as S/C is isomorphic to a subgroup of S_b. Also, C has b_0 orbits on $\Omega = \{1, \cdots, n\}$ of size d_0, and the preceding argument yields $k(C) \leq (2^{d_0-1})^{b_0}$. By (1.7b)

$$k(S) \leq 2^{(d_0-1)b_0} \cdot 2^{b-1} = 2^{n+b-b_0-1} \leq 2^{n-1}.$$

because $b_0 d_0 = n$ and $b \leq b_0$.

Hence we are reduced to the case that S is primitive. Let first $S \subseteq \mathrm{AGL}_m(p)$ be of *affine type* ($n = p^m$), and let $M = C_S(M)$ be a minimal (regular) normal subgroup of S. By Theorem 2.3c the principal block is the unique p-block of S. Therefore by the Brauer–Feit Theorem 2.4,

$$k(S) \leq 1 + \frac{1}{4}|P|^2$$

where P is a Sylow p-subgroup of S. Since P/M is isomorphic to a subgroup of a Sylow p-subgroup of $\mathrm{GL}_m(p)$, which has order $p^{m(m-1)/2}$, we have $k(S) \leq 1 + \frac{1}{4}p^{m(m+1)}$. We have to examine the cases where $1 + \frac{1}{4}p^{m(m+1)} > 2^{p^m-1}$, that is, where $p^{m(m+1)} \geq 2^{p^m+1}$. This happens only when $p = 2$ and $m \leq 4$. Here the cases $m = 1, 2, 3$ are easily treated. Note that every subgroup of $\mathrm{GL}_3(2)$ has at most 7 conjugacy classes. In the $m = 4$ case consider the stabilizer $C = C_S(w)$ for some $w \in M^\sharp$, so that C/M is isomorphic to a subgroup of a parabolic subgroup $R : \mathrm{GL}_3(2)$ of $\mathrm{GL}_4(2)$, with unipotent radical $R \cong Z_2^{(3)}$, and $|S : C| \leq 15$. Using (1.7b) we conclude that $k(S/M) \leq 15 \cdot 7 \cdot 8 < 2^{10}$ and $k(S) \leq 2^{14}$, as desired.

Let next S be of *diagonal type*. So the socle M of S is the direct product of $m > 1$ copies of a nonabelian simple group L, and $n = |L|^{m-1}$. Further S/M is isomoprphic to a subgroup of $\mathrm{Out}(L) \times S_m$. Clearly $k(L) \leq |L|/2$ and $|L| \geq 60$. From the classification of the finite simple groups and Lemma 9.2b one has the very crude estimate $|\mathrm{Out}(L)| \leq |L|$. (If L is an alternating or sporadic simple group, then $|\mathrm{Out}(L)| \leq 2$, except when $L = A_6$ where $|\mathrm{Out}(L)| = 4$.) Consequently

$$k(S) \leq k(L)^m |\mathrm{Out}(L)| \cdot k(S_m) \leq |L|^{m+1} \cdot 2^{-m} \cdot 2^{m-1} \leq |L|^{m+1},$$

which gives the result.

Let S be of *product type*. So $n = b^d$ for integers b, d greater than 1, and S is a subgroup of $T \operatorname{wr} S_d$ for some primitive subgroup T of S_b. By induction we know that every subgroup of T has at most 2^{b-1} conjugacy classes. Hence every subgroup of the base group $N = T^{(d)}$ has at most $2^{(b-1)d}$ conjugacy classes. Since $S/S \cap N$ is isomorphic to a subgroup of S_b, we have $k(S/S \cap N) \leq 2^{d-1}$. By (1.7b) $k(S) \leq 2^{(b-1)d} \cdot 2^{d-1} \leq 2^{bd-1}$. Of course $bd \leq b^d$.

So let S be *almost simple*, say $L \subseteq S \subseteq \operatorname{Aut}(L)$ for some simple nonabelian group L. If $L = A_n$ for $n \neq 6$, then $\operatorname{Aut}(L) = S_n$ and the result follows (see above). If $L = A_6$ or if L is a sporadic simple group, the desired conclusion follows by inspection of the Atlas. Use that n is larger than the degree of each faithful character.

So let L be a group of Lie type over $\mathbb{F}_r$, and $L \not\cong A_n$. Let $r = q^t$ for some prime q, and let $\ell = \operatorname{rk}(L)$. Clearly $n \geq P(L)$. We use Lemmas 9.2a, 9.2b and 9.2c, and we use (1.7b). The cases where $L = L_2(r)$ or L is a Suzuki or Ree group are easily treated using the information given above. The groups $G_2(4)$, $^3D_4(2)$ and $^2F_4(2)'$ are treated with the help of the Atlas. In the remaining cases we have

$$k(S) \leq (6r)^{\ell} |\operatorname{Out}(L)| \leq 2^{(r^{\ell+1}-1)/(r-1)-1} \leq 2^{n-1},$$

as desired. $\square$

9.4. Improvements

Recently [Maróti, 2005] has shown that, for $n \geq 3$, the Liebeck–Pyber bound can be improved to $k(S) \leq \sqrt{3}^{\,n-1}$. When S is solvable or $|S|$ not divisible by 3, 5 or 7, this stronger bound had been established previously by [Kovács–Robinson, 1993] and [Riese–Schmid, 2003], and this suffices for our purposes.

Lemma 9.4a. *Let p be any of the primes $3, 5$ or 7. If S is a primitive permutation group of order prime to p and of degree $n \geq 7$, then $k(S) \leq 2^{\frac{n}{2}}$.*

Proof. Of course we invoke the O'Nan–Scott Theorem 9.2d. Suppose first that S is of *affine type*. Then S has a minimal normal q-subgroup M for some prime $q \neq p$, with $C_S(M) = M$ and $|M| = n = q^m$ for some m. Also $S = M : H$ where H is an irreducible p'-subgroup of $\operatorname{GL}_m(q)$. Arguing as before we have to examine the cases where $1 + \frac{1}{4}q^{m(m+1)} > \sqrt{2}^{q^m}$, that is, where

$$q^{m(m+1)} \geq 2^{(q^m+4)/2}.$$

This implies that $q = 7$, $m = 1$ or $q = 3$, $m = 2, 3$ or $q = 2$, $m = 3, 4, 5$ (noting that $n = q^m \geq 7$). If $(q, m) = (7, 1)$, then $p = 3$ or 5 and $k(S) \leq n = 7$. If $(q, m) = (3, 2)$, then $p = 5$ or 7 and H may be any irreducible subgroup of $\mathrm{GL}_2(3)$. If $3 \nmid |H|$, then $k(S) \leq 1 + \frac{1}{4}3^4$ by (2.4), hence $k(S) \leq 21 \leq 2^{9/2}$. Otherwise $H = \mathrm{GL}_2(3)$ or $\mathrm{SL}_2(3)$, in which cases we get $k(S) = 11, 10$, respectively.

Let $(q, m) = (3, 3)$. Then again H may be any irreducible subgroup of $\mathrm{GL}_3(3)$. If $H = \mathrm{GL}_3(3)$, then $k(S) = k(\mathrm{GL}_3(3)) + k(\mathrm{AGL}_2(3)) = 24 + 11 = 35$ by (1.10b), because the stabilizer in H of any nontrivial linear character λ of M is isomorphic to $\mathrm{AGL}_2(3)$ and λ extends to $I_S(\lambda)$ [Atlas, p. 13]. For $H = \mathrm{SL}_3(3)$ we get $k(S) = 12 + 11 = 23$. In the remaining cases H is solvable of order dividing 72, and $|S| \leq 2^{3^3/2}$.

Let $(q, m) = (2, 3)$. If $H = \mathrm{GL}_3(2)$, then $p = 5$ and $k(S) = k(\mathrm{GL}_3(2)) + k(S_4) = 6 + 5 = 11$ by (1.10b) since the stabilizer $I_H(\lambda) \cong S_4$ for any nontrivial linear character λ of M. Otherwise H is cyclic of order 7 or is a Frobenius group of order 21, and we get $k(S) \leq 8$.

Let $(q, m) = (2, 4)$. Then we must have $p = 7$ and, by inspection of the possible irreducible $7'$-subgroups of $\mathrm{GL}_4(2)$, $k(S) \leq k(4^2 : \mathrm{GL}_2(4)) \leq 2 \cdot (15 + 4) \leq 2^{q^m/2}$. For $(q, m) = (2, 5)$, H must be cyclic of order 31 or a Frobenius group of order $31 \cdot 5$ (when $p \neq 5$), and the result follows.

Suppose next that S is of *product type*. Here $n = b^d$ for certain integers b, d greater than 1 and S is a subgroup of $T \mathrm{wr}\, S_\mathrm{d}$, T being a primitive subgroup of S_b. Let N denote the base group of $T \mathrm{wr}\, S_\mathrm{d}$. By Theorem 9.3 each subgroup of T has at most 2^{b-1} conjugacy classes. Thus $k(S \cap N) \leq 2^{(b-1)d}$. Since $S/S \cap N$ is isomorphic to a subgroup of S_d we have $k(S/S \cap N) \leq 2^{d-1}$, again by Theorem 9.3. The result follows if $bd - 1 \leq b^d/2$. Since $n = b^d \geq 7$, this is true unless $b = 2, d = 3$ or $b = 3, d = 2$. The primitive permutation p'-groups of degree 8 or 9 are solvable, hence are of affine type and already treated.

Suppose that S is of *diagonal type*. Here the socle M of S is the direct power of $m > 1$ copies of some simple p'-group L, $n = |L|^{m-1}$ and S/M is isomorphic to a subgroup of $\mathrm{Out}(L) \times S_m$. Hence

$$k(S) \leq k(L)^m \cdot |\mathrm{Out}(L)| \cdot 2^{m-1} \leq |L|^{m+1} \leq \sqrt{2}^{|L|^{m-1}},$$

because $k(L) \leq |L|/2$, $|\mathrm{Out}(L)| \leq |L|$ and $|L| \geq 60$.

So we may assume that S is *almost simple*, say $L \subseteq S \subseteq \mathrm{Aut}(L)$ for some simple p'-group L. The alternating groups are ruled out by assumption. Also, the case $p = 3$ is easy since then $L = \mathrm{Sz}(q)$ is a Suzuki group, and Lemmas 9.2a, 9.2c apply. The only sporadic groups meeting the requirements are M_{11}, M_{12} and J_3 for $p = 7$, for which the result holds [Atlas]. For groups of Lie type we argue as in the proof for Theorem 9.3. The exceptional groups in Lemma 9.2a, including the Ree groups ${}^2G_2(r)$, and the groups of rank 1 $(L = L_2(r))$ are treated as before. Letting L be defined over of $\mathbb{F}_r$ and $\ell = \mathrm{rk}(L)$, as usual, we therefore have $\ell \geq 2$ and $n \geq P(L) \geq (r^{\ell+1} - 1)/(r - 1)$, and it suffices to verify the inequality

$$(6r)^\ell |\mathrm{Out}(L)| \leq 2^{\frac{1}{2}(r^{\ell+1}-1)/(r-1)}.$$

Using the bound for $|\mathrm{Out}(L)|$ given in Lemma 9.2b this inequality holds unless $\ell = 2, r \leq 4$ or $\ell = 3, 4$ and $r = 2$. The corresponding p'-groups L are all in the Atlas, namely the $5'$-groups $A_2(3)$, ${}^2A_2(3) \cong G_2(2)'$, $G_2(3)$ and ${}^3D_4(2)$, and the $7'$-groups ${}^2A_3(2) \cong \Omega_6^-(2)$, $C_2(4) = \mathrm{PSp}_4(4)$, ${}^2A_2(4)$, ${}^2A_4(2)$ and ${}^2F_4(2)'$. The desired estimate holds in each case. (Except for $L = A_2(3)$ one can even replace $\sqrt{2}$ by $\sqrt[5]{2}$.) □

Proposition 9.4b. *Let again p be any of the primes $3, 5, 7$, and let $S \subseteq S_n$ be a p'-group. Suppose S has r orbits of size 2 and s orbits not of size 2. Then $k(S) \leq 2^r \cdot \sqrt{3}^{\,n-2r-s}$. In particular, $k(S) \leq \sqrt{3}^{\,n-1}$ for $n \geq 3$.*

Proof. Assume the proposition is false, and let S be a counterexample with n minimal. Then S is transitive. For if S has an orbit of size 2, then $k(S) \leq 2 \cdot 2^{r-1} \cdot \sqrt{3}^{\,n-2-2(r-1)-s}$ by induction, using (1.7b). If S has no orbit of size 2 but is not transitive, argue as in the proof for Theorem 9.3. By inspection $n \geq 5$.

Suppose we have shown that S is even primitive. Then Lemma 9.4a yields that $n \leq 6$. For $n \leq 4$ the result follows by inspection. For $n = 5$ we have $k(S) \leq 7$. There are just four primitive $(7'-)$ groups of degree $n = 6$, which are isomorphic (as groups) to S_5, A_5, S_6 or A_6, and $k(S) \leq \sqrt{3}^{\,n-1}$.

Consequently S is *not* primitive. Let T be a point stabilizer in S. We first claim that if H is a proper intermediate group, $T \subset H \subset S$, then $|S : H| = 2$ or $|H : T| = 2$. Let $|S : H| = b$ and $|H : T| = d$, and assume $b > 2$ and $d > 2$. Let $C = \mathrm{Core}_S(H)$ and $|C : C \cap H| = |CT : H| = d_0$,

$|S : CT| = b_0$. Then C has b_0 orbits on $\Omega = \{1, \cdots, n\}$, each of size d_0. Of course S/C is a transitive permutation group of degree b, $3 \le b < n$, and so $k(S/C) \le \sqrt{3}^{b-1}$. If $d_0 > 2$, then by assumption and (1.7b)

$$k(S) \le \sqrt{3}^{(d_0-1)b_0} \cdot \sqrt{3}^{b-1} \le \sqrt{3}^{n+b-b_0-1} \le \sqrt{3}^{n-1},$$

because $n = b_0 d_0$ and $b_0 \ge b$. Thus $d_0 = 2$ and $b_0 = \frac{n}{2}$. It follows that C is an elementary abelian 2-group, with $k(C) = |C| = 2^{\frac{n}{2}}$, and $CT \subset H$ $(d > d_0 = 2)$. Hence $b \le \frac{n}{4}$. Since $2^{\frac{1}{2}} \cdot \sqrt{3}^{\frac{1}{4}} < \sqrt{3}$, we get the desired result.

So letting H be a proper intermediate group, the following three possibilities have to be considered:

(i) T is maximal in H and $|S : H| = 2$;

(ii) H is maximal in S and $|H : T| = 2$;

(iii) There is a further intermediate subgroup X, such that $T \subset H \subset X \subset S$ say, and then H is maximal in X and X is maximal in S, and $|H : T| = 2 = |S : X|$.

Case (i): Let $d = |H : T|$, and let $N = \mathrm{Core}_H(T)$. Suppose first that $d \ge 7$. Observe that H/N is a primitive permutation group of degree d. Hence $k(H/N) \le \sqrt{2}^d$ by Lemma 9.4a. For any $x \in S \setminus H$ we have $N \cap N^x = 1$ and $k(N) = k(N^x) = k(N \cdot N^x/N) \le \sqrt{3}^{m-1}$ by induction $(d \ge 3)$. Thus by $k(H) \le 2 \cdot k(H/N) \cdot k(N) \le \sqrt{2}^{d+2} \cdot \sqrt{3}^{d-1}$. Since we are assuming that $k(S) > \sqrt{3}^{2d-1}$, we have $\sqrt{2}^{d+2} > \sqrt{3}^d$. But this is false for $d \ge 4$ (even). Thus $d \le 6$.

Note that $p = 7$ when S is not solvable. Let $d = 6$. Then by inspection $k(N) \le k(H/N) \le 11$ and $k(S) \le 2 \cdot k(H/N) \cdot k(N) \le 242 < \sqrt{3}^{11}$. If $d = 5$ then $k(H/N) \le 7$, $k(N) \le 7$ and $k(S) \le 98 < \sqrt{3}^9$. For $d \le 4$, H and hence S are solvable. Consider $d = 3$. Then $|S| \le 2|H/N|^2 \le 72$. Since $k(S_3 \mathrm{wr}\, S_2) = 9$ and $k(S) > \sqrt{3}^{2d-1} \ge 16$ by assumption, we have $|S| \le 36$, even $|S| = 36$ (for otherwise S were abelian). Then $|H| = 18$ and $|N| = 3$, and $k(S) \le 3 \cdot 3 \cdot 2 = 12$, a contradiction.

Let $d = 4$. Then by inspection $k(H/N) \le 5$, and $k(N/N \cap N^x) \le 5$ for $x \in S \setminus H$. This gives $k(S) \le 50$, However, we are assuming that $k(S) > \sqrt{3}^7$, hence $k(S) \ge 47$. This forces that $k(H/N) = k(H/N^x) = 5$. Since H/N and H/N^x are isomorphic to primitive permutation groups of degree 4, we see that $H/N \cong N \cong S_4$. We conclude that $S \cong S_4 \mathrm{wr}\, S_2$ and $k(S) = 20$.

Case (ii): Let $b = |S : H|$, and let $N = \mathrm{Core}_S(H)$. Here S/N is a primitive permutation group of degree b, and N is an elementary abelian 2-group of order at most 2^b. If $b \geq 7$ then $k(S/N) \leq \sqrt{2}^{\,m}$ by Lemma 9.4a. It follows that $k(S) \leq \sqrt{2}^{\,+2b}$. Since by assumption $k(S) > \sqrt{3}^{\,2b-1}$ this forces that $\sqrt{2}^{\,3b} > \sqrt{3}^{\,2b-1}$ and $b < 10$.

Let $b = 9$. If S/N is solvable, then S/N is of affine type and so $k(S/N) \leq 11$ as seen in the proof of Lemma 9.4a. This gives the result. If S/N is a nonsolvable (primitive) $5'$-subgroup of S_9, then it is isomorphic to a (subnormal) subgroup of $L_2(8) : 3$ and so even $k(S/N) \leq 9$ [Atlas]. There is no nonsolvable primitive $7'$-subgroup of S_9.

Let $b = 8$ or 7. If S/N is solvable (of affine type), then $k(S/N) \leq b$, which is as required. If S/N is a nonsolvable, primitive p'-subgroup of S_8, then $p = 5$ and S/N is isomorphic to $L_3(2)$ or to $L_3(2) : 2$ and $k(S/N) \leq 9$, or it is the affine group $2^3 : L_3(2)$ of degree 8. We know already that $k(2^3 : L_3(2)) = 11$. We get the desired contradiction here, as well as for $b = 7$ where $k(S/N) = k(L_2(7)) = 6$.

Let $b = 6$. Then S is solvable or $p = 7$ and $k(S/N) \leq 11$, yielding the result.

Let $b = 5$. If S/N is a nonsolvable $(7'$-$)$ subgroup of S_5, then it is S_5 or A_5 and so $k(S/N) = 7$ or 5. From the [Atlas] we infer that S is isomorphic to a subgroup of $2^5 : S_5 = Z_2 \mathrm{wr}\, S_5$, which has just 36 conjugacy classes. Since 2^5 is the natural permutation module for S_5 over $\mathbb{F}_2$, this may be computed by means of formula (1.10b) and shows that S contains $2^4 : A_5$. At any rate, $k(S) \leq 36 < \sqrt{3}^{\,9}$. The remainder is straightforward.

Case (iii): Let $|X : H| = m$ and $N = \mathrm{Core}_X(H)$, $M = \mathrm{Core}_X(T)$. Then $M \cap M^x = 1$ for any $x \in S \setminus X$, N/M is an elementary abelian 2-group of order at most 2^m, and X/N is a primitive permutation group of degree m (and order prime to p). Suppose that $m \geq 7$. Then $k(X/N) \leq \sqrt{2}^{\,m}$ by (9.4a), hence $k(X/M) \leq \sqrt{2}^{\,3m}$. Further $k(M) = k(M^x M/M) \leq \sqrt{3}^{\,2m-1}$ by induction. It follows that $k(S) \leq 2 \cdot \sqrt{2}^{\,3m} \cdot \sqrt{3}^{\,2m-1}$. Since $k(S) > \sqrt{3}^{\,4m-1}$, this forces that $m < 12$. If X/N is solvable (of affine type), one even obtains $m \leq 9$, and these cases are easily treated. So let X/N be not solvable in what follows. Note that $k(X/N) > 3^m/2^{m+1}$ by assumption.

Let $m = 11$. Then $p = 7$ and either $X/N \cong L_2(11)$ and $k(X/N) = 8$, or $X/N \cong M_{11}$ and $k(X/N) = 10$. We obtain the desired contradiction.

Let $m = 10$. Then $p = 7$ and X/N either is a (primitive) subgroup of $A_6.2^2$, in which case $k(X/N) \le 22$, or of S_5 (in its primitive action on the 2-subsets of $\{1, \cdots, 5\}$).

The possibilities $m = 9, 8, 7$ are ruled out using the information given in Case (ii). For $m = 6$ we have $X/N \cong S_6$ or A_6, but we have to argue more carefully. Choose $x \in S \smallsetminus X$. Then $N \cdot M^x$ is a subnormal subgroup of X. Thus either $MN^x \subseteq N$ and $k(M) = k(MM^x/M) \le 2^6$, or NM^x/N is S_6 or A_6, that is, NM^x is a subgroup of X of index 1 or 2. The latter case cannot happen since $1 = M \cap M^x = M \cap (M^x \cap N)$ and so $M \cong M(M^x \cap N)/(M^x \cap N)$ would be an elementary abelian 2-group (but M^x would not). Using (1.10b) we get $k(S) \le 2 \cdot 11 \cdot 2^6 \cdot 2^6 < \sqrt{3}^{24-1}$. The same argument works for $m = 5$. This completes the proof. $\square$

Examples 9.4c. The upper bound given in Lemma 9.4a can be improved considerably for large degree n. For instance, in [Gluck *et al.*, 2004] it has been shown that if S is a primitive permutation $5'$-group of degree $n > 20$, then $k(S) \le 2^{\frac{n}{5}}$. On the other hand, there are examples (of imprimitive groups) where one has proper lower bounds.

(i) Let $S = S_4 \mathrm{wr}\, Z_m$ for some integer $m \ge 2$, which is a transitive (but imprimitive) permutation group of degree $n = 4m$. Since the base group of the wreath product is a direct product of m copies of S_4 and has index m in S, by part (i) of Theorem 1.7a $k(S) \ge \frac{5^m}{m} = \frac{4}{n} 5^{\frac{n}{4}}$. By Proposition 8.5d, $k(S) \le 5^{\frac{n}{4}} + \frac{5n}{4} - 5$.

(ii) Let $S = Z_p \mathrm{wr}\, Z_p$ for some prime p, which is a (transitive) Sylow p-subgroup of S_n, $n = p^2$. By Proposition 8.5d, $k(S) = p^{p-1} + p^2 - 1 > 2^{\log_2 p\,(\sqrt{n}-1)}$.

(iii) Let S be a Sylow p-subgroup of S_n where $n = p^{m+1}$ for some prime p and some $m \ge 2$. One knows that $S \cong Z_p \mathrm{wr}\, Z_p \mathrm{wr}\, \cdots \mathrm{wr}\, Z_p$ with $m + 1$ occurrences of Z_p. By induction one gets that $k(S) \ge p^{\frac{1}{p-1}} \cdot b^{p^{m-1}}$ where $\log_p b = (p-2)/(p-1)$. So for instance $k(S) > 3^{\frac{n}{6}}$ when $p = 3$.

Chapter 10

The Final Stages of the Proof

In proving the $k(GV)$ theorem we may assume that G is irreducible on V (Proposition 3.1a). By the Robinson–Thompson Theorem 5.2b we may assume that there is no real vector in V for G. Then, by the Riese–Schmid Theorem 8.4, the pair (G, V) is nonreal induced. The proof is accomplished now using that the class numbers for the nonreal reduced pairs are rather small, and using the Liebeck–Pyber Theorem 9.3 and its improvement. This actually works except for one case ($p = 5$), where some additional considerations are needed.

10.1. Class Numbers for Nonreal Reduced Pairs

Throughout we let (H, W) be a nonreal reduced pair inducing the pair (G, V). Let E be the core of H.

Proposition 10.1. *Let* Y *be a subnormal subgroup of* H.

(i) *If* $Y \supseteq E$ *then* $k(YW) < |W|/2$ *unless* (H, W) *is of type* (Q_8) *for* $p = 5$ *or* $p = 7$. *For the type* (Q_8) *and* $p = 7$ *(where* $|W| = 49$) *we have* $k(YW) \leq k(HW) = 27$.

(ii) *Let* $p = 3$ *(where* (H, W) *is of type* (2^5_-) *and* $|W| = 81$) *and let* $Y \supseteq Z(E)$. *Then* $k(YW) \leq k(Z(E)W) = 42$.

(iii) *Let* $p = 5$ *(where* (H, W) *is of type* (Q_8) *and* $|W| = 25$) *and let* $Y \neq 1$. *Then* $k(YW) \leq k(HW) = 20$.

Proof. We make use of Proposition 3.1b and knowledge of the orbit structure of H on W, including knowledge of the point stabilizers (Secs. 6.1 and 7.1). The class numbers $k(H)$, $k(Y)$ are computed using the Clifford–Gallagher formula (1.10b). We carry out only a few cases. The assumption that Y is *subnormal* in H is crucial. Let $Z = Z(E)$ and $U = E/Z$.

Type (Q_8) : Let $p = 5$. Then $H \cong (Q_8 \circ Z_4).S_3$. There are $k(S_3) = 3$ irreducible characters of H lying over each of the two faithful irreducible characters of $Y_4 = Q_8 \circ Z_4$ (extendible to H since $M(S_3) = 1$), and over each of its two S_3-invariant linear characters. (We use the notation which will be introduced in Lemma 10.4a below.) Two further linear characters of Y_4 are stable under a fixed involution in S_3, each having two extensions to this inertia group and induce up to II. Thus $k(II) - 3 \cdot 2 + 3 \cdot 2 + 2 \cdot 2 - 16$. Of course $k(Q_8) = 5$, $k(Y_4) = 10$ and $k(H') = 7$ ($H' = Y_3 \cong \mathrm{SL}_2(3)$). For the subgroup $Y = Y_2$ of H with index 2 we have $k(Y_2) = 14$. Now use that H is transitive on $W^\sharp$ and that $C_H(w) \cong Z_4$ for any $w \in W^\sharp$. We obtain that $k(HW) = 20$ and $k(Y_2W) = 16$.

The cases $p = 7, 13$ are treated similarly.

Type (2_-^5) : Let $p = 3$. Then $H \cong (2_-^{1+4}) : \bar{H}$ where $\bar{H} = Z_5 : Z_4$. There are $k(\bar{H}) = 5$ irreducible characters of H over the unique faithful irreducible character of $E \cong 2_-^{1+4}$ (as $\mathrm{M}(\bar{H}) = 1$). Since Z_5 is irreducible on U, Z_4 fixes a unique singular point in U, and since the $|\bar{H} : Z_2| = 10$ nonsingular points have stabilizer Z_2 in $\bar{H}$, we get $k(H) = 5 \cdot 1 + 4 \cdot 1 + 2 \cdot 1 + k(\bar{H}) = 16$. Using that H is transitive on $W^\sharp$ and that $C_H(w) \cong Z_8$ for each $w \in W^\sharp$ we get $k(HW) = 16 + 8 = 24$. Of course $k(E) = 17$, and E has 5 orbits on $W^\sharp$ of size 16, whence $k(EW) = 17 + 5 \cdot 2 = 27$. Also $Z = Z(E)$ has 40 orbits on $W^\sharp$ of size 2, so $k(ZW) = 42$.

Observe that if Y is generated by a noncentral involution in E, then $k(YW) = 54$. So the assumption $Y \supseteq Z(E)$ cannot be omitted.

Type (3_+^3) : Let $p = 7$ and $H \cong \left(3_+^{1+2} : \mathrm{Sp}_2(3)\right) \times Z_2$. There are three H-orbits on $W^\sharp$ with point stabilizers Z_6, $Z_2 \times Z_3^{(2)}$ and $\mathrm{Sp}_2(3)$. Hence $k(H) = 2(2 \cdot 7 + 1) = 30$ and $k(HW) = 30 + 6 + 18 + 7 = 61$ (whereas $|W| = 7^3$). Similar computation for the subnormal subgroups Y containing $E \cong 3_+^{1+2}$.

Type ($2.\mathbf{A_5}$) : Let $p = 31$. Then $H \cong 2.A_5 \times Z_{15}$ and $k(H) = 9 \cdot 15 = 135$. There are two orbits of H on $W^\sharp$ with point stabilizers Z_3, Z_5, hence $k(HW) = 135 + 3 + 5 = 143$ (whereas $|W| = 31^2$). The group $E = 2.A_5$ has 8 regular orbits on $W^\sharp$, hence even $k(EW) = 9 + 8 = 17$.

Type ($2.\mathbf{A_6}$) : Here $H \cong 2^\pm S_6 \times Z_3$ and $p = 7$, and we have three H-orbits on $W^\sharp$ with stabilizers Z_3, Z_6 and $Z_3 \mathrm{wr}\, S_2$. Apply Proposition 3.1b.

Type ($\mathbf{Sp_4(3)}$) : Let $p = 7$. Then $H \cong \mathrm{Sp}_4(3) \times Z_3$ and $k(H) = 3 \cdot 34 = 102$ (Atlas). There are two orbits on $W^\sharp$ with point stabilizers $\mathrm{SL}_2(3) \times Z_3$ and $(3_+^{1+2} : Q_8) \times Z_3$, which have class numbers 21 and 33, respectively. Hence $k(HW) = 156$ (whereas $|W| = 7^4$). The remainder is done similarly. $\quad\square$

10.2.   Counting Invariant Conjugacy Classes

By assumption we have an imprimitivity decomposition

$$V = W_1 \oplus \cdots \oplus W_n$$

for some integer $n \geq 1$, where the pairs $(N_G(W_i)/C_G(W_i), W_i)$ are isomorphic to (H, W) and permuted transitively by G. From the classification of the nonreal reduced pairs in Chapters 6 and 7 it follows that V is an absolutely irreducible $\mathbb{F}_p G$-module, and that $p \in \{3, 5, 7, 11, 13, 19, 31\}$.

Let us write $G_i = N_G(W_i)$ and $H_i = G_i/C_G(W_i)$ for each i, and let $N = \bigcap_{i=1}^n G_i$ be the normal core. Let $E \cong E_i$ be the cores of $(H, W) \cong (H_i, W_i)$. Identify (H, W) with (H_1, W_1) say (as reduced pairs). Recall that distinct normal subgroups of H are not isomorphic (and hence are characteristic in H). Let $S = G/N$ act (transitively) on the set $\Omega = \{1, \cdots, n\}$ like on the $\{W_i\}$. Because of Proposition 10.1 we may assume that $n \geq 2$. The degree n of the permutation group S is also called the degree of (G, V). Since G is faithful on V, we may identify G with a subgroup of the wreath product $H \,\mathrm{wr}\, S$ (Lemma 8.2a). Let $T_i = C_N(W_i)$ and $N_i = N/T_i$ for $i \in \Omega$, and denote by $N_0 \cong N_i$ their common image in $H = H_0$. Let $R_i = \bigcap_{j \neq i} T_j = C_N(\bigoplus_{j \neq i} W_j)$ so that

$$R = R_1 \times \cdots \times R_n$$

is a subgroup of N whose direct factors are permuted transitively by G (or S). Let $R_0 \cong R_i \cong RC_G(W_i)/C_G(W_i)$ denote the common image in H. We know from Theorem 8.5c that $R_0 \supseteq E$ except possibly when $p = 3$ or $p = 5$. In the exceptional cases at least $R_0 \supseteq Z(E)$, and always $N_0 \supseteq E$. Of course R is normal in G, and both N_0 and R_0 are characteristic normal subgroups of $H = H_0$ ($R_0 \subseteq N_0$). We let $Q = N/R$ and $Q_0 = N_0/R_0$.

We need some bounds for the number $k_g(N) = |C_{C\ell(N)}(g)|$ of conjugacy classes of N fixed by some element $g \in G$. Let us call a vector $v \in V$ an n-vector provided its support is Ω, that is, $v = w_1 + \cdots + w_n$ with all components $w_i \in W_i$ being nonzero.

Proposition 10.2a. *Suppose $S = \langle Ng \rangle$ is cyclic ($g \in G$). Then:*

(i) $k_g(N) \leq \min\{|N_0|, k(R_0) \cdot k_g(Q)\}$.

(ii) $k_g(NV) \leq \min\{|N_0 W|, k(R_0 W) \cdot k_g(Q)\}$.

(iii) $k_g(NV) \leq k_g(N) + \sum_j k(C_N(v_j))$ *where $\{v_j\}$ is a set of representatives of the N-orbits on $V^{\sharp}$ consisting of n-vectors.*

Proof. For later use we describe exactly the assumptions needed for the argument. Of course g normalizes N and R, acts on Q and permutes the $\{R_i\}$ (and $\{H_i\}$) cyclically. The nth power g^n normalizes each R_i (since $g^n \in N$ or, what is essential, since g^n induces the identity permutation on the set $\{R_i\}$).

There is a corresponding embedding of GV into the wreath product $(HW)\mathrm{wr}\,S$, having base group $BV = H_1 W_1 \times \cdots \times H_n W_n$. Here $RV = R_1 W_1 \times \cdots \times R_n W_n$ is a normal subgroup of NV, with quotient group $Q = (NV)/(RV)$, and g leaves NV and RV invariant and permutes the $\{R_i W_i\}$ cyclically. Therefore (ii) is immediate from (i).

Let us prove (i). For each $y \in N$, $C_N(gy)$ maps onto a subgroup of $C_Q(gy)$ with kernel $C_R(gy)$. Hence $|C_N(gy)| \leq |C_Q(gy)| \cdot |C_R(y)|$. By (1.7c) therefore

$$k_g(N) = \frac{1}{|N|} \sum_{y \in N} |C_N(gy)| \leq \frac{1}{|N|} \sum_{y \in N} |C_Q(gy)| \cdot |C_R(gy)|.$$

Now $C_N(gy) \cap C_N(W_1) = 1$ for each $y \in N$, because gy permutes the $\{W_i\}$ cyclically and y leaves each W_i invariant. Hence $C_N(gy)$ maps injectively into $N_1 = N/C_N(W_1) \cong N_0$ and so $|C_N(gy)| \leq |N_0|$, for all $y \in N$. Consequently $k_g(N) \leq |N_0|$.

Let $\alpha = \frac{1}{|N|} \sum_{y \in N} |C_Q(gy)|$. Let $\{t_j\}$ be a transversal to R in N, and let $\bar{g} = Rg$. Since $C_Q(gt_j x) = C_Q(gt_j)$ for each $x \in R$, and since $\frac{|R|}{|N|} = \frac{1}{|Q|}$, we get

$$\alpha = \frac{1}{|Q|} \sum_{\bar{y} \in Q} |C_Q(\bar{g}\bar{y})| = k_g(Q),$$

again by Eq. (1.7c). Let β be the maximum of the $\frac{1}{|R|} \sum_{x \in R} |C_R(gt_j x)|$, with t_j varying over the transversal, the maximum being obtained for $t_j = t_0$ say. The estimate in the preceding paragraph shows that $k_g(N) \leq \alpha \cdot \beta$. So we have to show that $\beta \leq k(R_0)$. Let $x = x_1 \cdots x_n$ and $h = h_1 \cdots h_n$ be elements of R (with $x_i, h_i \in R_i$). Assume g, and hence gt_0 and $g_0 = gt_0(x/x_1)$, map onto the cycle $(12...n)$ in S. Then h is centralized by $gt_0 x = g_0 x_1$ if and only if $h_1^{g_0 x_1} = h_2, \cdots, h_{n-1}^{g_0 x_1} = h_n$ and $h_1 = h_1^{(g_0 x_1)^n}$. Now $(g_0 x_1)^n = g_0^n x_1^{g_0^{n-1}} x_1^{g_0^{n-2}} \cdots x_1$ and so $h_1^{(g_0 x_1)^n} = h_1^{g_0^n x_1}$, because h_1 is centralized by R_i for $i \neq 1$. We conclude that $|C_R(gt_0 x)| = |C_R(g_0 x_1)| \leq |C_{R_1}(g_0^n x_1)|$. Just using that g_0^n normalizes R_1, as before

$$\frac{1}{|R_1|} \sum_{x_1 \in R_1} |C_{R_1}(g_0^n x_1)| = k_{(g_0^n)}(R_1) \leq k(R_1) = k(R_0).$$

Thus $\frac{1}{|R_1|} \sum_{x_1 \in R_1} |C_R(gt_0 x_1 x_2 \cdots x_n)| \leq k(R_0)$ for all $\frac{|R|}{|R_1|} = |R_0|^{n-1}$ elements $(x_2, \cdots, x_n) \in R_2 \times \cdots \times R_n$. Consequently $\beta = \frac{1}{|R|} \sum_{x \in R} |C_R(gt_0 x)| \leq k(R_0)$.

It remains to prove (iii). $C\ell(NV)$ and $\mathrm{Irr}(NV)$ are isomorphic $\langle g \rangle$-sets by Brauer's permutation lemma (Theorem 1.4b). The partition $V = \biguplus_j v_j N$ into N-orbits corresponds to a partition $\mathrm{Irr}(NV) = \biguplus_j \mathrm{Irr}(NV|\lambda_j)$, with $C_N(v_j) = I_N(\lambda_j)$ when the $\lambda_j \in \mathrm{Irr}(V)$ are chosen properly (Proposition 3.1b). These partitions are preserved by g since $\mathrm{Irr}(NV|\lambda_j)^g = \mathrm{Irr}(NV|\lambda_j^g)$ for each j. Thus $k_g(NV)$ is the sum of the fixed points under g in $\mathrm{Irr}(NV|\lambda_j)$, the sum taken over those λ_j corresponding to $\langle g \rangle$-invariant N-orbits on V. One such summand is $k_g(N)$, the number of fixed points under $\langle g \rangle$ of $\mathrm{Irr}(NV|1_V) = \mathrm{Irr}(N)$. Since g permutes the $\{W_i\}$ cyclically, only those N-orbits on $V^{\sharp}$ can be $\langle g \rangle$-invariant which are represented by n-vectors. Hence the result.   □

We would like to have a substitute for Proposition 10.2a in the general case. Let $g \in G$. If $B = \prod_J B_J$ is a decomposition of the base group according to the cycle decomposition of the permutation $\bar{g} = Ng$ in S, the natural projection map $N \to \prod_J B_J$ is a $\langle g \rangle$-homomorphism, and it is *injective*. However, letting N_J be the image of N in B_J, the induced map $C\ell(N) \to \prod_J C\ell(N_J)$ need not be injective. (An enlightening example is $S_3 \Delta_{Z_2} S_3$.) So the map on the fixed point sets under $\langle g \rangle$ need not be injective likewise. The following lemma is used only in the very last step ($p = 5$) of the proof of the $k(GV)$ conjecture (and for NV in place of N).

Lemma 10.2b. *Let $g \in G$. According to the the the cycle decomposition of Ω under the permutation $\bar{g} = Ng$ in S, with orbits J, let N_J be the image of N in $B_J = \prod_{i \in J} H_i$. Then $R_J = \prod_{i \in J} R_i$ is a normal subgroup of N_J, and both R_J and N_J are stable under g. For each J let k_J be the maximum of the $|C_{C\ell(Y)}(gt)|$ where Y varies over all subgroups of N_J containing R_J which are normalized by gt for some $t \in N$. Then $k_g(N) \leq \prod_J k_J$.*

Proof. Let $J_1, \cdots, J_r$ be the distinct $\langle \bar{g} \rangle$-orbits on Ω. We argue by induction on r. For $r = 1$ the assertion is trivial. So let $r > 1$. Let M be the image of N under the $\langle g \rangle$-projection of B onto $\prod_{i=1}^{r-1} B_{J_i}$ with respect to B_{J_r}. Since the maps $N \twoheadrightarrow N_{J_i}$, $1 \leq i \leq r - 1$, factor through $N \twoheadrightarrow M$, the inductive hypothesis applies showing that $k_g(M) \leq \prod_{i=1}^{r-1} k_{J_i}$.

Fix $y_m \in M$ such that $y_m^M \in C_{C\ell(M)}(g)$. Of course $y_m^M = y_m^N$. Let Y be the image of $C_N(y_m)$ under the $\langle g \rangle$-projection of B onto B_{J_r}. Then

$Y \supseteq R_{J_r}$. There exists $t \in N$ such that $y_m^{gt} = y_m$. Since gt centralizes y_m and normalizes N, it normalizes $C_N(y_m)$ and Y (as the projection is compatible with the action of gt). Let y^N be a $\langle g \rangle$-invariant conjugacy class of N which maps onto y_m^M. We may replace $y \in N$ by an N-conjugate, if necessary, such that y maps onto y_m. Then $y \in C_N(y_m)$, and we may write uniquely $y = y_m y_r = y_r y_m$ where y_r is the image of y in Y. There exists $t_0 \in N$ such that $y^{gt} = y^{t_0}$, and we have

$$y_m y_r^{gt} = y^{gt} = y^{t_0} = y_m^{t_0} y_r^{t_0}.$$

We conclude that $y_m = y^{t_0}$ and $y_r^{gt} = y_r^{t_0}$ (by uniqueness), hence $t_0 \in C_N(y_m)$ and $y_r^Y \in C_{Cl(Y)}(gt)$.

Now let $\widetilde{y}^N$ be another $\langle g \rangle$-invariant conjugacy class of N mapping onto y_m^M, and choose $\widetilde{y}$ such that y_m is its component in M. Write uniquely $\widetilde{y} = y_m \widetilde{y}_r = \widetilde{y}_r y_m$ with $\widetilde{y}_r \in Y$ and $\widetilde{y}_r^Y \in C_{C\ell(Y)}(gt)$, as before. If $\widetilde{y}_r^Y = y_r^Y$, then

$$\widetilde{y}^{C_N(y_m)} = y_m \widetilde{y}_r^Y = y_m y_r^Y = y^{C_N(y_m)}$$

and so $\widetilde{y}^N = y^N$. The result follows since, by definition, $k_{gt}(Y) \leq k_{J_r}$. $\square$

10.3. Nonreal Induced Pairs

Theorem 10.3 (Riese–Schmid). *Let (G, V) be properly induced ($n \geq 2$) from the nonreal reduced pair (H, W). Then $k(GV) \leq \frac{1}{2}|V|$, except possibly when $p = 5$.*

Proof. We use the notation introduced in Sec. 10.2. In addition let $Y_n = T_n = N/C_N(W_n)$ and let

$$Y_i = C_N(W_{i+1} \oplus \cdots \oplus W_n)/C_N(W_i \oplus W_{i+1} \oplus \cdots \oplus W_n)$$

for $i = 1, \cdots, n-1$. Since $C_N(W_{i+1} \oplus \cdots \oplus W_n) \supseteq R_i$ and $R_i \cap C_N(W_i \oplus \cdots \oplus W_n) = 1$, Y_i may be identified with a subnormal subgroup of H_i containing R_i. So W_i is a faithful $\mathbb{F}_p Y_i$-module.

Suppose first that $p \neq 3$ (and $p \neq 5$). Then $Y_i \supseteq R_i \supseteq E_i$ by Theorem 8.5c. Hence $k(Y_i W_i) < \frac{1}{2}|W|$ by Proposition 10.1 (i) unless (H, W) is of type (Q_8) with $p = 7$. In this exceptional case we have $k(Y_i W_i) \leq 27 = \frac{27}{49}|W|$. Repeated application of (1.7b) yields that

$$k(NV) \leq \prod_{i=1}^{n} k(Y_i W_i) < \left(\frac{1}{2}\right)^n |V|$$

if (H, W) is not of type (Q_8) with $p = 7$. In the exceptional case $k(NV) \leq (\frac{27}{49})^n |V|$. Now by (1.7b) once again $k(GV) \leq k(NV) \cdot k(S)$, and $k(S) \leq 2^{n-1}$ by the Liebeck–Pyber Theorem 9.3. Moreover, in the exceptional case S is a $7'$-group and so $k(S) \leq \sqrt{3}^{n-1}$ for $n \geq 3$ by Proposition 9.4b. This gives the assertion except when (H, W) is of type (Q_8) with $p = 7$ and $n = 2, 3$. Note that $(\frac{27}{49})^n \sqrt{3}^{n-1} \leq \frac{1}{2}$ for $n > 3$. This is less than 0.502 for $n = 3$, and $(\frac{27}{49})^2 \cdot 2 \leq 0.61$.

We have to improve the last bounds (for $n = 2, 3$). Here $H \cong 2^- S_4 \times Z_3$ ($p = 7$). By Theorem 8.5c, $R_0 \supseteq E \cong Q_8$. Let $n = 2$, and let $S = \langle Ng \rangle$ be of order 2. Then $Q = N/R$ is a quotient group of $S_3 \times Z_3$. Hence $k(Q) \leq 3 \cdot 3 = 9$, and if $k(Q) = 9$ then $R_0 \cong Q_8$ and $k(R_0 W) = 11$. It is easy to check that $k(Q) \cdot k(R_0 W) \leq 99$ in each case. We know that $k(NV) \leq (\frac{27}{49})^2 |V| = 729$. By Proposition 10.2a (ii) at most $k_g(NV) \leq 99$ of the at most 729 conjugacy classes of NV are $\langle g \rangle$-invariant. By Theorem 1.4b at most 99 of the at most 729 irreducible characters of NV are stable in GV. By (1.10b) therefore

$$k(GV) \leq 2 \cdot 99 + (729 - 99)/2 = 513,$$

which is less than $\frac{1}{4}|V|$. For $n = 3$ we have to examine the cases where $S \cong A_3$ or S_3. Use that $k(NV) \leq 27^3$ and that $|H| \cdot |W| = 3 \cdot 48 \cdot 27$ conjugacy classes of NV are fixed by a 3-cycle in S by Proposition 10.2a.

Let $p = 3$. Then $|W| = 3^4$ and $k(Y_i W_i) \leq 42$ by Proposition 10.1. By Proposition 9.4b, $k(S) \leq \sqrt{3}^{n-1}$ for $n \geq 3$. Hence

$$k(GV) \leq k(NV) \cdot k(S) \leq (\frac{42}{81})^n \sqrt{3}^{n-1} |V| \leq \frac{1}{2}|V|,$$

except when $n = 2$, in which case $k(GV) \leq k(NV) \cdot 2 \leq (\frac{42}{81})^2 \cdot 2 < 0.6|V|$. Again this bound is easily improved. The worst case happens when $R_0 \cong Z(H)$ has order 2, in which case $Q \cong H/Z(H)$. Then $k(R_0 W) = 42$ and $k(Q) = 12$. Hence by Proposition 10.2a at most $42 \cdot 12 = 504$ of the $k(NV) \leq 42^2$ irreducible characters of NV are invariant in GV. Application of (1.10b) gives $k(GV) \leq 2 \cdot 504 + (42^2 - 504)/2 = 1.638$, which is less that $\frac{1}{4}|V|$ ($|V| = 81^2$). The proof is complete. □

10.4. Characteristic 5

It remains to examine the case where (H, W) is the nonreal reduced pair of type (Q_8) with $p = 5$. Having established Theorem 10.3 in 2001, published in [Riese–Schmid, 2003], it took almost two years to handle this case too. The thorough analysis made in [Gluck *et al.*, 2004] yields in this case the estimate $k(GV) \leq \frac{1}{2}|V|$ as before ($n \geq 2$). In what follows we argue, using some ideas from [Keller, 2006], on the basis of Proposition 10.2a and Lemma 10.2b. This makes the approach considerably shorter (though the result is weaker). We use the notation introduced in Sec. 10.2.

For convenience of the reader we give some additional information on the group H, which is a 5-complement in $\mathrm{GL}_2(5)$ (uniquely determined up to conjugacy). As already seen in Sec. 3.2, H is transitive on the nonzero vectors of the standard module $W = \mathbb{F}_5^{(2)}$. Also, H is the standard holomorph of $Q_8 \circ Z_4$. Hence $H/Z(H) \cong S_4$, but the two covering groups $2^{\pm}S_4$ are not involved in H (nor a Singer cycle).

Lemma 10.4a. *Let (H, W) be of type (Q_8) with $p = 5$. There are up to conjugacy just 11 subnormal subgroups Y_a of $H = Y_0$ containing the centre of the core E of H (listed below). The numbering is such that whenever Y_a contains some conjugate of Y_b then $a \leq b$.*

Y_a	$k(Y_a)$	$k(Y_a W)$	$k(Y_a/Y_{11})$
$Y_0 = H$	16	20	10
$Y_1 = O_{2,3}(H)$	14	16	8
$Y_2 = H' \cong \mathrm{SL}_2(3)$	7	8	4
$Y_3 = O_2(H) \cong Q_8 \circ Z_4$	10	16	8
$Y_4 = E = H''$	5	8	4
$Y_5 \cong D_8$	5	14	4
$Y_6 \cong Z_4 \times Z_2$	8	14	4
$Y_7 = Z(H) \cong Z_4$	4	10	2
$Y_8 \cong Z_4$	4	10	2
$Y_9 \cong Z_2 \times Z_2$	4	16	2
$Y_{10} = Z(E)$	2	14	1

The statements are easily verified (arguing as for Proposition 10.1). The normal subgroups of H are determined by their order. The subnormal but not normal subgroups of H are of type Y_5, Y_6, Y_8 and Y_9, each having 3 conjugates under H. For each subgroup Y of H we have $k(YW) \leq 20$, except for two conjugacy classes of abelian groups $Y = C_H(w)$ respectively $C_H(w) \times Z(H)$ for some $w \in W^\sharp$, where $k(YW) = |W|$. If $Y \neq H$ is a nonabelian subgroup, then $k(YW) \leq 16$ except when $Y \cong Z_4 \mathrm{wr}\, Z_2$ is a Sylow 2-subgroup of H (in which case $k(YW) = 20$).

It is also easy to compute $k(Y_a/Y_b)$ whenever Y_b is a normal subgroup of Y_a. Knowing that $|C_{Y_a}(w)| = 1, 2$ or 4 for each $w \in W^\sharp$, which may be deduced from the first row of the table, application of Proposition 3.1b enables us to compute from $k(Y_aW) - k(Y_a)$ the number of Y_a-orbits on $W^\sharp$. The subgroups of H of order greater than or equal to 8 have at most 5 orbits on $W^\sharp$. (By direct computation or using [GAP] one can establish the complete orbit structure for each subgroup of H.)

Lemma 10.4b. *Let (H, W) be of type (Q_8) with $p = 5$, and let $n = 2$. Then $k(GV) \leq \frac{1}{2}|V|$.*

Proof. In this case $S = \langle Ng \rangle$ has order 2. Also $G_1 = N = G_2$ so that $N = H_1 \Delta_Q H_2$ is a fibre-product of the $H_i \cong H$ amalgamating $Q = N/R$ ($R = R_1 \times R_2$). We also have an extension $R_0 \rightarrowtail H \twoheadrightarrow Q$. As in the proof of Theorem 10.3 we have $k(NV) \leq k(N_1 W_1) \cdot k(R_2 W_2) \leq 20^2$ ($R_2 = C_N(W_1)$ and $N_1 = N/R_2 \cong H$).

Suppose $G = H \operatorname{wr} S$, that is, $R_0 = H = Y_0$. Then $GV = (HW) \operatorname{wr} S$ and so by Proposition 8.5e and Lemma 10.4a

$$k(GV) = 2 \cdot k(HW) + k(HW)(k(HW) - 1)/2 = 230.$$

Suppose $R_0 = Y_1, Y_2, Y_3, Y_4$ or Y_7. Application of Proposition 10.2a and of Lemma 10.4a yields that

$$k_g(NV) \leq k(R_0W) \cdot k(Q) = 32, 32, 48, 48, 50,$$

respectively. In the worst case, in view of Theorem 1.4b, at most 50 of the at most 20^2 irreducible characters of NV are stable in GV. On the basis of the Clifford–Gallagher formula (1.10b) we therefore obtain the estimate $k(GV) \leq 2 \cdot 50 + (400 - 50)/2 = 275$.

It remains to examine the case where $R_0 = Y_{10}$. Then $k_g(N) \leq 2 \cdot 10 = 20$ by part (i) of Proposition 10.2a, and part (ii) gives $k_g(NV) \leq k(R_0W) \cdot k(Q) = 14 \cdot 10$. We improve this latter bound by applying part (iii) of this proposition. We know that N_1 is transitive on $W_1^\sharp$ and that $C_N(w_1)/R_2$ is cyclic of order 4 for each $w_1 \in W_1^\sharp$. Since $R_1 = C_N(W_2)$ is fixed point free on $W_1^\sharp$, $C_N(w_1) \cap R_1 = 1$ and so $C_N(w_1)$ is faithful on W_2 (of order 8). Similarly, $C_N(v) = C_N(w_1) \cap C_N(w_2)$ intersects R_2 trivially for each $w_2 \in W_2^\sharp$, whence $|C_N(v)| = 1, 2$ or 4. We also know from Lemma 10.4a that $C_N(w_1)$ has at most 5 orbits on $W_2^\sharp$. (Either $C_N(w_1)$ is cyclic with

3 regular orbits on $W_2^\sharp$, or it corresponds to Y_9 having 4 orbits, or it is of type $C_H(w) \times Y_{10}$ for some $w \in W_2^\sharp$, in which case one has 5 orbits.) Thus

$$k_g(NV) \le k_y(N) + 5 \cdot 4 \le 40.$$

Noting that $k(NV) \le 20 \cdot 14 = 280$ here, formula (1.10b) yields that $k(GV) \le 2 \cdot 40 + (280 - 40)/2 = 200.$ $\square$

The last estimate can be easily improved. In fact, in Lemma 10.4b the upper bound $k(GV) \le 152$ holds unless $G = H \,\mathrm{wr}\, S_2$.

Proposition 10.4c. *Let (H, W) be of type (Q_8) with $p = 5$, inducing the nonreal pair (G, V) as before ($n \ge 2$). Then for any $g \in G \smallsetminus \bigcup_{i=1}^{n} G_i$ we have $k_g(NV) < |V|^{\frac{2}{3}}$.*

Proof. By hypothesis $\bar{g} = Ng$ is a permutation in $S = G/N$ without fixed points on Ω. According to the cycle decomposition of $\bar{g}$ we obtain a decomposition of the base group BV of $(HW)\mathrm{wr}\,S$. By virtue of Lemma 10.2b we may replace $\bar{g}$ by one of its cycles, in the following sense:

Suppose $g \in (HW)\mathrm{wr}\,S$ permutes $W_1, \cdots, W_n$ cyclically and leaves NV and RV invariant. Let Y be any subgroup of NV containing RV which is normalized by g. Then prove that $|C_{\mathrm{C}\ell(Y)}(g)| \le |V|^{\frac{2}{3}}$.

The crucial point is that $Y \supseteq RV$. In particular $V = O_5(Y)$ and Y/V is a $5'$-group. Note that RV is normal in Y as $Y \subseteq NV$. By the Schur–Zassenhaus theorem we find a 5-complement $\widetilde{N}$ in Y and, replacing g by gv for some $v \in V$, if necessary, we may assume that g normalizes $\widetilde{N}$. Then $\widetilde{R} = \widetilde{N} \cap RV$ is a g-invariant normal subgroup of $\widetilde{N}$ which is isomorphic to R. In fact $\widetilde{R}_i = C_{\widetilde{N}}(\bigoplus_{j \ne i} W_j)$ is isomorphic to R_i for each i.

By abuse of notation we write $\widetilde{N} = N$, $\widetilde{R} = R$ and $\widetilde{R}_i = R_i$. Then, as before, R is a g-invariant normal subgroup of N mapping onto the normal subgroup $R_0 \cong R_i$ of $H = Y_0$, with $R_0 \supseteq Y_{10}$ in the notation of Lemma 10.4a. Let $T_i = C_N(W_i)$ and $N_i = N/T_i$ for each i, as before, which are permuted cyclically by g. In contrast to our previous experience the N_i can be, at first, map onto subgroups of H which are not normal, possibly not even subnormal. But they all contain R_0. So we may define again $Q_0 = N_0/R_0$ (where $N_0 = N_1$ according to our convention).

In addition $Q = N/R$ is a g-invariant quotient group as before. We do not know whether g^n is an element of the (new) group N (or NV), but g^n

normalizes each N_i and each R_i. Consequently Proposition 10.2a applies. In particular

$$k_g(NV) \le |N_0 W| = |N_0| \cdot |W| \le 96 \cdot 25 = 2.400.$$

This is less than $|V|^{\frac{1}{2}} = |W|^{\frac{n}{2}}$ if $n \ge 5$. For $n = 4$ we have $\lfloor |V|^{\frac{2}{3}} \rfloor = 5.343$, and $|V|^{\frac{2}{3}} = 625$ for $n = 3$. Therefore we only have to consider the cases $n = 2, 3$.

n = 2: Then $\lfloor |V|^{\frac{2}{3}} \rfloor = 73$. Keep in mind that N_0 may be any subgroup of $H = Y_0$ containing $R_0 \supseteq Y_{10}$ (unlike the situation in Lemma 10.4b). Noting that $T_1 = R_2$, $T_2 = R_1$ and $R = R_1 \times R_2$ we see that $Q_0 = Q$ and that

$$N = N_1 \Delta_Q N_2$$

is a fibre-product. We know from Lemma 10.4a that $k(R_0 W) \le 20$, and that $k_g(NV) \le k(Q) \cdot k(R_0 W)$ by Proposition 10.2a. If $k(R_0 W) = 20$, then $R_0 = H$, $Q = 1$ and $k_g(NV) = 20$. We may thus assume that $k(R_0 W) \le 16$, and that $k(Q) > 4$. Inspection of Lemma 10.4a yields that R_0 must be contained in $Y_3 \cong Q_8 \circ Z_4$. We distinguish two cases.

Case 1: Q is a $3'$-group

Then N_0 is a 2-group, hence contained in a Syow 2-subgroup $P \cong Z_4 \mathrm{wr}\, S_2$ of H. Note that $P' = Y_8$ (up to conjugacy in H). Since $R_0 \supseteq Y_{10}$ is normal in H, we have $k(Q) \cdot k(R_0 W) < 73$ unless $R_0 = Y_{10}$ and either $N_0 = P$ or $N_0 = Y_3$ (Lemma 10.4a). Let us consider the (worse) case where $N_0 = P$. Then $k(Q) = 10$, $|R_0| = 2$ and $k(R_0 W) = 14$. Hence $k_g(N) \le k(Q) \cdot k(R_0) = 20$ by part (i) of Proposition 10.2a, and $k_g(NV) \le 140$ by part (ii). This will be improved by using part (iii) of this proposition.

One checks that $N_0 = P$ has two orbits on $W^\sharp$, with point stabilizers of order 2 and 4. Let $v = w_1 + w_2$ be a 2-vector in V ($w_i \in W_i^\sharp$ for $i = 1, 2$). Then $C_{N_1}(w_1) = C_N(w_1)/R_2$ has order 2 or 4. Since $R_1 = C_N(W_2)$ is fixed point free on $W_1^\sharp$, $C_N(w_1)$ is faithful on W_2 (of order 4 or 8). Similarly, $C_N(v) \cap R_2 = 1$ and so $|C_N(v)|$ is a divisor of 2 or of 4. If $C_N(w_1)$ is cyclic, it has 6 regular orbits on $W_2^\sharp$ when $|C_N(w_1)| = 4$, and 3 regular orbits otherwise. If $C_N(w_1)$ is not cyclic and of order 4, it is of type Y_9 and has 4 regular orbits and 4 orbits of size 2. If $|C_N(w_1)| = 8$ we have at most 8 orbits (Lemma 10.4a). We conclude that there are at most 5 orbits of N on 2-vectors of V with point stabilizers $C_N(v)$ of order dividing 4, and at most 8 such orbits with stabilizers of order dividing 2. Hence

$$k_g(NV) \le k_g(N) + 5 \cdot 4 + 8 \cdot 2 \le 56,$$

as desired. The case where $N_0 = Y_3$ is treated similarly: The number of N-orbits on V consisting of 2-vectors is at most $3 \cdot 8 = 24$, with point stabilizers of order 2 or 1. This yields that $k_g(NV) \leq k_g(N) + 24 \cdot 2 \leq 64$ in this case.

Case 2 : $|Q|$ is divisible by 3

Then $|N_0| \geq 24$. There are two conjugacy classes of subgroups in H which act regularly on $W^\sharp$, namely $Y_3 \cong \mathrm{SL}_2(3)$, which is normal in H, and the 3-Sylow normalizers $Y \cong S_3 \times Z_4$ in H. Since $k(Q) > 4$ and $R_0 \supseteq Y_{10}$, we only have to treat the second case, with $N_0 = Y$ and $R_0 = Y_{10}$. Then $k(Q) = 6$ and $k(R_0W) = 14$, and $k_g(N) \leq 12$, $k_g(NV) \leq 6 \cdot 14$ by Proposition 10.2a. We have to improve the latter bound. For any 2-vector $v = w_1 + w_2$ the stabilizer $C_N(w_1) = R_2$ has 12 regular orbits on $W_2^\sharp$, so that we get $k_g(NV) \leq k_g(N) + 12 \cdot 1 \leq 24$.

It remains to examine the cases where $N_0 = Y_0$ or Y_1 (Lemma 10.4a). Let us consider the worse case $N_0 = Y_0 = H$. Then we must have $R_0 = Y_4$, Y_7 or Y_{10}. If $R_0 = Y_4$, then $k(R_0W) = 8$ and $k(Q) = 6$. If $R_0 = Y_7$, then $k(R_0W) = 10$ and $k(Q) = 5$. Thus it remains to examine the case where $R_0 = Y_{10}$. Here $k(Q) = 10$ and $k_g(N) \leq k(Q) \cdot k(R_0) = 20$ by part (i) of Proposition 10.2a, and $k_g(NV) \leq k(Q) \cdot k(R_0W) = 140$ by part (ii). Again we improve the latter bound by using part (iii) of this proposition. Let $v = w_1 + w_2$ be a 2-vector in V. Then $w_1N = W_1^\sharp$ and $C_N(w_1)/R_2$ is cyclic of order 4 (Lemma 10.4a). $C_N(w_1)$ is faithful on $W_2^\sharp$ since it intersects $R_1 = C_N(W_2)$ trivially (being fixed point free on $W_1^\sharp$). Similarly $C_N(v) \cap R_2 = 1$ and so $|C_N(v)| = 1, 2$ or 4. As mentioned above $C_N(w_1)$ (being of order 8) has at most 5 orbits on $W_2^\sharp$. At any rate,

$$k_g(NV) \leq k_g(N) + (1 \cdot 5) \cdot 4 \leq 40.$$

The case $N_0 = Y_1$ is treated similarly.

n = 3: By Proposition 10.2a we may assume that $25|N_0| > 625 = |V|^{\frac{2}{3}}$. Thus $|N_0| > 24$ and so $N_0 = Y_0$ or Y_1 by Lemma 10.4a, or N_0 is a Sylow 2-subgroup of $H = Y_0$. Note that g permutes the $T_i = C_N(W_i)$ cyclically ($i = 1, 2, 3$). Let $T = T_1T_2$, and let T_0 be the image of T/T_1 ($\cong T/T_2$) in N_0. Using that $R_3 = T_1 \cap T_2$ we see that $T/R_3 \cong T_0 \times T_0$. From $R = R_1 \times R_2 \times R_3$ we infer that the T_i are pairwise distinct, and from $T_1R = T_1R_1$ and $R_1 \subseteq T_2$ we get

$$T_1R \cap T_2R = (T_1R \cap T_2)R = (T_1 \cap T_2)R_1R = R.$$

Thus $T/R \cong Q_0 \times Q_0$, and

$$Q = (N_1/R_1)\Delta_{N/T}(N_2/R_2) \cong Q_0 \Delta_{N_0/T_0} Q_0$$

is a fibre-product, identifying the R_i with their images in $N_i = N/T_i$. We have a group extension $T_0/R_0 \rightarrowtail Q \twoheadrightarrow Q_0$. Thus $k(Q) \le k(T_0/R_0) \cdot k(Q_0)$ by (1.7b).

We assert that $k_g(Q) \le |Q_0|$. For any $y \in N$, the element gy does not centralize a nontrivial element in $T_1R/R = T_1R_1/R$. Hence $C_Q(gy)$ maps injectively into $N/T_1R \cong N_0/R_0 = Q_0$, and by (1.7c)

$$k_g(Q) = \frac{1}{|Q|} \sum_{y \in Q} |C_Q(gy)| \le |Q_0|,$$

as asserted. By Proposition 10.2a, $k_g(NV) \le k_g(Q) \cdot k(R_0W)$. Hence we may assume that $|Q_0| \ge 625/k(R_0W) \ge 25$. This forces that $N_0 = H = Y_0$ and that $R_0 = Y_{10}$.

Thus $k(Q_0) = 10$ and $k(R_0W) = 14$. Further T_0 is a normal subgroup of H, and by Proposition 10.2a we may assume that $k(Q_0) \cdot k(T_0/R_0) \ge k(Q) > 625/k(R_0W)$. It follows that $k(T_0/R_0) \ge 5$. This implies that $T_0 = Y_0, Y_1$ or Y_3 (Lemma 10.4a).

Now $T_1/R_3 \cong T_0$ has 1 or 3 orbits on $W_2^\sharp$ (depending on whether $T_0 = Y_0, Y_1$ or Y_3). Let $v = w_1 + w_2 + w_3$ be a 3-vector in V ($w_i \in W_i^\sharp$ for $i = 1, 2, 3$). Then $w_1 N = W_1^\sharp$ and $C_N(w_1)/T_1$ is cyclic of order 4, and $C_N(w_1)$ has at most 3 orbits on $W_2^\sharp$. From the structure of N_0/T_0 we infer that $C_N(w_1) \cap T_2$ properly contains R_3 (which has 12 regular orbits on $W_3^\sharp$). It follows that $C_N(w_1+w_2) = C_N(w_1) \cap C_N(w_2)$ has at most 8 orbits on $W_3^\sharp$. Clearly $C_N(v)/C_{T_1}(v)$ is isomorphic to a subgroup of $C_N(w_1)/T_1$. Similarly, $C_{T_1}(v) \subseteq C_{T_1}(w_2)$ and so $C_{T_1}(v)/C_{R_3}(v)$ is isomorphic to a subgroup of $C_{T_1}(w_2)/R_3$. Since R_3 is fixed point free on $W_3^\sharp$ and $R_3 \subseteq C_N(w_1 + w_2)$, therefore $C_N(v)$ is a 2-group of order dividing $1 \cdot 4 \cdot 4 = 16$, even $|C_N(v)| \le 8$ unless $T_0 = H$. In this latter case $C_N(w_1)$ is transitive on $W_2^\sharp$. At any rate, since $(3 \cdot 8) \cdot 8 > (1 \cdot 8) \cdot 16$, application of part (iii) of Proposition 10.2a yields that

$$k_g(NV) \le k_g(N) + (1 \cdot 3 \cdot 8) \cdot 8 \le 288.$$

Here we use that $k_g(Q) \le |Q_0| = 48$ and so $k_g(N) \le k_g(Q) \cdot k(R_0) \le 96$, in view of part (i) of that proposition. □

Theorem 10.4d (Gluck–Magaard–Riese–Schmid). *Let (G,V) be induced from the nonreal reduced pair (H,W) in characteristic $p = 5$. Then we have $k(GV) < |V|$.*

Proof. We argue by induction on the degree n of (G,V), the degree of the permutation group $S = G/N$. We may assume that $n \geq 3$ (Lemma 10.4b). We may also assume that $k(XV) \leq |V|$ for each proper subgroup X of G. For otherwise repeated application of (1.7b) yields that there exist a subnormal subgroup Y of X and an irreducible Y-constituent U of V such that $k((Y/C_Y(U)) \cdot U) > |U|$. (Use that all modules are completely reducible. Start with an irreducible X-submodule U_1 of V, let $X_1 = C_X(U_1)$, $V_1 = V/U_1$ and note that

$$k(XV) \leq k((X/X_1)U_1) \cdot k(X_1V_1).$$

If $k((X/X_1)U_1) > |U_1|$ take $U = U_1$. Otherwise pick an irreducible X_1-submodule U_2 of V_1, and let $X_2 = C_{X_1}(U_2)$, $V_2 = V_1/U_2$, etc. .) By Theorem 5.2b there is no real vector in U for Y. By Theorem 8.4, $(Y/C_Y(U),U)$ is induced from a nonreal reduced pair (in characteristic 5). But (H,W) is the unique, up to isomorphism, nonreal reduced pair in characteristic $p = 5$ by the classification of these pairs. Clearly the degree of $(Y/C_Y(U),U)$ is less than n. Thus by the inductive hypothesis $k((Y/C_Y(U)) \cdot U) < |U|$, a contradiction.

By Proposition 10.4c for any $g \in G \setminus \bigcup_{i=1}^n G_i$ we have $k_g(NV) \leq |V|^{\frac{2}{3}}$. Let r be the number of conjugacy classes of $S = G/N = (GV)/(NV)$ which are contained in $\bigcup_{i=1}^n G_i/N$. We have $k(S) > r \geq 1$ (Jordan). Application of Theorem 1.7d yields that

$$k(GV) \leq k(G_1V) + (k(S) - r)|V|^{\frac{2}{3}}.$$

By (1.7b) $k(G_1V) \leq k(H_1W_1) \cdot k(C_G(W_1) \cdot (W_2 \oplus \cdots \oplus W_n))$. Clearly $|C_G(W_1)| < |G|$, hence $k(C_G(W_1) \cdot V) \leq |V|$. On the other hand $C_G(W_1) \cdot V = W_1 \times C_G(W_1) \cdot (W_2 \oplus \cdots \oplus W_n)$, and so $k(C_G(W_1) \cdot (W_2 \oplus \cdots \oplus W_n)) \leq |W|^{n-1}$. It follows that $k(G_1V) \leq k(HW) \cdot |W|^{n-1} = 20 \cdot |W|^{n-1}$. Since S is a $5'$-group and $n \geq 3$, $k(S) \leq \sqrt{3}^{n-1}$ by Proposition 9.4b. Consequently

$$k(GV) \leq 20 \cdot |W|^{n-1} + (\sqrt{3}^{n-1} - 1)|W|^{\frac{2n}{3}} = (0.8 + c_n)|V| < |V|$$

for each $n \geq 3$, because $c_3 = 0.08$ and $c_n < 0.6^n$ in general. We just use that $|W| = 25$. $\square$

10.5. Summary

Let G be a p'-subgroup of $\mathrm{GL}(V) = \mathrm{GL}_m(p)$ for some prime p and some integer $m \geq 1$ ($|V| = p^m$). With Theorem 10.4d the proof of the $k(GV)$ theorem is completed. Let us recall the basic steps:

• We may assume that G is irreducible on V (Proposition 3.1a).

• We may assume that there is no real vector in V for G (Theorem 5.2b).

• The minimal counterexamples (G, V), with respect to $|V|$, which admit no real vector, can be described via Clifford theory and lead to the so-called nonreal reduced pairs (Theorem 5.4).

• The nonreal reduced pairs have been classified (Theorems 6.1 and 7.1).

• If G is irreducible on V and if there is no real vector in V for G, then (G, V) is induced from a nonreal reduced pair (H, W), and G is not far from being a wreath product $H \mathrm{wr}\, S$ for some permutation group S (Theorems 8.4 and 8.5c).

• The cases where (G, V) is nonreal induced, in the above sense, can be treated using upper bounds for the class numbers of permutation groups (Theorems 9.3 and 10.3, 10.4d).

Theorem 10.5a. *We have $k(GV) \leq |V|$, where equality can hold only if there is a strongly real vector in V for G.*

Proof. We have proved that $k(GV) \leq |V|$ (see above). Suppose we have $k(GV) = |V|$. Then $k(G_i V_i) = |V_i|$ for each irreducible submodule V_i of V and $G_i = G/C_G(V_i)$ (Proposition 3.1a), and G is the direct product of the G_i. Combining Theorems 10.3, 10.4d and 8.4 we obtain that, for each i, there is a real vector $v_i \in V_i$ for G_i, which by Theorem 5.2b must be *strongly* real. Now $v = \sum_i v_i$ is strongly real for G. □

Theorem 10.5b. *We have $k(G) \leq |V| - 1 = p^m - 1$, and equality holds if and only if G is a Singer cycle in $\mathrm{GL}_m(p)$.*

Proof. Note that $k(G) < k(GV)$ as G is a proper quotient group of $GV = V : G$. If $k(G) = |V| - 1$, then G must be abelian and act transitively on $V^\sharp$ by Proposition 3.1b. Hence G is a Singer cycle in $\mathrm{GL}(V)$. The converse is known from Sec. 3.2. □

Chapter 11

Possibilities for $k(GV) = |V|$

It appears to be rather intricate to characterize the pairs (G, V) where $k(GV) = |V|$ (in the usual coprime situation). By Proposition 3.1a it suffices to consider the case that G is irreducible on V. Assuming $k(GV) = |V|$ we shall establish certain congruences which indicate that, at least for large characteristics, G should be a Singer cycle in $\mathrm{GL}(V)$. Unfortunately Clifford theory seems to fail in attacking this question.

11.1. Preliminaries

Let p be a prime, G a finite p'-group, and let V be a finite faithful $\mathbb{F}_p G$-module. Assume (without loss of generality) that G is irreducible on V.

Proposition 11.1a. *Suppose that $k(GV) = |V|$. Then we have the following partial results:*

(i) *If G has a regular orbit on V, then G is abelian.*

(ii) *If G is transitive on $V^\sharp$, then either G is a Singer cycle in $\mathrm{GL}(V)$ or $|V| = 2^3$ and G is the Frobenius group of order 21, or $|V| = 3^2$ and G is semidihedral of order 16.*

(iii) *If there is $v \in V$ such that $H = C_G(v)$ is abelian, then $k(HV) = |V|$ and H acts on each irreducible H-submodule of $[V, H]$ as a Singer cycle.*

(iv) *There is a strongly real vector in V for G.*

Proof. This has been established in Theorems 1.5d, 3.4d and 10.5a. □

We know that $k(GV) = |V|$ if G is a Singer cycle, and also in the two further cases described in (ii) above. (For the dihedral subgroups G of $\mathrm{GL}_2(3) = \mathrm{GL}(V)$ the equality also holds.)

Recall that Knörr's generalized character δ_V of G is given by $\delta_V(x) = |V : C_V(x)|$; cf. Eq. (3.3b). Let $e = p \cdot \exp(G)$, $K = \mathbb{Q}(e^{2\pi i/e})$ and R be the ring of integers of K. Let Γ be the subgroup of $\mathrm{Gal}(K|\mathbb{Q})$ fixing each p'-root of unity in K. So $\Gamma \cong \mathrm{Gal}(\mathbb{Q}(\varepsilon_p)|\mathbb{Q})$ where $\varepsilon_p = e^{2\pi i/p}$. Let $\mathfrak{p}|p$ be a prime ideal of R above p. Then $\mathfrak{p}_0 = \mathfrak{p} \cap \mathbb{Q}(\varepsilon_p)$ is the unique (totally ramified) prime ideal of $R_0 = \mathbb{Z}[\varepsilon_p] = R \cap \mathbb{Q}(\varepsilon_p)$ above p. We have $\mathfrak{p}_0 = (1 - \varepsilon_p)R_0$.

Lemma 11.1b. *Suppose that $k(GV) = |V|$. Let $v \in V$ be strongly real for G, and let $p \neq 3$. Then $H = C_G(v)$ maps into $\mathrm{SL}(V)$, and there is an integer-valued generalized character ψ of H such that $\psi(1) = 1$ and $\psi^2 = \delta_V$ (on H). We have $\langle \psi\chi, \theta \rangle_H = \pm 1$ for all $\chi \in \mathrm{Irr}(G)$ and $\theta \in \mathrm{Irr}(H)$.*

Proof. If H does not map into $\mathrm{SL}(V)$, clearly p is odd. Combining Theorems 5.2a and 3.3d yields that then $k(GV) \leq \frac{p+3}{2}|V|$. But this is impossible by assumption. By Lemma 5.2a there exists ψ as asserted. From Theorem 3.3c it follows that $\langle \psi\chi, \psi\chi \rangle_H = k(H)$ for all $\chi \in \mathrm{Irr}(G)$. Let $\theta \in \mathrm{Irr}(H)$. Then

$$|H|\langle \psi\chi, \theta \rangle_H = \sum_{h \in H} \psi(h)\chi(h)\theta(h^{-1}) \equiv \chi(1)\theta(1) \,(\mathrm{mod}\,\mathfrak{p}),$$

because $\psi(h)^2 = |V : C_V(h)|$ is a proper power of p for $h \in H^\sharp$, as H is faithful on V, and $\psi(1) = 1$. On the other hand, $\chi(1)\theta(1)$ is not divisible by p (Theorem 1.3b). Thus $\langle \psi\chi, \theta \rangle_H \neq 0$, and the result follows. $\square$

Keep the assumptions of the above lemma. As in Theorem 3.3d define the class function $\Psi = \Psi^v$ on $X = GV$ by letting $\Psi(x) = |C_V(h)|\psi(h)$ if $x \in X$ is conjugate to hv for some $h \in H$, and letting $\Psi(x) = 0$ otherwise.

Lemma 11.1c. *For every $\chi \in \mathrm{Irr}(X)$ the multiplicity $\langle \Psi, \chi \rangle = \pm 1$ if $p = 2$ or if V is in the kernel of χ, and otherwise it is a $2p$th root of unity. Moreover,*

$$|H|\langle \Psi, \chi \rangle = \chi(v^{-1}) + p\alpha$$

for some $\alpha \in \mathbb{Z}[\varepsilon_p]$.

Proof. We make use of Theorem 3.3d, and its proof. By its very construction as an induced class function, Ψ is a $\mathbb{Z}[\varepsilon_p]$-linear combination of characters of $X = GV$. Let us write $f_\chi = \langle \Psi, \chi \rangle$ for $\chi \in \mathrm{Irr}(X)$. So $f_\chi \in \mathbb{Z}[\varepsilon_p]$. Using Frobenius reciprocity one gets

$$|H|f_\chi = \sum_{h \in H} \psi(h)\chi(h^{-1}v^{-1}) = \chi(v^{-1}) + p\alpha$$

for some $\alpha \in R$, because $\psi(h)$ is divisible by p for all $h \in H^\sharp$. Observe that $\chi(v^{-1}) \in R_0 = \mathbb{Z}[\varepsilon_p]$ and $\chi(v^{-1}) \equiv \chi(1) \not\equiv 0 \,(\mathrm{mod}\,\mathfrak{p}_0)$ since p does not divide $\chi(1)$ (Theorem 1.3b). We see that $\alpha \in \mathbb{Q}(\varepsilon_p) \cap R = R_0$ and, of course, that $f_\chi \neq 0$. As in Theorem 3.3d one also computes that $\langle \Psi, \Psi \rangle = \frac{1}{|H|}\sum_{h \in H}|C_V(h)|\psi(h)^2 = |V|$.

Recall that for $\chi \in \mathrm{Irr}(X)$, $\sigma \in \Gamma$ and $h \in H$ we have $\chi^\sigma(hv) = \chi(hv)^\sigma = \chi(hv^s)$ if σ maps ε_p to ε_p^s. It follows that

$$f_\chi^\sigma = \langle \Psi, \chi^\sigma \rangle \neq 0.$$

Let N denote the norm for $\mathbb{Q}(\varepsilon_p)|\mathbb{Q}$. We know that $\mathrm{N}(f_\chi) = \prod_{\sigma \in \Gamma} f_\chi^\sigma$ is a nonzero integer. By the arithmetic–geometric mean inequality (1.5b)

$$\frac{1}{p-1} \sum_{\sigma \in \Gamma} |f_\chi^\sigma|^2 \geq |\mathrm{N}(f_\chi)|^{\frac{2}{p-1}},$$

and we have equality if and only if all $|f_\chi^\sigma|$, $\sigma \in \Gamma$, agree.

Now by hypothesis $k(X) = |V|$, that is, $|\mathrm{Irr}(X)| = |V|$. Observe that

$$(p-1)|V| = (p-1)\langle \Psi, \Psi \rangle = \sum_{\chi \in \mathrm{Irr}(X)} \sum_{\sigma \in \Gamma} |f_\chi^\sigma|^2.$$

Thus, for each $\chi \in \mathrm{Irr}(X)$, we have equality above with $|\mathrm{N}(f_\chi)| = 1$, whence $|f_\chi^\sigma| = 1$ for every $\sigma \in \Gamma$. Being a cyclotomic integer this implies that f_χ is a root of unity (in $R_0 = \mathbb{Z}[\varepsilon_p]$). Hence $f_\chi = \langle \Psi, \chi \rangle = \pm 1$ when $p = 2$, and otherwise it is a $2p$th root of unity. If V is in the kernel of χ, then

$$f_\chi = \langle \Psi, \chi \rangle = \frac{1}{|H|} \sum_{h \in H} \psi(h)\chi(h^{-1}) = \langle \psi, \chi \rangle_H = \pm 1$$

by the preceding lemma. This completes the proof. $\qquad\square$

11.2. Some Congruences

We keep the assumptions made in the preceding section. In particular V is an irreducible, faithful, coprime $\mathbb{F}_p G$-module, $X = GV$ and $k(X) = |V|$. The following congruences are established in [Schmid, 2005] and indicate that G should be a Singer cycle when p is large enough. For $p = 2, 3$ the congruences tell us nothing. For $p = 2$ the group G has odd order and so the Burnside congruence (1.5a) applies.

Theorem 11.2a. *For any irreducible character χ of $X = GV$ we have $\chi(1) \equiv \pm 1 \pmod{p}$, and $k(G) \equiv |G| \equiv \pm 1 \pmod{p}$.*

Proof. We may assume that $p \geq 5$. As above let $v \in V$ be strongly real for G, and let $H = C_G(v)$. Let ψ and Ψ be the generalized characters of H and G studied in Lemmas 11.1b and 11.1c, respectively.

Let $\chi \in \mathrm{Irr}(X)$ be an irreducible character of X. Then $\langle \Psi, \chi \rangle = \pm 1$ if $p = 2$ or if V is in the kernel of χ, and $\langle \Psi, \chi \rangle$ is a $2p$th root of unity otherwise (11.1c). Letting $R_0 = \mathbb{Z}[\varepsilon_p]$ and $\mathfrak{p}_0 = (1 - \varepsilon_p)R_0$ be as before, we have $\langle \Psi, \chi \rangle \equiv \pm 1 \pmod{\mathfrak{p}_0}$, at any rate. We have also seen that $|H|\langle \Psi, \chi \rangle = \chi(v^{-1}) + p\alpha$ for some $\alpha \in R_0$. We conclude that

$$\pm|H| \equiv |H|\langle \Psi, \chi \rangle \equiv \chi(1) + p\alpha \equiv \chi(1) \pmod{\mathfrak{p}_0}.$$

It follows that $\chi(1) \equiv \pm|H| \pmod{p}$. Taking for χ the 1-character of X we get that $|H| \equiv \pm 1 \pmod{p}$, and this in turn shows that $\chi(1) \equiv \pm 1 \pmod{p}$ in general. Picking $\chi = \chi_{v,1}$ as described in Proposition 3.1b, which is an irreducible character of X of degree $|G : H|$, we get $|G : H| \equiv \pm 1 \pmod{p}$. Thus $|G| = |H| \cdot |G : H| \equiv \pm 1 \pmod{p}$. Finally

$$|G| = \sum_{\chi \in \mathrm{Irr}(G)} \chi(1)^2 \equiv k(G) \pmod{p},$$

completing the proof. $\square$

Theorem 11.2b. *Let $H = C_G(v)$ for any vector $v \in V$. Then $k(H) \equiv |H| \equiv \pm 1 \pmod{p}$ and $\theta(1) \equiv \pm 1 \pmod{p}$ for every $\theta \in \mathrm{Irr}(H)$.*

Proof. Let $\chi = \chi_{v,1}$ in the notation of Proposition 3.1b. So χ is an irreducible character of $X = GV$ with degree $|G : H|$. Thus $|G : H| = \chi(1) \equiv \pm 1 \pmod{p}$ by the preceding theorem. Similarly, for every $\theta \in \mathrm{Irr}(H)$ the degree $\chi_{v,\varsigma}(1) = \chi(1) \cdot \theta(1) \equiv \pm 1 \pmod{p}$ likewise. Hence $\theta(1) \equiv \pm 1 \pmod{p}$ and, therefore,

$$|H| = \sum_{\theta \in \mathrm{Irr}(H)} \theta(1)^2 \equiv k(H) \pmod{p}.$$

Use finally that $|H| = |G|/|G : H|$. $\square$

11.3. Reduced Pairs

Though Clifford reduction seems to fail in attacking the present problem, it is of some interest to examine the reduced pairs (G, V). We use the notation introduced in Sec. 5.5. Hence V is a faithful FG-module where $F = \mathbb{F}_r$ for some power r of the prime p, the core E of G is absolutely irreducible on V, and $G_0 = N_{\mathrm{GL}(V)}(E)$. Also $Z = C_{G_0}(E) \cong F^\star$, and χ is the Brauer character of G (and of G_0) afforded by V.

Proposition 11.3a. *Suppose (G, V) is a reduced pair of quasisimple type. Then $k(GV) < |V|$.*

Proof. Assume $k(GV) = |V|$. Then by Proposition 11.1a, (i) there is no regular G-orbit on V. We may appeal to Theorem 7.2a. The minimal point stabilizers $H = C_{G_0}(v)$ listed there are abelian in most cases, in which case part (iii) of Proposition 11.1a applies. If H is nonabelian, the desired contradiction follows from Theorem 11.2b. For instance, the case $r = 7$, $H \cong D_{12}$ is ruled out as there is $\zeta \in \mathrm{Irr}(H)$ with $\zeta(1) = 2$.

Hence it remains to examine the permutation pairs (G, V), where the core E is an alternating group A_{d+1} and V is the deleted permutation module over F of dimension d, with $p \geq d + 2$ (Example 5.1a). Since there is no regular vector in V for G, $r = p = d + 2$ or $d + 3$. At first, G can be any subgroup of $G_0 = S \times Z$ containing E, where $S \cong S_{d+1}$ acts on V in the natural way. We know that there is a vector $u \in V$ such that $C_{G_0}(u)$ is cyclic of order dividing $p - 1$ (Example 5.1a). From $k(GV) = |V|$ and Theorem 11.2b we infer that $|C_G(u)| = |C_{G_0}(u)| = p - 1$. This forces that $p - 1 = d + 1$.

We next pick a strongly real vector $v \in V$ for G_0 as in Example 5.1a, namely $v = dw_0 - \sum_{i=1}^{d} w_i$ (where $\{w_i\}_{i=0}^{d}$ is a permutation basis). Then $H_0 = C_{G_0}(v) \cong S_d$ and $\mathrm{Res}_{H_0}^{G_0}(V)$ is the natural permutation module. Since the transpositions in H_0 act with determinant -1 on V and since $p \geq 7$, $H = C_G(v) \cong A_d$ by assumption and Lemma 11.1b. So $|H| = \frac{1}{2}d! = \frac{1}{2}(p-2)!$, and from Theorem 11.2b we get that

$$(p - 2)! \equiv \pm 2 \ (\mathrm{mod}\, p).$$

It follows that $(p - 1)! \equiv \mp 2 \ (\mathrm{mod}\, p)$. On the other hand, $(p - 1)! \equiv -1 \ (\mathrm{mod}\, p)$. This forces that $p = 3$, a contradiction. $\qquad \square$

Remark. Recently [Guralnick–Tiep, 2005] have shown that the inequality $k(GV) \leq \frac{1}{2}|V|$ holds for each reduced pair (G, V) of quasisimple type.

Proposition 11.3b. *Suppose (G, V) is reduced of extraspecial type. Then $k(GV) < |V|$ except possibly when $r = 3$.*

Proof. Assume that $k(GV) = |V|$ and that $r \neq 3$. Here E is a q-group of extraspecial type for some prime $q \neq p$, and $|E/Z(E)| = q^{2m}$ for some integer $m \geq 1$. By Proposition 11.1a, (i) there is no regular G-orbit on V. Similarly, there is no $v \in V$ such that $C_G(v) \neq 1$ is an elementary abelian 2-group. For then $r = 3$ by part (iii) of Proposition 11.1a, which has been excluded. There is also no strongly real vector $v \in V$ such that $H = C_G(v) \neq 1$ is cyclic. For then $k(HV) = |V|$, and from Proposition 3.1a it follows that $HV = H_1 V_1 \times \cdots \times H_n V_n$ where either $H_i = 1$ (and $|V_i| = r$) or $H_i \neq 1$ is a Singer cycle on V_i. Since H is cyclic, the orders of the H_i are pairwise prime to each other. Since V is self-dual as an FH-module, this implies that each V_i is self-dual as an FH_i-module (and $H = H_i$ for some i unless $p = 2$). From Lemma 4.1c it follows that $|H_i| = 2$ and $|V_i| = 3 = r$ for some i, against our assumption.

Probably the case $r = 3$ is not exceptional. But excluding $r = 3$ enables us to appeal to Theorem 6.3b. We get that $q = 3$, $m \leq 3$ or $q = 2$, $m \leq 5$, and r and p are suitably bounded above (Comments 6.3c).

Let first $p = 2$. Then we make use of Theorem 6.4 (including its proof). We have $E \cong 3_+^{1+2m}$ for $m = 2$ or 3 and $GZ = X \circ Z$ where $X = E : H$ for some point stabilizer $H = C_G(v)$, where v is strongly real for G. If $m = 2$ then H is cyclic of order 5, which is impossible. So we have $m = 3$, in which case H either is cylic of order 7 or is a Frobenius group of order 21. The former case cannot happen. From Theorem 6.4 we also know that the Weil character $\chi = \xi$ takes the value $\chi(h) = -1$ on the elements $h \in H$ of order 7, and $\chi(y) = 0, 3$ or -9 for the elements $y \in H$ of order 3. In addition $\chi(y) = 0$ for the noncentral elements y of E.

The dimensions of the eigenspaces on V of the noncentral elements of X of prime order are not greater than 12. Since

$$2|X|r^{14} = 2 \cdot 21 \cdot 3^7 r^{14} < r^{27}$$

for $r \geq 4$, there is a regular G-orbit on V, contradicting Proposition 11.1a.

Let next p be odd. If $q = 3$ then $p \geq 5$, and we make use of Theorem 6.5 (and its proof). If $m = 1$, then either there is $v \in V$ such that $C_G(v)$ is an elementary abelian 2-group, or (G, V) is nonreal reduced of type (3_+^3) and $k(GV) \leq \frac{1}{2}|V|$ by Proposition 10.1. If $m = 2$, then we find v as before, or $r = p = 7$ in which case χ is an irreducible character of G of degree $\chi(1) = 3^2 \not\equiv \pm 1 \pmod 7$, contradicting Theorem 11.2a. Let $m = 3$. We must have $r = 19, 13$ or 7 (Comments 6.3c). Here χ is an irreducible character of G of degree $\chi(1) = 3^3$, so that Theorem 11.2a applies for $r = p = 19$. For $r = 13$ or 7 we have $G \neq G_0$, and G maps onto a proper irreducible subgroup $\bar{G}$ of $\mathrm{Sp}_6(3)$. Checking the possible groups $\bar{G}$ and their irreducible character degrees gives the result.

Let $q = 2$ (and $p \neq 2$). There is no problem with $m = 1$ (Proposition 6.6a). For $m = 2$ we have $r \leq 83$ (Comments 6.3c). If $E \cong 2_+^{1+4}$, by Proposition 6.6b we have $p \neq 3$, and there is a strongly real $v \in V$ for G such that $C_G(v) \cong D_{12}$ when $r = 7$ and $C_G(v) \cong D_8$ otherwise (as point stabilizers cannot be cyclic or elementary abelian 2-groups). But then $C_G(v)$ has an irreducible character of degree 2, in contrast to Theorem 11.2b. For $E \cong 2_-^{1+4}$ either we get nonreal reduced pairs of type (2_-^5) or $r \geq 11$ and we find v as before. For $p = 3$, G maps into a $3'$-subgroup of $O_4^-(2) \cong S_5$, and we get a regular orbit. If $E \cong 2_0^{1+4}$, then we find a strongly real v for G such that $C_G(v)$ is an elementary abelian 2-group for $p = 3$ and, otherwise, is isomorphic to S_3 or A_3 or to other explicit given groups. In all the $p \neq 3$ cases Theorems 11.2a and 11.2b apply.

Let $E \cong 2_-^{1+6}$. Here we have the (crude) bound $r \leq 139$ (Comments 6.3c). By Proposition 6.6c we have $p \neq 3$, and if $r = 7 = p$ then there is $v \in V$ such that $C_G(v) \cong S_4$. Of course $24 \not\equiv \pm 1 \pmod 7$. For $r \geq 11$ there is a strongly real vector $v \in V$ for G such that $H = C_G(v)$ has a normal Sylow 3-subgroup $S \cong 3_+^{1+2}$ with index 2 or 4, and Theorem 11.2b applies.

Let $E \cong 2_+^{1+2m}$ with $m = 3, 4, 5$. For $p = 3$, by considering the irreducible $3'$-subgroups of $O_{2m}^+(2)$, one gets an elementary abelian 2-vector in V for G. Otherwise by Proposition 6.7a there is a strongly real vector $v \in V$ such that $C_{G_0}(v) \cong \mathrm{GL}_m(2)$. Similar statements when $E \cong 2_0^{1+2m}$ (Proposition 6.7b). Apply Theorems 11.2a and 11.2b.

Let finally $E \cong 2_-^{1+2m}$ with $m = 4, 5$. As before one rules out the case $p = 3$. Otherwise by Proposition 6.7c there is a strongly real vector $v \in V$ for G such that $C_{G_0}(v) \cong \mathrm{GL}_m(2)$, or $m = 4$ and $C_{G_0}(v) \cong G_2(2)'$. We have $r \leq 71$ for $m = 4$ and $r \leq 23$ for $m = 5$ (Comments 6.3c). Again Theorems 11.2a and 11.2b apply. $\square$

Remark. The proof is less laborious once better upper bounds for r are available, and these can be obtained by computations with [GAP]. In our investigations in Chapter 6 we attempted to avoid computer calculations.

Chapter 12

Some Consequences for Block Theory

There are various long-standing conjectures in modular representation theory. Presumably the most outstanding conjecture, and the most difficult one, is Brauer's $k(B)$ problem discussed in Chapter 2 of this monograph. In that chapter we already proved that the $k(GV)$ theorem implies Brauer's $k(B)$ conjecture for p-solvable groups. We shall recover this in terms of Brauer correspondence and blocks with normal defect groups.

12.1. Brauer Correspondence

Let p be a prime, and let B be a p-block of the finite group G with defect group D. As usual $k_0(B)$ denotes the number of (ordinary) irreducible characters belonging to B which are of height zero in the block. The *Brauer correspondent* b of B for $N_G(D)$ refers to Brauer's first main theorem on blocks (mentioned in Sec. 2.5) and is that block of $N_G(D)$ with defect group D for which $b^G = B$; sometimes b is called the *germ* of B (with respect to D). Observe that $D = O_p(N_G(D))$ by Theorem 2.3c.

Let us recall some conjectures in modular representation theory.

• Alperin–McKay conjecture: $k_0(B) = k_0(b)$.

• Olsson conjecture: $k_0(B) \leq |D : D'|$.

• Brauer's height zero conjecture: $k_0(B) = k(B)$ *if and only if D is abelian.*

• Broué conjecture: $k(B) = k(b)$ *if D is abelian.*

The Broué conjecture would follow from the Alperin–McKay conjecture and one-half of Brauer's height conjecture. It would also be a consequence of Alperin's weight conjecture [Alperin, 1987], or of the overall conjectures by [Dade, 1992]. From the $k(GV)$ theorem it follows that the above conjectures hold at least *locally*:

Theorem 12.1a. *We have $k(b) \leq |D|$ and $k_0(b) \leq |D : D'|$. Further $k_0(b) = k(b)$ if and only if the defect group D is abelian.*

So if the Broué conjecture holds, Brauer's $k(B)$ problem would be settled for abelian defect groups. The Alperin–McKay conjecture is known to be true for p-solvable groups [Okayama–Wajima, 1980], and for some classes of finite groups G (including the symmetric groups and certain groups of Lie type). Brauer's height conjecture has been also settled for p-solvable groups [Gluck–Wolf, 1984]. Hence we have the following.

Corollary 12.1b. *Suppose G is p-solvable. Then we have $k(B) \leq |D|$ and $k_0(B) \leq |D : D'|$. Also, $k_0(B) = k(B)$ if and only if D is abelian.*

12.2. Clifford Theory of Blocks

It has been shown by [Reynolds, 1963] that Brauer's height zero conjecture holds for blocks with normal defect groups. In order to prove Theorem 12.1a we have to appeal to his work. This is Clifford theory of blocks, dealing with *root blocks*. If B is a p-block of G with defect group D, there is a block b of $H = DC_G(D)$ with $b^G = B$ (Sec. 2.5). Each such b is a called a root of B in H. It follows from the first main theorem on blocks that these roots form a $N_G(D)$-conjugacy class of blocks of H. Also, D is the unique defect group of such a root b by Theorem 2.3c. The *inertial index* of B is defined as the index of the inertia group in $N_G(D)$ of b (or its *canonical character*, see below) over H, and it is crucial that this is a p'-number.

    As in Chapter 2 we let $X_{p'}$ denote the set of p'-elements of a finite group X, and $\chi_{p'} = \operatorname{Res}^X_{X_{p'}}(\chi)$ for a character χ of X.

Lemma 12.2a. *Suppose $H = DC_G(D)$ for some p-subgroup D of the finite group G, and let b be a p-block of H with defect group D. Then b contains a unique irreducible character θ having D in its kernel, and $\theta_{p'}$ is the unique irreducible Brauer character in b. This canonical character θ belongs to a block of H/D with defect zero and is of height zero in b. We have $k(b) = k(D)$; indeed $\operatorname{Irr}(b)$ consists of the characters θ_ζ for $\zeta \in \operatorname{Irr}(D)$, defined by $\theta_\zeta(y) = \theta(y)\zeta(y_p)$ if the p-part $y_p \in D$ and zero otherwise.*

For a proof we refer to [F, V.4.7].

Lemma 12.2b. *Let D, H, b and θ be as above. Let $T = I(\theta)$ be the inertia group in $N_G(D)$ of θ. Then the block $B = b^G$ of G has defect group D if and only if $|T : H|$ is not divisible by p.*

Proof. By the first main theorem on blocks (and transitivity of block induction) we may assume that $G = N_G(D)$. Hence $T = I_G(\theta)$. Since $\theta_{p'}$ is the unique irrreducible Brauer character in b, T is the inertia group in G of the block b, that is, T is the stabilizer of the block idempotent e_b to b (by conjugation). Thus the central character ω_b of b is stable in G.

We know that $D \subseteq D_0$ for some defect group D_0 of B (Sec. 2.5). By Lemma 2.3b there exists a conjugacy class c_0 of G with $c_0 \subseteq H$, $a_b(c_0) \neq 0$ and $\omega_B(\widehat{c_0}) \neq 0$ (in characteristic p), such that D_0 is a defect group of c_0 (see also Theorem 2.2b). Let $x \in c_0$ with $D_0 \subseteq C_G(x)$, and let $c = x^H$. Then $c_0 = \bigcup_i c^{t_i}$ where the t_i range over a right transversal of $N_G(c)$ in G. As in the proof for Theorem 2.3c,

$$0 \neq \omega_B(\widehat{c_0}) = \omega_b(\widehat{c_0}) = \sum_i \omega_b(\widehat{c}^{\,t_i}) = |G : N_G(c)| \omega_b(\widehat{c}).$$

Since $N_G(c) = C_G(x)H$ and D_0 is a Sylow p-subgroup of $C_G(x)$, it follows that p does not divide $|G : D_0 H|$. Thus $D_0 \subseteq H$ if and only if $|G : H|$ is not divisible by p, and then Lemma 2.3b implies that c and b have a defect group in common. $\qquad\square$

Lemma 12.2c. *Let N be a normal subgroup of the finite group X such that $G = X/N$ is a p'-group, and let $Z = \langle \varepsilon \rangle$ be generated by a primitive $\exp(N)$th root of unity ε. Suppose $\theta \in \mathrm{Irr}(N)$ is G-stable and $\theta_{p'} \in \mathrm{IBr}(N)$. Then the Clifford obstructions $\mu_G(\theta) = \mu_G(\theta_{p'})$ agree in $\mathrm{H}^2(G, Z_{p'})$.*

Proof. Let $K = \mathbb{Q}(\varepsilon)$ and $R = \mathbb{Z}_{(p)}[\varepsilon]$, and let $\mathfrak{p}$ be a prime of R above p and $F = R/\mathfrak{p}$. This R is a principal ideal domain. By Proposition 1.9a we may identify the Clifford obstruction $\mu_G(\theta) = \mu_{KG}(\theta)$ with an element of $\mathrm{H}^2(G, Z) = \mathrm{H}^2(G, Z_{p'}) \times \mathrm{H}^2(G, Z_p)$. Since G is a p'-group by hypothesis, $\mathrm{H}^2(G, Z_p) = 0$. Clearly the Brauer character $\theta_{p'}$ is G-invariant too.

By Theorem 1.1d there is a KN-module W affording θ. Now

$$\mathrm{End}_{KG}\big(\mathrm{Ind}_N^G(W)\big) = \bigoplus_{g \in G} K\tau_g$$

is a crossed product for some units τ_g sending $W \otimes g$ to W ($\tau_1 = 1$). By the preceding paragraph we may choose these units such that the factor set $\tau(g, h) = \tau_{gh}^{-1}\tau_g\tau_h \in Z_{p'}$ for all $g, h \in G$. Let $\{t_g\}_{g \in G}$ be a transversal to N in X, with $t_1 = 1$, and let $t(g, h) = t_{gh}^{-1}t_g t_h$ be the corresponding factor set. There are unique $\alpha_g \in \mathrm{GL}(W)$ such that

$$(w \otimes t_g)\tau_g = (w)\alpha_g$$

for all $w \in W$, $g \in G$. For $y \in N$ we have $(wy)\alpha_g = (wy \otimes \iota_g)\tau_g = (w \otimes t_g y)\tau_g = (w \otimes t_g y^{t_g})\tau_g = (w \otimes t_g)\tau_g y^{t_g} = (w)\alpha_g y^{t_g}$ (where we used that the tensor product is over KN and that τ_g is a KX-automorphism). One similarly shows that $\alpha(g, h) = \alpha_{gh}^{-1}\alpha_g\alpha_h$ is the map $w \mapsto \tau(g, h)wt(g, h)$ on W.

Let $\{w_j\}$ be a K-basis of W. Let U be the R-submodule of W generated by all $(w_j y)\alpha_g$, $y \in N$, $g \in G$. Since U is finitely generated and torsion-free, it is a free R-module. Since U contains a basis of W and is contained in W, the rank of U is $\theta(1) = \dim_K W$. From $(w)\alpha_g y^{t_g} = (wy)\tau_g$ for $w \in W$ we infer that U is stable under N. From

$$(w)\alpha_g \tau_h = w\alpha(g, h)\alpha_{gh} = \big(\tau(g, h)wt(g, h)\big)\alpha_{gh}$$

and $\tau(g, h) \in Z_{p'} \subseteq R^\star$ (unit group) we see that $(U)\alpha_g \subseteq U$ for all $g \in G$. Since $\alpha_g \alpha_{g^{-1}} = \alpha(g, g^{-1})$ is the map $u \mapsto \tau(g, g^{-1})ut(g, g^{-1})$ on U, which is invertible, we may view the α_g, via restriction, as elements of $\mathrm{GL}(U) = \mathrm{GL}_R(U)$. This in turn shows that U is a G-stable RN-lattice affording θ, yielding a projective R-representation of G with the same factor set τ as before.

The FN-module $V = U/\mathfrak{p}U$ affords $\theta_{p'}$. Identifying $Z_{p'}$ with the group $Z_{p'}(1 + \mathfrak{p})/(1 + \mathfrak{p})$ we get that $\mu_G(\theta_{p'}) = \mu_{FG}(\theta_{p'})$ agrees with $\mu_{KG}(\theta)$ in $\mathrm{H}^2(G, Z_{p'})$, as desired. $\qquad\square$

Proposition 12.2d (Reynolds). *Suppose B is a p-block of G with normal defect group, D. Then there exists a group G_0 having a normal Sylow p-subgroup $D_0 \cong D$ and a p-block B_0 of G_0 such that the irreducible characters and Brauer characters of B_0 and B are in 1-1 correspondence preserving heights, and B_0, B have the same decomposition and Cartan matrices.*

Proof. Let b be a root of B in $H = DC_G(D)$. By Theorem 2.3c, B is the unique block of G covering b. Let θ be the canonical character of b, and let $T = I_G(\theta)$. Observe that $D = O_p(H)$ is the unique defect group of b and that T contains the inertia groups of all irreducible characters or Brauer characters in b. By Lemma 12.2b, p does not divide the inertial index $|T : H|$ of B. We have $B = b^G = (b^T)^G$, and it follows from Theorem 1.8b that the blocks B and b^T behave (via character induction) as asserted for B, B_0 (cf. Theorem 2.6b). Thus we may assume that $T = G$. Then b is the unique block of H covered by B.

Let $N = C_G(D)$. Recall that the block idempotents to B and b are in the group algebra for N (Theorem 2.3c). Let b_N be the block of N having the same block idempotent as b. Then B is the unique block of G covering b_N, and b_N is the unique block of N covered by B. Note that $Z(D) = N \cap D$ is the unique defect group of b_N.

Let $\bar{G} = G/H$, and let $\bar{G}(\theta)$ be the representation group of θ. Since $\bar{G}$ is a p'-group, $\bar{G}(\theta)$ may be understood as a central extension of $\bar{G}$ with a cyclic group $Z_{p'}$ of order $\exp{(N)}_{p'}$ (Lemma 12.2c). Let $G(\theta) = \bar{G}(\theta) \Delta_{\bar{G}} G$ be the extended representation group. By Theorem 1.9c there is $\widehat{\theta} \in \mathrm{Irr}(G(\theta))$ extending θ when viewed as a character of $\mathrm{Ker}(G(\theta) \twoheadrightarrow \bar{G}(\theta)) \cong N$. Letting $\widehat{\theta}^{-1}$ be the unique linear constituent of $\widehat{\theta}$ on $\mathrm{Ker}(G(\theta) \twoheadrightarrow G) \cong Z_{p'}$ we have a 1-1 correspondence $\chi \leftrightarrow \zeta$ between $\mathrm{Irr}(G|\theta)$ and $\mathrm{Irr}(\bar{G}(\theta))|\widehat{\theta})$, given by $\chi = \widehat{\theta} \otimes \zeta$ (Theorem 1.9c).

By Schur–Zassenhaus there is a complement $\bar{Y} = Y/N$ to H/N in G/N. The group Y acts on D in the natural way (with kernel N). Let $\widehat{G} = D : Y$ and $\widetilde{G} = D : \bar{Y}$ be the semidirect products. Then $\widehat{G}$ maps onto $\widetilde{G}$, and $\widehat{G}$ maps onto G by replacing the direct product $N \times D$ by the central product $H = N \circ D$ over $Z(D)$. The characters of $\widehat{G}$ having the "diagonal" of $Z(D) \times Z(D)$ (within $N \times D$) in the kernel correspond to the characters of G.

The group of the proposition is nothing but the extended representation group $G_0 = G(\theta)$. This G_0 has a normal Sylow p-subgroup $D_0 \cong D$ with $C_{G_0}(D_0) = Z(D_0) \times Z_{p'}$. By Theorem 2.3c there is a unique p-block B_0 of G_0 covering $\widehat{\theta}$, and this has defect group D_0. In particular $\mathrm{Irr}(B_0) = \mathrm{Irr}(G_0|\widetilde{\theta})$. By Lemma 12.2a, θ is of defect zero in $N/Z(D)$, so that $\theta(1)_p = |N/Z(D)|_p$. Also, $\theta_{p'}$ is the unique irreducible Brauer character in b_N, and $\mathrm{Irr}(b_N)$ consists of the characters θ_λ, for each (linear) $\lambda \in \mathrm{Irr}(Z(D))$, satisfying $\theta_\lambda(y) = \theta(y)\lambda(y_p)$ if $y_p \in Z(N)$ and $= 0$ otherwise. We have $(\theta_\lambda)_{p'} = \theta_{p'}$ for each λ. Hence the above Clifford correspondence gives, by restriction to p'-elements, a bijection from $\mathrm{IBr}(B)$ onto $\mathrm{IBr}(B_0) = \mathrm{IBr}(G_0|\widetilde{\theta})$. Note that $(\widehat{\theta})_{p'}$ is irreducible and $(\widetilde{\theta})_{p'} = \widetilde{\theta}$.

Fix $\lambda \in \mathrm{Irr}(Z(D))$. Regard $\theta_\lambda \in \mathrm{Irr}(b_N)$ as a character of $(N \times D)/D \cong N$. The inertia group $T_\lambda = I_{\widehat{G}}(\lambda) = I_{\widehat{G}}(\theta_\lambda)$ contains $N \times D$. By Lemma 12.2c, $\mu_{\bar{T}_\lambda}(\theta_\lambda)$ is the restriction to $\bar{T}_\lambda = T_\lambda/(N \times D)$ of $\mu_{\bar{G}}(\theta)$. Hence we may regard $\bar{T}_\lambda(\theta_\lambda)$ as a subgroup of $\bar{G}(\theta)$. Let G_λ and S_λ be the subgroups of $G(\theta)$ and G_0, respectively, mapping onto $\bar{T}_\lambda$, so that $|G(\theta) : G_\lambda| = |G_0 : S_\lambda| = |G : T_\lambda|$. By Lemma 12.2c we can further pick a character $\widehat{\theta}_\lambda$ of G_λ extending θ_λ (in the usual sense) such that

$$(\widehat{\theta}_\lambda)_{p'} = \mathrm{Res}_{G_\lambda}^{G(\theta)}(\widehat{\theta}_{p'}).$$

Hence on the p'-group $Z_{p'}$, embedded into G_λ, $\widehat{\theta}_\lambda$ is a multiple of $\widetilde{\theta}^{-1}$. By Theorem 1.9c we have a 1-1 correspondence $\psi \leftrightarrow \varphi$ between $\mathrm{Irr}(T_\lambda|\theta_\lambda)$ and $\mathrm{Irr}(G_0|\lambda^{-1}\widetilde{\theta})$, given by $\psi = \widehat{\theta}_\lambda \otimes \varphi$. Here we use that each $\psi \in \mathrm{Irr}(T_\lambda|\theta_\lambda)$ lies over λ and that the diagonal of $Z(D) \times Z(D)$ in $\widehat{G}$ must be in the kernel of ψ when inflated to a subgroup of this group. It follows that $S_\lambda = I_{G_0}(\lambda^{-1}\widetilde{\theta})$ is the inertia group of $\lambda^{-1}\widetilde{\theta}$, viewed as a linear character of $Z(D_0) \times Z_{p'}$. By Theorem 1.8b, $\chi = \mathrm{Ind}_{T_\lambda}^G(\psi)$ and $\zeta = \mathrm{Ind}_{S_\lambda}^{G_0}(\varphi)$ are irreducible. By Frobenius reciprocity $\chi \in \mathrm{Irr}(B)$, as B is the unique block of G covering b_N, and $\zeta \in \mathrm{Irr}(B_0) = \mathrm{Irr}(G_0|\theta)$. Furthermore

$$\chi_{p'} = \mathrm{Ind}_{T_\lambda}^G(\psi)_{p'} = \mathrm{Ind}_{T_\lambda}^G(\psi_{p'}) = (\widehat{\theta})_{p'} \otimes \zeta_{p'}.$$

The correspondence $\chi \leftrightarrow \zeta$ between $\mathrm{Irr}(B)$ and $\mathrm{Irr}(B_0)$ preserves heights, and B, B_0 have the same decomposition and Cartan matrices. □

12.3. Blocks with Normal Defect Groups

In order to prove Theorem 12.1a we have to consider blocks with normal defect groups. The following lifting property for blocks has been established in [Külshammer, 1987].

Lemma 12.3a. *Let D be a normal p-subgroup of G. Then the natural map $G \twoheadrightarrow G/D'$ induces a bijection beteween the p-blocks of G and that of G/D'.*

Proof. Let $A = FG$ where F is a field of characteristic p. Let J be the kernel of the natural map $FG \twoheadrightarrow F[G/D']$. Thus J is the (left) ideal of $A = FG$ generated by all $t - 1$, $t \in D'$. But D' is generated by all commutators $[x, y]$ for $x, y \in D$, and

$$[x, y] - 1 = x^{-1}y^{-1}\big((x - 1)(y - 1) - (y - 1)(x - 1)\big) \in J(FD)^2 \subseteq J(A)^2,$$

and $xy - 1 = (x-1)(y-1) + (x-1) + (y-1) \equiv (x-1) + (y-1) \pmod{J(FD)^2}$. Consequently $J \subseteq J(A)^2$. (Cf. [Jennings, 1941] for a more precise result; in fact one may replace D' by the Frattini subgroup of D.) For any central idempotent e of $A = FG$, $J + e$ is a central idempotent of $A/J \cong F[G/D']$. If $J + e = J + f$ for a central idempotent f of A, then $e - ef = (e - 1)f$ is a central idempotent of A contained in J, hence is zero as J is nilpotent. Hence $e = ef$ and, similarly, $f = ef$, hence $e = f$.

Now let $\bar{e}$ be a central idempotent of A/J. By the usual lifting properties of idempotents there is an idempotent e of A such that $\bar{e} = J + e$ [F, I.12.3]. It remains to show that e is a *central* idempotent of A. Since $\bar{e}$ is central (and $\bar{e}(1 - \bar{e}) = 0$),

$$J + eA(1 - e) = J + e(1 - e)A = J.$$

Hence $eA(1 - e) \subseteq J \subseteq J(A)^2$ and $eA(1 - e) \subseteq J(A)^2(1 - e)$. Suppose we know already that $eA(1 - e) \subseteq eJ(A)^n(1 - e)$ for some integer $n \geq 2$. Then $eA(1 - e) \subseteq eJ(A)^{n-1}eJ(A)(1 - e) + eJ(A)^{n-1}(1 - e)J(A)(1 - e) \subseteq eJ(A)^{n+1}(1 - e)$. Since $J(A)$ is nilpotent, this shows that $eA(1 - e) = 0$. Similarly $(1 - e)Ae = 0$. For any element $a \in A$ we therefore have $ea = eae + ea(1 - e) = eae$ and, analogously, $ae = eae$. Consequently $ea = ae$. Thus e is a central idempotent of $A = FG$, completing the proof. $\qquad \square$

Theorem 12.3b. *Suppose B is a p-block of G with normal defect group D. Then $k(B) \leq |D|$ and $k_0(B) \leq |D : D'|$.*

Proof. By Proposition 12.2d we may assume that D is a (normal) Sylow p-subgroup of G. It follows that G is p-solvable. Thus $k(B) \leq |D|$ by Theorems 2.6c/10.5a. By Lemma 12.3a there is a unique p-block B' of G/D' corresponding to B. Clearly D/D' is the unique defect group of B'.

Let $\chi \in \mathrm{Irr}(B')$. Then χ is an irreducible character of G (via inflation) with kernel $\mathrm{Ker}(\chi) \supseteq D'$. Let θ be an irreducible (linear) constituent of $\mathrm{Res}_D^G(\chi)$, and let $T = I_G(\theta)$. By Theorem 1.8b,

$$\chi(1) = e_\chi |G : T|\theta(1) = e_\chi |G : T|$$

where the ramification index e_χ is a divisor of $|T/D|$. Hence p does not divide $\chi(1)$. Therefore χ is of height zero.

On the other hand, suppose $\chi \in \mathrm{Irr}(B)$ is of height zero. Since the defect group D of B is a Sylow p-subgroup of G, $\chi(1)$ is not divisible by p. Let θ be an irreducible constituent of $\mathrm{Res}_D^G(\chi)$. By Theorem 1.8b, $\theta(1)$ is a divisor of $\chi(1)$. Hence $\theta(1)$ is not divisible by p. Since D is a p-group, $\theta(1)$ is a power of p. Hence $\theta(1) = 1$ and so χ has D' in its kernel. By definition χ, as a character of G/D', belongs to B'.

We have proved that $k_0(B) = k_0(B') = k(B')$. From Theorem 10.5a it follows that $k(B') \leq |D/D'|$. $\qquad \square$

Remark. The proof that $k(B) = k_0(B)$ if and only if D is abelian [Reynolds, 1963] is quite similar.

Chapter 13

The Non-Coprime Situation

As already mentioned in the preface, the inequality $k(GV) \le |V|$ does not hold if one drops the assumption of coprimeness (of $|G|$ and $|V|$), in general.

Examples 13.1. Let p be a prime and $m \ge 2$ be an integer.

(i) Suppose G is a Sylow p-subgroup of $\mathrm{GL}_m(p)$, consisting of the matrices with ones in the main diagonal and zeros above it, and let $V = \mathbb{F}_p^{(m)}$ be the standard module. The structure of G is well understood [Huppert, 1967, III. §16]. One knows that G has (elementary) abelian normal subgroups of order $p^{m^2/4}$ if m is even and of order $p^{(m^2-1)/4}$ otherwise. In [Higman, 1960] it is proved that $k(G) < (m-1)! p^{\lfloor \frac{m}{4} \rfloor}$. Following Higman we show that $k(G) > p^{\frac{1}{13} m^2}$. Since $k(G) \ge k(G/G') = p^{m-1}$, we may assume that $m \ge 12$. For positive integers $r \ge s$ and $r \ge t$, with $2r + s + t = m$, consider the elements of G consisting of block matrices of the form

$$\begin{pmatrix} I_s & & & \\ A & I_r & & \\ & J & I_r & \\ & & B & I_t \end{pmatrix}.$$

Here A is a $r \times s$ matrix having nonzero entries on the diagonal leading from the bottom right corner and zeros below this diagonal, J an $r \times r$ matrix with nonzero entries on its subsidiary diagonal and zeros otherwise, and B is a $t \times r$ matrix having nonzero entries on the diagonal leading from the top left-hand corner and zeros below this diagonal. Such (s, r, r, t) partitioned matrices are conjugate in G only if they are equal. Now consider those matrices where $r = \lceil \frac{m}{3} \rceil$ and $s = \lfloor \frac{m}{6} \rfloor$. The number of freely disposable positions in matrices of the above form is $c_m = r(s+t) - s(s+1)/2 - t(t+1)/2 + 2$, so that $k(G) \ge p^{c_m}$. One verifies that $c_m > \frac{m^2}{13}$.

(ii) $G = \mathrm{GL}_m(p)$ acts irreducibly on $V = \mathbb{F}_p^{(m)}$. We claim that the affine group $GV = \mathrm{AGL}_m(p)$ has class number $k(GV) = k(G) + k\big(\mathrm{AGL}_{m-1}(p)\big) > |V|$ (for $m \ge 2$). It is known that $p^m - p^{m-1} < k(G) \le p^m - 1$ [Green, 1955]. For $\lambda \ne 1_V$ in $\mathrm{Irr}(V)$ we have $I_G(\lambda) \cong \mathrm{AGL}_{m-1}(p)$. Now argue by induction using (1.10b) and the fact that G is transitive on $\mathrm{Irr}(V)^\sharp$. Of course $k\big(\mathrm{AGL}_1(p)\big) = p$.

(iii) $G = \mathrm{GL}_2(p) \mathrm{wr}\, Z_m$ acts faithfully and irreducibly on $V = \mathbb{F}_p^{(2m)}$ (in the natural way). We have $GV = \left(\mathbb{F}_p^{(2)} : \mathrm{GL}_2(p)\right) \mathrm{wr}\, Z_m$ and

$$k(GV) \geq \frac{1}{m}\left(\frac{p^2 + p - 1}{p^2}\right)^m |V|$$

by Theorem 1.7a, because the affine group $\mathrm{AGL}_2(p) = \mathbb{F}_p^{(2)} : \mathrm{GL}_2(p)$ has class number $k(\mathrm{GL}_2(p)) + k(\mathrm{AGL}_1(p)) = (p^2 - 1) + p$. (For $p = 2$ the group $GV \cong S_4 \mathrm{wr}\, Z_m$ has been already studied in Example 9.4c.) On the other hand, $k(G) \leq (p^2 - 1)^m + (p^2 - 1)m - (p^2 - 1) < p^{2m} = |V|$ by Proposition 8.5d.

Proposition 13.2 (Liebeck–Pyber). *If G is an irreducible subgroup of $\mathrm{GL}_m(p)$, then $k(G) \leq p^{cm}$ for some constant $c \leq 10$.*

This is Theorem 4 in [Liebeck–Pyber, 1997]. One might conjecture that here $k(G) \leq p^m - 1$, like in the coprime situation (Theorem 10.5b). The recent work by [Guralnick–Tiep, 2005] gives some evidence that this might be true at least when G is almost quasisimple.

We turn to the computation of $k(GV)$. Inspection of the proof for Theorem 2.6c shows that if $k(GV) \leq |V|$ holds in some more general situations, the $k(B)$ conjecture holds for a corresponding wider class of p-constrained groups. First two general observations.

Lemma 13.3. *Suppose the finite group X has an abelian normal subgroup V. Let $G = X/V$, and let $GV = V : G$ be the semidirect product (with G acting as before). Then $k(X) \leq k(GV)$.*

Proof. This is immediate from formula (1.10b) since linear characters of V can be extended to their inertia groups provided these split over V. $\square$

Lemma 13.4. *Let V be a faithful $\mathbb{F}_p G$-module, and let $O_p(G) = 1$. Let $\widetilde{V}$ be the direct sum of the composition factors of V (in a fixed composition series). Then $k(GV) \leq k(G\widetilde{V})$.*

Proof. Note that G is faithful on $\widetilde{V}$ since $O_p(G) = 1$ (and since G is faithful on V). Let $\{g_j\}$ be a set of representatives for the conjugacy classes of G. By Lemma 3.1c and (1.4a)

$$k(GV) = \sum_j |C\ell(C_G(g_j)|V/[V, g_j])| = \sum_j \frac{1}{|C_G(g_j)|} \sum_{h \in C_G(g_j)} |C_{V/[V, g_j]}(h)|.$$

Hence $k(GV) = \frac{1}{|G|} \sum_{g \in G} \sum_{h \in C_G(g)} |C_{V/[V,g]}(h)|$. There is a corresponding identity for $k(G\widetilde{V})$. Hence it suffices to show that

$$\dim C_{V/[V,g]}(h) \leq \dim C_{\widetilde{V}/[\widetilde{V},g]}(h)$$

for all elements $g \in G$ and $h \in C_G(g)$. Let $H = \langle g, h \rangle$. For any short exact sequence $0 \to U \to V \to W \to 0$ of H-modules, $U/[U,g] \to V/[V,g] \to W/[W,g] \to 0$ is an induced exact sequence of $\langle h \rangle$-modules. This in turn gives rise to the exact sequence $0 \to (W/[W,g])^* \to (V/[V,g])^* \to (U/[U,g])^*$ for the dual $\langle h \rangle$-modules, and then $0 \to C_{(W/[W,g])^*}(h) \to C_{(V/[V,g])^*}(h) \to C_{(U/[U,g])^*}(h)$ is exact. By Theorem 1.4b the cyclic group $\langle h \rangle$ has on any module M the same number of fixed points as on its dual module $M^* = \mathrm{Irr}(M)$. Consequently

$$\dim C_{V/[V,g]}(h) \leq \dim C_{U/[U,g]}(h) + \dim C_{W/[W,g]}(h).$$

Now apply this (inductively) to the given G-composition series of V. $\quad\square$

Proposition 13.5 (Kovács–Robinson). *Suppose the finite group X has a normal elementary abelian p-subgroup V of order p^m. Assume $G = X/V$ is faithful and completely reducible on V. If G is p-solvable, then there is a constant c, not depending on p or m, such that $k(X) \leq c^m |V|$.*

This has been proved in [Kovács–Robinson, 1993, Theorem 4.1]. Independence of the constant c from the prime p comes from the Fong–Swan theorem (mentioned in Sec. 2.7). In [Liebeck–Pyber, 1997] it has been shown that one actually may take $c = 103$. But by assumption and Theorem 2.3c, X has a unique p-block. Since the $k(B)$ problem is solved for p-solvable groups, application of the Fong–Swan theorem, lifting the Brauer characters afforded by the irreducible summands of V to p-rational characters of G, and of a result by Schur [I, 14.19] therefore yields that

$$k(X) \leq |X|_p < p^{mp/(p-1)^2} |V|$$

in this case, and $k(X) \leq |V|$ if $m < p - 1$. Of course the constant $c_p = p^{p/(p-1)^2}$ depends on the prime p, but $c_p < 2$ for $p \neq 2, 3$ and $c_p \searrow 1$ when p tends to infinity. One might conjecture that $k(X) \leq |V|$ in this p-solvable situation, provided $p \geq 5$ is large enough. (By Example 13.1, (iii) this is false for $p = 2, 3$.) For arbitrary G a (weak) upper bound is given in [Liebeck–Pyber, 1997].

Proposition 13.6 (Robinson). *Suppose the finite group X has a normal abelian subgroup V such that $C_X(V)$ is a p-group. For any integer $s \geq 0$ let $k_s(X)$ be the number of irreducible characters of X such that $p^s\chi(1)_p = |X|_p$. If $k(X) \leq |V|$, then $\sum_{s=0}^{\infty} k_s(X)/p^{2s} \leq 1/|V|$, and equality only holds when V is a Sylow p-subgroup of X.*

Proof. Let $|V| = p^m$ and $|X|_p = p^a$. Then $p^m\chi(1)_p \leq p^a$ for all $\chi \in \mathrm{Irr}(X)$ by Theorem 1.3b. Hence $k_s(X) = 0$ for $s < m$, and

$$\sum_{s=0}^{\infty} k_s(X)/p^{2s} \leq k(X)/p^{2m} \leq |V|/p^{2m} = 1/|V|.$$

Assume the equality is attained. Then $\sum_{s=m}^{a} k_s(X)p^{2s-2m} = 1$ and so $k_s(X) = 0$ for $s > m$. Thus $p^a = p^m\chi(1)_p = p^m$, picking $\chi = 1_X$. $\square$

In this proposition the group X again has a unique p-block, and $k_s(X)$ is the number of irreducible characters of X with height p^{a-s} in this block ($|X|_p = p^a$). This is motivated from [Brauer–Feit, 1959], counting characters with prescribed height (see also [F, V.§9]).

Epilogue: There are lots of open questions in this area, and work continues. [Guralnick and Tiep, 2005] have studied thoroughly the case where $G = X/V$ is almost quasisimple and faithful, irreducible on V. They proved that $k(X) \leq \frac{1}{2}|V|$ provided G does not involve A_n for $5 \leq n \leq 16$ or a group of Lie type of (untwisted) rank at most 6 or a classical group with V related to the standard module. The exceptional cases have still to be examined. In a forthcoming paper D. Gluck has treated permutation pairs (Example 5.1a) in the non-coprime situation. As I understand, he is also investigating the question whether $k(G) \leq p^m - 1$ when G is an irreducible subgroup of $GL_m(p)$.

Although the status of the classification of the finite simple groups is satisfactory [Aschbacher, 2004], a proof of the $k(GV)$ theorem independent of that would be greatly appreciated. Likewise challenging is Brauer's $k(B)$ problem. [Solomon, 2001] has the dream of returning in 100 years to ask about the meaning of the sporadic simple groups. My dream is to return in 100 years to ask, among others, about this mysterious Brauer problem.

Appendix A

Cohomology of Finite Groups

Let G be a *finite* group and let V be a (right) G-module. The (Tate) cohomology is a $\mathbb{Z}$-sequence of abelian groups $\mathrm{H}^n(G,V)$ defined by $\mathrm{H}^0(G,V) = V^G/Vtr_G$, where $V^G = C_V(G)$ is the fixed module and $tr_G = \sum_{g \in G} g$ in the group algebra $\mathbb{Z}G$, which all vanish if $V = \mathrm{Ind}_1^G(U)$ for some abelian group U, together with *connecting homomorphisms* declared functorially for short exact sequences of G-modules. The existence is guaranteed via projective resolutions of the trivial G-module $\mathbb{Z}$. For an introduction see [Hilton–Stammbach, 1971], which will be quoted as [HS].

In this book basically 1- and 2-cohomology is used, where we have the following familiar interpretations. Suppose $V \rightarrowtail X \twoheadrightarrow G$ is an extension of the G-module V, so that X is a group with normal subgroup V and quotient group $G = X/V$, and X induces by conjugation the given G-module structure on V. Choosing a transversal $\{t_g\}_{g \in G}$ to V in X we get a factor set (2-cocycle) $\tau \in Z^2(G,V)$ by defining $\tau(g,h) = t_{gh}^{-1}t_g t_h$. This $Z^2(G,V)$ is an (abelian) group by pointwise multiplication. Passing to other transversals gives a congruence relation on $Z^2(G,V)$, and $\mathrm{H}^2(G,V)$ is the corresponding quotient group. Conversely, each $\tau \in Z^2(G,V)$ defines such a group extension $X = X(\tau)$, with underlying set $G \times V$ and multiplication $(g,v) \cdot (h,w) = (gh, vh + w + \tau(g,h))$. Thus $\mathrm{H}^2(G,U)$ is in natural 1-1 correspondence with the (equivalence classes) of these group extensions, the zero element describing the semidirect product $GV = V : G$.

In the semidirect product $X = GV$ there may exist another subgroup $G_\alpha = \{gv_g |\ g \in G\}$ *complementing* V in X ($G_\alpha \cap V = 1$ and $X = G_\alpha V$). Then $\alpha : g \mapsto v_g$ is a crossed homomorphism (1-cocycle) of G in V, and with respect to pointwise multiplication we get a group $Z^1(G,V)$ acting regularly on the complements. Conjugacy of the complements in X (under V) yields a congruence relation, and $\mathrm{H}^1(G,V)$ is the corresponding quotient group. So $\mathrm{H}^1(G,V) = 0$ if and only if all complements are conjugate.

Given any extension $V \rightarrowtail X \twoheadrightarrow G$ of the G-module V, $Z^1(G,V)$ may be identified with the group of all automorphisms of X centralizing V and G, and $\mathrm{H}^1(G,V)$ accords with the quotient modulo those inner

213

automorphisms of X coming from V. If $V = V^G$ is a trivial G-module (X a *central extension* of G), then $\mathrm{H}^1(G, V) = \mathrm{Hom}(G, V)$.

Some general results on cohomology can be proved by describing them in cohomological dimension zero, say, and extending them functorially by *dimension–shifting*. This is due to the fact that there is a 1-induced G-module $\widehat{V} = \mathrm{Ind}_1^G\big(\mathrm{Res}_1^G(V)\big)$ having V as quotient and as submodule. For $n \geq 1$ the groups $\mathrm{H}_n(G, V) = \mathrm{H}^{-(n+1)}(G, V)$ are called *homology groups*.

(A1) The exponent of $\mathrm{H}^n(G, V)$ is a divisor of $|G|$. This is obvious for $n = 0$, hence is true for all n. For $n = 1, 2$ it leads, together with Sylow's theorem (and Feit–Thompson), to the *Schur–Zassenhaus* theorem.

(A2) *Inflation, restriction:* In dimension zero the restriction map Res_H^G for a subgroup H of G is defined by $v + V tr_G \mapsto v + V tr_H$ ($v \in V^G$), is induced from $H \rightarrowtail G \to V$ on 1-cocycles, and is extended to all dimensions by dimension–shifting. Observe that $\mathrm{Res}_H^G(\widehat{V})$ is a 1-induced H-module by Mackey decomposition. Let H be normal in G. Then V^H is a G/H-module in the obvious way, and so is $\mathrm{H}^0(H, V)$, and $\mathrm{H}^n(H, V)$ for each n. The inflation map is $v + V^H tr_{G/H} \mapsto v + V tr_G$ ($v \in V^G$) in dimension 0, is induced from $G \twoheadrightarrow G/H \to V^H \rightarrowtail V$ on 1-cocycles, and is extended by dimension–shifting. There is a natural exact sequence

$$0 \to \mathrm{H}^1(G/H, V^H) \overset{\mathrm{Inf}}{\to} \mathrm{H}^1(G, V) \overset{\mathrm{Res}}{\to} \mathrm{H}^1(H, V)^{G/H} \to \mathrm{H}^2(G/H, V^H) \overset{\mathrm{Inf}}{\to} K$$

where K is the kernel of $\mathrm{Res} : \mathrm{H}^2(G, V) \to \mathrm{H}^2(H, V)$ [HS, VI.81]. One can even describe the image of $\mathrm{H}^2(G/H, V^H)$ in K as the kernel of a certain transgression map $K \to \mathrm{H}^1\big(G/H, \mathrm{H}^1(H, V)\big)$, *e.g.*, using spectral sequences.

(A3) *Shapiro's lemma:* Let $V = \mathrm{Ind}_H^G(U)$ for some subgroup H of G and some H-module U. Then restriction to H and H-projection of V onto U gives rise to natural isomorphisms $\mathrm{H}^n(G, V) \cong \mathrm{H}^n(H, U)$. This is easily verified for $n = 0$, hence holds everywhere (see also [HS], pp. 164 and 224).

(A4) *Restriction, corestriction:* Let H be a subgroup of G. The corestriction map Cor_H^G is induced from the *relative trace* $tr_{H \backslash G} : V^H \to V^G$ (with respect to any right transversal to H in G) in dimension zero, hence in general. Composition $\mathrm{Cor} \circ \mathrm{Res}$ is multiplication with $|G : H|$ on $\mathrm{H}^n(G, V)$ (see also [HS, VI.16.4]). It follows that if p is a prime not dividing $|G : H|$, then the restriction map $\mathrm{H}^n(G, V) \to \mathrm{H}^n(H, V)$ is injective on the p-components (which gives *Gaschütz's* splitting theorem).

(A5) $\widehat{\mathbb{Z}} = \mathbb{Z}G$ is the regular module (group algebra) over $\mathbb{Z}$, which maps onto $\mathbb{Z}$ with kernel I_G, the so-called augmentation ideal, and contains $\mathbb{Z}tr_G \cong \mathbb{Z}$. One gets $H_1(G, \mathbb{Z}) \cong I_G/I_G^2 \cong G/G'$ [HS, VI .4.1]. The *Schur multiplier* of G is defined as $M(G) = H_2(G, \mathbb{Z})$ or, equivalently, by the *Hopf–Schur* formula $M(G) = R \cap \mathcal{F}'/[R, \mathcal{F}]$ in any free presentation $R \rightarrowtail \mathcal{F} \twoheadrightarrow G$ of G [HS, VI.§9]. Suppose Z is any trivial G-module. Then, by the universal coefficient theorem [HS, VI.15.2], there is a natural exact sequence

$$0 \to \mathrm{Ext}(G/G', Z) \overset{\mathrm{Inf}}{\rightarrowtail} H^2(G, Z) \overset{\rho}{\to} \mathrm{Hom}(M(G), Z) \to 0.$$

Here $\mathrm{Ext}(G/G', Z)$ governs the *abelian* extensions of G/G' by Z, described by *symmetric* 2-cocycles, and ρ sends the cohomology class of a central extension $Z \rightarrowtail X \twoheadrightarrow G$ to the (transgression) map $M(G) \to Z$, which has the image $Z \cap X'$. This may be derived on the basis of the Hopf–Schur formula. Using that $\mathrm{Ext}(G/G', Z) = 0$ if Z is a divisible group we see that $H^2(G, \mathbb{C}^\star)$ is the dual group to $M(G)$.

The central extension $Z \rightarrowtail X \twoheadrightarrow G$, with cohomology class $\mathbf{x}$ say, is called *proper* if $Z \subseteq X'$ ($\rho_{\mathbf{x}}$ epimorphic), and a *covering group* (or Schur cover) if in addition $M(G) \cong Z$ (via $\rho_{\mathbf{x}}$). If $G = G'$ is perfect, there is a unique (universal) covering group $\widehat{G}$ of G, up to isomorphism; every automorphism of G can be lifted to $\widehat{G}$.

(A6) The Schur multipliers of all quasisimple groups are known (*Schur, Steinberg, Griess* and others). In particular: $M(A_n) = Z_2$ for $n \neq 6, 7$ ($M(A_6) = M(A_7) = Z_6$); $M(\mathrm{GL}_m(2)) = 1$ for $m \neq 3, 4$; $M(\mathrm{Sp}_{2m}(q)) = 1$ for all m and all odd prime powers q except when $(m, q) = (1, 9)$; $M(\mathrm{Sp}_{2m}(2)) = 1$ for $m \geq 4$; $M(\Omega_{2m}^{\pm}(2)) = 1$ for $m \neq 2, 3$ and with the exception $M(\Omega_8^+(2)) = Z_2 \times Z_2$; $M(\mathrm{SU}_m(q)) = 1$ for $m \geq 3$ and $(m, q) \neq (4, 2), (4, 3), (6, 2)$.

(A7) *Module extensions:* Let $U \rightarrowtail V \twoheadrightarrow W$ be an extension of G-modules. We are only dealing with $\mathbb{Z}$-split extensions. Even more, we shall be concerned mostly with FG-modules over a finite field F. These extensions are classified by $\mathrm{Ext}_{FG}(W, U) = H^1(G, \mathrm{Hom}_F(W, U))$. Here $\mathrm{Hom}_F(W, U) \cong W^* \otimes_F U$ as FG-modules (via diagonal actions), and $W^* = \mathrm{Hom}_F(W, F)$ is the *dual module* to W. From projective resolutions of the trivial FG-module F it is immediate that only irreducible modules belonging to the principal block can have nontrivial cohomology. If γ is a group or field automorphism and V^γ is the *twisted* FG-module thus obtained, $H^n(G, V^\gamma) \cong H^n(G, V)$ via γ. (All Ext-groups vanish if $\mathrm{char}(F) \nmid |G|$; *Maschke's* theorem.)

Remark. Let V be an FG-module and $T^2(V) = V \otimes_F V$. Identify $\Lambda^2(V) = \text{Alt}^2(V)$ with the alternating square, and let $\text{Sym}^2(V)$ be the symmetric square (fixed space of $v \otimes w \mapsto w \otimes v$). Let $G = \text{SL}(V)$. If $\text{char}(F) \neq 2$, then both $\Lambda^2(V)$ and $\text{Sym}^2(V)$ are irreducible FG-modules and $T^2(V) = \Lambda^2(V) \oplus \text{Sym}^2(V)$, as in (1.1). Otherwise $\Lambda^2(V)$ is a submodule of $\text{Sym}^2(V)$ with quotient module isomorphic to the twist V^γ of V under the Frobenius automorphism γ of F, and $T^2(V)$ is a uniserial FG-module with composition factors $\Lambda^2(V), V^\gamma, \Lambda^2(V)$.

(A8) Let $V = \mathbb{F}_p^{(2m)}$ for some prime p and some $m \geq 1$. Then $\text{M}(V) = \Lambda^2(V)$ by the *Künneth* theorem [HS, V.3.1]. Letting $R \rightarrowtail \mathcal{F} \twoheadrightarrow V$ be a free presentation of (minimal) rank $2m$, we have a natural exact sequence $\text{M}(V) \rightarrowtail R/[R, \mathcal{F}]R^p \twoheadrightarrow V$. This sequence splits, and splits naturally through the pth power map when p is odd. The dual of this sequence agrees with the universal coefficient sequence $V^* \rightarrowtail \text{H}^2(V, \mathbb{F}_p) \twoheadrightarrow \Lambda^2(V)^*$. Now regard $V \cong V^*$ as the standard module for $G = \text{Sp}_{2m}(p)$. Since G is absolutely irreducible on V, $\text{Hom}_G(V \otimes V^*, \mathbb{F}_p) = \text{End}_G(V) = \mathbb{F}_p$. In the odd case $\text{H}^2(V, \mathbb{F}_p) \cong V^* \oplus \Lambda^2(V)^*$ as a G-module, and there is a unique alternating form $\neq 0$, up to scalar multiples, which is G-invariant. This defines the extraspecial group p_+^{1+2m} (since the group of exponent p^2 does not admit the symplectic group G). For $p = 2$ we have a similar result replacing the symplectic group by $\text{O}_{2m}^\pm(2)$, defining thus the groups $2_\pm^{1+2m}$.

(A9) The cohomology of certain classical groups on their standard modules and related modules is known [Griess, 1973], [Sah, 1977], [Bell, 1978]. It vanishes by (A2) whenever a nontrivial scalar multiplication appears. Also, $\text{H}^1(\text{SL}_m(q), \mathbb{F}_q^{(m)}) = 0$ for $(m, q) \neq (2, 2^r)$ with $r > 1$ and $(m, q) \neq (3, 2)$, $\text{H}^2(\text{SL}_m(q), \mathbb{F}_q^{(m)}) = 0$ for $(m, q) \neq (2, 2^r)$ with $r > 2$, $(m, q) \neq (3, 3^r)$ with $r > 1$, and $(m, q) \neq (3, 2), (4, 2), (5, 2), (3, 5)$; $\text{H}^n(\text{Sp}_{2m}(2), \mathbb{F}_2^{(2m)}) = Z_2 = \text{H}^n(\text{Sp}_{2m}(2) \times Z_2, \mathbb{F}_2^{(2m)})$ for $m > 1$ and $n = 1, 2$, but $\text{H}^2(\text{Sp}_4(2)', \mathbb{F}_2^{(4)}) = 0$; further $\text{H}^1(\text{O}_{2m}^\pm(2), \mathbb{F}_2^{(2m)}) = 0$ except for $m = 3$ and positive type, and $\text{H}^2(\text{O}_{2m}^\pm(2), \mathbb{F}_2^{(2m)}) = Z_2$ for $m > 2$ and the same result for $\Omega_{2m}^\pm(2)$ except for $\Omega_6^+(2)$ where the 2-cohomology vanishes (as it does for $m = 1, 2$). – The computer was needed to compute the 2-cohomology of some orthogonal groups on the standard module, e.g., for $\Omega_{10}^+(2)$ and $\Omega_{12}^-(2)$ (Derek Holt).

(A10) The nonzero element, if any, of the 2-cohomology of the orthogonal groups $\text{O}_{2m}^\pm(2)$, $\Omega_{2m}^\pm(2)$ and the groups $\text{Sp}_{2m}(2)$, $\text{Sp}_{2m}(2) \times Z_2$, on the standard modules is represented by the automorphism group of the corresponding 2-groups of extraspecial type [Griess, 1973].

Appendix B

Some Parabolic Subgroups

The nonabelian finite simple groups fall into the following four classes: The alternating groups A_n ($n \geq 5$), the finite classical groups, the exceptional groups of Lie type, and the 26 sporadic groups. In this appendix we are concerned only (for convenience) with the classical groups $\mathrm{Sp}_{2m}(q)$ and $\mathrm{SO}_{2m}^+(q)$, $\mathrm{SO}_{2m}^-(q)$ (q a prime power, $m \geq 1$). In Lie terminology the symplectic group is the simply connected group of type C_m whereas the orthogonal groups under consideration may be identified as those groups of type D_m (plus type) and ^2D_m (negative type) which are neither simply connected nor adjoint. In the orthogonal case there is a subgroup $\Omega_{2m}^\pm(q)$ of index 2 in $\mathrm{SO}_{2m}^\pm(q)$, which is the kernel of the Dickson invariant when q is even, and this is the unique subgroup with index 2 unless we have plus type and $m = 2 = q$.

Let U be the standard module for these classical groups, and write $G = G(U)$ for the related isometry group (introduced above; we write $G(U) = \mathrm{O}_{2m}^\pm(q)$ in the even orthogonal case). Let $F = \mathbb{F}_q$. Let τ denote either the symplectic form on U or the symmetric form to the quadratic form Q in the orthogonal case.

A nondegenerate symplectic form requires even dimension, and then the form on U is uniquely determined up to isomorphism (and so is the group). In the orthogonal case $Q : U \to F$ satisfies $Q(cu) = c^2 Q(v)$ for $c \in F$, $u \in U$, and $\tau(u,v) = Q(u+v) - Q(u) - Q(v)$ defines the associated (symmetric) bilinear form on U. The form Q is nondegenerate if τ is nondegenerate. If q is odd, the theory of quadratic forms is that of symmetric bilinear forms. If q is even then τ necessarily is symplectic. Actually we are interested in quadratic forms only in characteristic 2, hence the assumption that $\dim_F U = 2m$ is even. There are up to isomorphism two distinct nondegenerate quadratic forms $Q = Q_{2m}^+$ (Witt index m) and $Q = Q_{2m}^-$ on V (index $m - 1$). This means that U possesses maximal totally singular subspaces of dimension m resp. of dimension $m - 1$; in the symplectic case we have index m. Reference for all this is [Dieudonné, 1971].

(B1) *Suppose U has index m. Then there exist totally singular subspaces W, W^* of dimension m such that $U = W \oplus W^*$. Let $P = N_G(W)$ and let $L = N_G(W, W^*)$ be the normalizer of both subspaces, $R = C_G(W, U/W)$ be the centralizer of both W and U/W. Then $P = R : L$ where $L \cong \mathrm{GL}(W)$ and W is the standard module of L and W^* is the dual module. Also $R \cong \mathrm{Alt}^2(W)$ as module for L in the orthogonal case, and $R \cong \mathrm{Sym}^2(W)$ in the symplectic case.*

Proof. By hypothesis there is basis $\{e_1, \cdots, e_m\} \cup \{e_1^*, \cdots, e_m^*\}$ of U such that $\tau(e_i, e_i^*) = 1$ for all i and all other scalar products of the e_i, e_j^* vanish, and that in addition $Q(e_i) = 0 = Q(e_j^*)$ in the orthogonal case. Then the $\langle e_i, e_j^* \rangle$ are *hyperbolic planes* in U, and $W = \langle e_1, \cdots, e_m \rangle$ and $W^* = \langle e_1^*, \cdots, e_m^* \rangle$ are maximal totally singular subspaces.

Let $x \in \mathrm{GL}(W)$ and $y \in \mathrm{GL}(W^*)$. Then $\begin{pmatrix} x & 0 \\ 0 & y \end{pmatrix}$ is an isometry of the symplectic space U if and only if $\tau(e_i x, e_j^* y) = \tau(e_i, e_j^*) = \delta_{ij}$ for all i, j. This means that $y = (x^{-1})^t$ has to be the inverse transpose of x. In the orthogonal case, observe that $\begin{pmatrix} x & 0 \\ 0 & x^{-t} \end{pmatrix}$ preserves W and W^* and that $Q(w + w^*) = 0$ for $w \in W$ and $w^* \in W^*$ if and only if $\tau(w, w^*) = 0$. So

$$L = \{ \begin{pmatrix} x & 0 \\ 0 & x^{-t} \end{pmatrix} \mid x \in \mathrm{GL}(W) \}$$

may be viewed as a subgroup of G in each case, and it is the stabilizer in G of W and W^*. Of course, $L \cong \mathrm{GL}(W)$ and W is the standard module, W^* its dual.

It is clear that $L \cap R = 1$. Let $x \in P$. We find $y \in L$ such that xy centralizes W. But $U/W \cong W^*$ and so xy centralizes U/W as well, that is, $xy \in R$. Thus $P = R : L$.

By (A7) it is clear that R is an FL- submodule of

$$\mathrm{Hom}_F(U/W, W) \cong \mathrm{Hom}_F(W^*, W) \cong T^2(W).$$

Let $\alpha : U/W \to W$ be a linear map. View α as an endomorphism of U via inflation. Write $e_i^* \alpha = \sum_j a_{ij} e_j$. We have to examine when $1 + \alpha$ is in G. In the symplectic case this requires that

$$0 = \tau(e_i^*, e_k^*) = \tau(e_i^* + \sum_j a_{ij} e_j, \; e_k^* + \sum_j a_{kj} e_j) = a_{ki} \tau(e_i^*, e_i) + a_{ik} \tau(e_k, e_k^*)$$

for all i, k. Thus the matrix (a_{ik}) has to be symmetric. It is well known that $T^2(W)$ is naturally isomorphic to $M_m(F)$ (sending $e_i \otimes e_k$ to the elementary matrix e_{ik}), in which case $\text{Sym}^2(W)$ maps onto the symmetric matrices and $\text{Alt}^2(W)$ onto the skew symmetric matrices.

In the orthogonal case we compute

$$0 = Q(e_i^*) = Q(e_i^* + \sum_j a_{ij}e_j) = \tau(e_i^*, \sum_j a_{ij}e_j) = a_{ii}$$

and, for $i \neq k$, $0 = Q(e_i^* + e_k^*) = Q\big(e_i^* + e_k^* + \sum_j(a_{ij} + a_{kj})e_j\big) = a_{ik} + a_{ki}$. Thus in this case the (necessary and sufficient) condition is that the matrix (a_{ik}) is skew symmetric. The proof is complete. $\qquad\square$

(B2) *Suppose* $W \neq 0$ *is a totally singular subspace of* U, *and let* $P = N_G(W)$. *There exists a totally singular subspace* W^* *of* U *such that* $U' = W \oplus W^*$ *is nondegenerate. Let* $U'' = (U')^\perp$, $L = N_G(W, W^*, U'')$ *and* $R = C_G(W, W^\perp/W, U/W^\perp)$. *Then* $P = R : L$ *and* $L = L_w \times G(U'')$, *where* $L_w \cong \text{GL}(W)$ *acts naturally on* W *and* W^* *is the dual* FL_w-*module, and where* $G(U'')$ *centralizes* $R_w = C_R(W^\perp)$. *Moreover,* $R_w \subseteq Z(R)$, $R/R_w \cong W \otimes_F U''$ *as an* $L = L_w \times G(U'')$-*module, and* $R_w \cong \text{Sym}^2(W)$ *as an* L_w-*module in the symplectic case and* $R_w \cong \text{Alt}^2(W)$ *otherwise.*

Proof. By Witt's theorem we may assume that W has the basis $\{e_1, \cdots, e_n\}$ taken from the standard basis above $(n \leq m)$. Let then $W^* = \langle e_1^*, \cdots, e_n^* \rangle$. So (B1) applies to $U' = W \oplus W^*$, and $U = U' \perp U''$ is an orthogonal sum of nondegenerate spaces. Hence $L = L_w \times \text{GL}(U'')$ is as asserted.

It is obvious that $L \cap R = 1$. As a P-module $U/W^\perp \cong W^*$ (through the given nondegenerate pairing). Thus to $x \in P$ we find $y \in L_w$ such that xy centralizes W and $U/W^\perp$. We have $W^\perp = W \oplus U''$. Hence each element in $W^\perp/W$ can be written as $W + u''$ for some unique $u'' \in U''$. We get a linear map $\alpha : U'' \to U''$ satisfying $(W + u'')\alpha = W + u''xy$. It is easy to see that $\alpha \in G(U'')$. This proves that $P = R : L$.

The quotient R/R_w acts faithfully on $W^\perp$, centralizing both W and $W^\perp/W \cong U''$, whence may be identified with $\text{Hom}_F(U'', W)$. In fact, for each linear map $\varphi : W^\perp/W \to W$ consider its inflation to $W^\perp$, and then $1 + \varphi$ gets an isometry of $W^\perp$, which extends to an element of G by Witt's theorem. Thus $R/R_w \cong \text{Hom}_F(U'', W)$ as an $L = G(U'') \times L_w$-*module*. Now use that U'' is a self-dual $G(U'')$-module.

Since R_w centralizes $W^\perp \cong U/W$ and R centralizes $U/W^\perp$ and W, we have $[U, R, R_w] = 1 = [R_w, U, R]$. It follows that $[R, R_w, U] = 1$ (by the 3-subgroups lemma) and so $R_w \subseteq Z(R)$. Now R_w centralizes U'' and acts on the nondegenerate space $U' = W \oplus W^*$ and centralizes W, with $U'/W \cong W^*$. Apply (B1) to U'. $\qquad\qquad\qquad\qquad\qquad\qquad\qquad\qquad\qquad$ $\square$

Remark. From (A7) we infer that L_w acts irreducibly on R_w unless $p = 2$ and $G(U) = \mathrm{Sp}_{2m}(q)$ is a symplectic group (in which case R_w is indecomposable with 2 composition factors). Also, $R/R_w \cong W \otimes_F U''$ is an irreducible $L_w \times G(U'')$-module unless $G(U'') \cong O_2^+(2)$ or $SO_2^+(q)$ for odd q, where $G(U'')$ is not irreducible on the standard module.

(B3) *Let W, P, R, L be as in (B2). Then P is a maximal parabolic subgroup in $G = G(U)$, with unipotent radical R and Levi complement L, except when $G(U) \cong SO_{2m}^+(q)$ and $\dim_F W = m$, in which case P is a maximal parabolic in $\Omega_{2m}^+(q)$. The maximal parabolics are all of of this form, and are determined up to conjugacy by the dimension of W (which may vary from 1 to the Witt index).*

Proof. It follows from the construction, and from (B2), that $P = N_G(R)$ is a proper subgroup of G and that $R = O_p(P)$. The order of P is determined by $n = \dim_F W$. Indeed $\dim_F U'' = 2m - 2n$ and so

$$|R| = |R/R_w| \cdot |R_w| = q^{n(2m-2n)+n(n\pm 1)/2},$$

where the positive sign holds in the symplectic case. One similarly computes $|L| = |\mathrm{GL}(W)| \cdot |G(U'')|$. This shows that P contains a Sylow p-subgroup of G, even a p-Sylow normalizer, except when $n = m$ and $G \cong SO_{2m}^+(q)$. So P is a parabolic in G or, in the exceptional case, in $\Omega_{2m}^+(q)$ (where there are two classes which fuse in $O_{2m}^+(q)$).

If P were not maximal in G (resp. in $\Omega_{2m}^+(q)$), there were a maximal parabolic $P_0 = R_0 : L_0$ properly containing P. Then at any rate $R_0 \subset R$ and $U_0 = C_U(R_0) \supseteq W$. We even have $W \subseteq U_0 \cap U_0^\perp = U_1$, and U_1 is P_0-invariant. Either U_1 is totally singular or q is even, U is orthogonal, and the subspace U_1' of U_1 of singular vectors has codimension 1. But then U_1' is P_0-invariant and $U_1' \supseteq W$. So P_0 stabilizes a totally singular subspace $W_0 \supseteq W$ of U. Since P_0 is a maximal subgroup, P_0 is the normalizer of W_0, and $W_0 \neq W$ as $P_0 \neq P$. We get a contradiction by comparing the orders of R and R_0.

Witt's theorem, and a survey of the Dynkin diagram, give the final uniqueness statement. $\qquad\qquad\qquad\qquad\qquad\qquad\qquad\qquad\qquad\qquad\qquad$ $\square$

Appendix C

Weil Characters

Let G be any of the classical groups $\mathrm{SL}_m(q)$, $\mathrm{Sp}_{2m}(q)$ or $\mathrm{SU}_m(q)$ for some $m \geq 1$ and some power $q = p^f$ of a prime p, and let U be the natural G-module over $\mathbb{F}_q$ (resp. $\mathbb{F}_{q^2}$ in the unitary case). In the symplectic case assume that q is odd. Let G_u be the associated *universal* group, that is, $G_u = \mathrm{GL}_m(q)$, $\mathrm{CSp}_{2m}(q)$ or $\mathrm{GU}_m(q)$, respectively. Then $G_u/Z(G_u)$ is the *adjoint* group, which may be identified with the normal subgroup of $\mathrm{Aut}(G)$ generated by the inner and diagonal automorphisms. Stimulated by [Weil, 1964] one associates to G_u and G certain distinguished characters related to extraspecial groups (or *Heisenberg groups*). It turns out that, with few exceptions, these *Weil characters* are faithful of minimal degree.

If Z is a trivial (additive) U-module, every biadditive map $\tau : U \times U \to Z$ is a factor set in $Z^2(U, Z)$, and if $Z \rightarrowtail E \twoheadrightarrow U$ is an extension to τ, the commutator map on $E = E(\tau)$ induces the alternating (symplectic) form given by $[u, u'] = \tau(u, u') - \tau(u', u)$.

(C1) *Let $G = \mathrm{SL}_m(q)$ and $G_u = \mathrm{GL}_m(q)$ for some integer $m \geq 2$. The permutation character of G_u on U decomposes as $\pi_U = 2 \cdot 1_{G_u} + \sum_{j=0}^{q-2} \xi_j$, where the ξ_j are pairwise distinct (complex) irreducible characters of G_u satisfying $\xi_0(1) = (q^m - q)/(q-1)$ and $\xi_j(1) = (q^m - 1)/(q-1)$ for $j > 0$. For $m \geq 3$, and excluding $(m, q) = (3, 2), (3, 4), (4, 2), (4, 3)$, these characters remain irreducible on G, and we have*

$$R_0(L_m(q)) = (q^m - q)/(q - 1).$$

Each faithful projective irreducible character of $L_m(q) = \mathrm{PSL}_m(q)$ of degree at most $\frac{q^m - 1}{q - 1}$ then is such a Weil character ξ_j.

Proof. Let U^* be the dual $\mathbb{F}_q G_u$-module to U, and let $W = U \oplus U^*$. Let further $\mathbb{F}_q \rightarrowtail E \twoheadrightarrow W$ be the associated Heisenberg group. Thus $E = E(\tau)$ is the set of pairs (w, z) in $W \times \mathbb{F}_q$ with $(w, z) \cdot (w', z') = (w + w', z + z' + \tau(w, w'))$ where the factor set $\tau : W \times W \to \mathbb{F}_q$ is $\mathbb{F}_q$-bilinear, given by $\tau((u, \lambda), (u', \lambda')) = \lambda(u')$. For $g \in G_u$ we define $(u, \lambda, z)^g = (u^g, \lambda^{g^*}, z)$ where g^* is the inverse transpose of g. It is immediate that in this manner

221

G_u preserves the factor set τ and so G_u acts on E (faithfully) centralizing $Z(E) \cong \mathbb{F}_q$. Even more, the subgroup $A = U \times \{0\} \times \{0\}$ of E is G_u-invariant and is the standard module, and $A^* = \{0\} \times U^* \times \{0\}$ is the dual module.

Every nontrivial linear character λ of $Z(E) \cong \mathbb{F}_q$ gives rise to an irreducible character χ_0 of E by inducing up to E the character $1_{A^*} \times \lambda$ of $A^* \times Z(E)$. Inducing up to $E : G_u$ the linear character $1_{A^*:G_u} \times \lambda$ of $(A^* : G_u) \times Z(E)$ we obtain an irreducible character χ of $E : G_u$ extending χ_0. Let ξ be the restriction to G_u of this irreducible character. Using that A is a set of right coset representatives for $(A^* : G_u) \times Z(E)$ in $E : G_u$, with G_u acting transitively on $A^\sharp$, and using Mackey decomposition, we see that $\xi = \pi$ is the permutation character of G_u on the set $A \cong U$. Of course G_u is transitive on $U^\sharp$.

$Z(G_u) \cong \mathbb{F}_q^*$ has $(q^m - 1)/(q - 1)$ regular orbits (cycles) on $U^\sharp$. Hence we have $q - 1$ distinct constituents ξ'_j of $\pi_{U^\sharp}$ lying over the distinct linear characters λ_j of $Z(G_u)$, and all these ξ'_j have degree $(q^m - 1)/(q - 1)$. The character ξ'_0 lying above the 1-character of $Z(G_u)$ is of the form $\xi'_0 = 1_{G_u} + \xi_0$ with $\xi_0(1) = (q^m - q)/(q - 1)$. Let $\xi'_j = \xi_j$ for $j > 1$.

The G_u-sets U, U^* are isomorphic through the inverse transpose automorphism. Thus $W \cong U \times U$ as G_u-sets and $\xi^2 = \pi_W$. Now G_u has exactly $q + 3$ orbits on the set $U \times U$ (being transitive on the set of linear independent pairs of vectors (u, u') and on the sets (u, tu) for each $t \in \mathbb{F}_q^*$). From $\xi = 2 \cdot 1_{G_u} + \sum_{i=1}^{q-1} \xi_i$ and $\langle \xi, \xi \rangle = \langle \xi^2, 1_{G_u} \rangle = q + 3$ we conclude that all ξ_i are irreducible and pairwise distinct.

The restrictions to $G = \mathrm{SL}(W) = \mathrm{SL}_m(q)$ of these characters need not be irreducible (nor distinct) for small m. We refer to [Tiep–Zalesskii, 1996, Theorem 3.1] for the final statements in (C1). One has $R_0(L_m(q)) = 3, 6, 7, 26$ for $(m, q) = (3, 2), (3, 4), (4, 2), (4, 3)$, respectively. $\square$

(C2) Let $G = \mathrm{Sp}_{2m}(q)$ and $G_u = \mathrm{CSp}_{2m}(q)$ for $m \geq 2$ and odd q. The permutation character of G on U is of the form $\pi_U = \xi \cdot \bar{\xi} = \xi^\diamond \cdot \overline{\xi^\diamond}$ for G_u-conjugate "generic" Weil characters $\xi \neq \xi^\diamond$. We have $\xi = \xi_1 + \xi_2$ and $\xi^\diamond = \xi_1^\diamond + \xi_2^\diamond$ with irreducible (Weil) characters $\xi_1, \xi_1^\diamond$ and $\xi_2, \xi_2^\diamond$ of degree $(q^m - 1)/2$ and $(q^m + 1)/2$, respectively. Moreover,

$$R_0(S_{2m}(q)) = (q^m - 1)/2,$$

and every faithful projective irreducible character of $S_{2m}(q) = \mathrm{PSp}_{2m}(q)$ of degree less than $\frac{1}{2(q+1)}(q^m - 1)(q^m - q)$ is one of $\xi_1, \xi_2, \xi_1^\diamond, \xi_2^\diamond$.

The final statement is a result due to [Tiep–Zalesskii, 1996, Theorem 5.2]. The other assertions follow from Theorems 4.3c and 4.5a. Recall, in particular, that ξ and ξ° are obtained by restriction to $\mathrm{Sp}_{2m}(q)$ of the faithful irreducible characters of degree q^m of the *symplectic holomorph* $E : \mathrm{Sp}_{2m}(q)$, where $E = p_{+}^{1+2fm}$ and where $U = E/Z(E)$ may be identified with the standard module.

(C3) *Let* $G = \mathrm{SU}_m(q)$ *and* $G_u = \mathrm{GU}_m(q)$, *with* $m \geq 3$. *For convenience exclude* $(m,q) = (3,2),(4,2),(6,2)$ *and* $(4,3)$. *Let* X *be the (standard) holomorph of* $E = p_{+}^{1+2fm}$ *for odd* p *and of* $E = 2_0^{1+2fm}$ *otherwise, with* $X/E \cong \mathrm{Sp}_{2m}(q)$ *(say). There is a distinguished subgroup* Y *of* X *such that* $Y \cap E = Z(E)$ *and such that* $YE/E \cong G_u.2$ *is a field extension subgroup of* $\mathrm{Sp}_{2m}(q)$. *We have a unique decomposition* $Y = Y_0 \times Z(E)$ *for odd* q, *where we write* $Y_0 = G_u.2 \supseteq G_u$, *and where* $Y' = G_u$ *otherwise. In both cases* $U = E/Z(E)$ *is the standard module for* G_u, *and the "generic" Weil character of* G_u *is defined as*

$$\xi_u = \mathrm{Res}_{G_u}^{X}(\chi) \cdot \mu,$$

where χ *is any faithful irreducible character of* X *of degree* q^m *and where* μ *is the linear character of* G_u *of order* $\gcd(q+1,2)$. *This* ξ_u *is well-defined, rational-valued and satisfies* $\xi_u^2 = \pi_U$. *Fixing a linear character* λ_u *of* $Z(G_u)$ *of order* $q+1$ *(which exists) let* ξ_j *be the constituent of* ξ_u *lying above* λ_u^j $(0 \leq j \leq q)$. *Then* $\xi_0(1) = (q^m + (-1)^m q)/(q+1)$ *and* $\xi_j(1) = \xi_0(1) - (-1)^m$ *for* $j > 0$, *and the* ξ_j *are pairwise distinct irreducible (Weil) characters of* G_u *and remain so when restricted to* G. *We have*

$$R_0(\mathrm{U}_m(q)) = \begin{cases} (q^m - q)/(q+1) & \text{if } m \text{ is odd} \\ (q^m - 1)/(q+1) & \text{otherwise,} \end{cases}$$

and each faithful projective irreducible character of $\mathrm{U}_m(q) = \mathrm{PSU}_m(q)$ *of degree at most* $\frac{q^m+1}{q+1}$ *for odd* m, *and* $\frac{q^m+q}{q+1}$ *otherwise, is one of the* ξ_j.

Proof. For existence (and uniqueness) of $G_u.2$ in $\mathrm{Sp}_{2m}(q)$ see [Aschbacher, 1984]; in the odd case it is an (almost) maximal subgroup, and determined up to conjugacy. Recall that $G_u/G \cong Z(G_u)$ are cyclic of order $q+1$, inverted by $G_u.2$ since the outer field automorphism of order 2 sends any z to $z^q = z^{-1}$. Let $Z_u = Z(G_u) = \langle z_u \rangle$, which acts on U as a group of scalar multiplications (with norm $z_u \cdot z_u^q = 1$). By assumption and (A6) the group $G = \mathrm{SU}_m(q)$ is perfect and $\mathrm{M}(G) = 1$. Using (A2), (A5) one shows that also $\mathrm{M}(G_u) = 1 = \mathrm{M}(G_u.2)$, and that $\mathrm{H}^n(G_u.2, U) = 0$ for all n.

Let α be an automorphism of X permuting the faithful linear characters of $Z(X) = Z(E)$ transitively (Theorems 4.3c and 4.3d). By the cohomological triviality stated above, we find Y in X as claimed, and α-invariant. In the even case use that α is trivial on X/E. If $|Z(E)| = p$ is odd, p does not divide $|G_u.2/G| = 2(q+1)$. In the even case $G_u.2/G$ is dihedral (and $q+1$ odd). Thus $Y = G_u.2 \times Z(E)$ when q is odd, and $Y' = G_u$ otherwise (A5), where G_u is α-invariant and α gets inner on G_u in each case. So the faithful irreducible characters of X of degree q^m agree on G_u, defining ξ_u, and ξ_u is rational-valued. As in Theorem 4.5a each element of G_u is good for U (use z_u). Hence $\xi_u^2 = \pi_U$ by Theorem 4.4.

Let $\varepsilon = e^{2\pi i/(q+1)}$, and choose λ_u such that $\lambda_u(z_u) = \varepsilon$. Now Z_u has on $U^\sharp$ just $(q^m - 1)/(q+1)$ regular orbits. It follows from Theorems 4.4 and 1.6c that $\xi_u(z_u) = (-1)^m$ (see also [I, 13.32]). Let $\rho = \sum_{j=0}^{q} \lambda_u^j$ denote the regular character of Z_u. We know that $\mathrm{Res}_{Z_u}^{G_u}(\xi_u^2) = \lambda_u^0 + \frac{q^m-1}{q+1} \cdot \rho$. Hence $\mathrm{Res}_{Z_u}^{G_u}(\xi)$ is not a multiple of ρ. Using that $\xi_u(z_u) = (-1)^m$ and $\rho(z_u) = 0$, and using elementary properties of sums of $(q+1)$th roots of unity we get

$$\mathrm{Res}_{Z_u}^{G_u}(\xi_u) = \begin{cases} a \cdot \rho + \lambda_u^0 & \text{if } m \text{ is even,} \\ a \cdot \rho + \sum_{j=1}^{q} \lambda_u^j & \text{otherwise} \end{cases}$$

for some nonnegative integer a. Of course $a = \frac{q^m-1}{q+1}$ when m is even, and $a = \frac{q^m-q}{q+1}$ otherwise. Let ξ_j denote the constituent of ξ_u lying over the linear character λ_u^j of Z_u. Then $\xi_0(1) = a+1$ when m is even, and $\xi_0(1) = a$ otherwise, and $\xi_j(1) = \xi_0(1) - (-1)^m$ for $j > 0$. So the ξ_j ($\neq 0$) are pairwise "disjoint", and $\xi_u = \sum_{j=0}^{q} \xi_j$. One knows that G has just $q+1$ orbits on the set U [Aschbacher, 1986, (22.4)]. Hence $\langle \xi_U, \xi_u \rangle_G = \langle \xi_u^2, 1_{G_u} \rangle_G = \langle \pi_u, 1_G \rangle = q+1$. This shows that the ξ_j are irreducible and pairwise distinct even as characters of $G = \mathrm{SU}_m(q)$. The final statements in (C3) are due to [Tiep–Zalesskii, 1996]. □

Remark. The values of the character ξ_u and of the irreducible Weil characters ξ_j on $G_u = \mathrm{GU}_m(q)$ have been computed in [Gérardin, 1977] and [Tiep–Zalesskii, 1996]. Suppose ξ_j lies over λ_u^j, as before, and let again z_u be the generator of $Z_u = Z(G_u)$ with $\lambda_u(z_u) = \varepsilon = e^{2\pi i/(q+1)}$. Let $g \in G_u$, and let $d_k = d_k(g)$ be the dimension of $C_U(z_u^{-k}g)$ over $\mathbb{F}_{q^2}$. Then $\xi_u(g) = (-1)^m(-q)^{d_0}$ and, for $j \in \{0, \cdots, q\}$,

$$\xi_j(g) = \frac{(-1)^m}{q+1} \sum_{k=0}^{q} (-q)^{d_k} \varepsilon^{kj}.$$

Bibliography

Alperin, J. L. (1987), Weights for finite groups, *Proc. Symp. Pure Math.* **47**, 369-379.

Aschbacher, M. (1984), On the maximal subgroups of the finite classical groups, *Invent. math.* **76**, 469-514.

Aschbacher, M. (1986), *Finite Group Theory*, Cambridge University Press, Cambridge.

Aschbacher, M. (2004), The status of the classification of the finite simple groups, *Notices Amer. Math. Soc.* **51**, 736–740.

Bell, G. W. (1978), On the cohomology of the finite special linear groups, I, II, *J. Algebra* **54**, 216-238 and 239-259.

Bourbaki, N. (1958), *Algèbre, Chap. 8*, Hermann, Paris.

Brauer, R. (1956), Number theoretical investigations on groups of finite order, *Proc. Int. Symp. Alg. Number Theory*, Tokyo Nikko, pp. 55–62.

Brauer, R. and Feit, W. (1959), On the number of irreducible characters of finite groups in a given block, *Proc. Nat. Acad. Sci. U.S.A.* **45**, 361-365.

Brauer, R. (1968), On blocks and sections in finite groups, II, *Amer. J. Math.* **90**, 895-825.

Conway, J. H. *et al.* (1985), *Atlas of Finite Groups* [Atlas], Clarendon, Oxford.

Dade, E. C. (1974), Character values and Clifford extensions for finite groups, *Proc. London Math. Soc.* (3) **29**, 216–236.

Dade, E. C. (1992), Counting characters in blocks, I, *Invent. math.* **109**, 187-210.

Dickson, L. E. (1901), *Linear Groups*, Teubner, Leipzig.

Dieudonné, J. A. (1971), *La Géométrie des Groupes Classique*, Springer, Berlin.

Dixon, J. D. and Mortimer, B. (1996), *Permutation Groups*, Springer, Berlin.

Erdős, P. (1942), On an elementary proof of some asymptotic formulas in the theory of partitions, *Annals Math.* (2) **43**, 437–450.

Feit, W. (1982), *The Representation Theory of Finite Groups* [F], North-Holland, Amsterdam.

Fong, P. (1962), Solvable groups and modular representation theory, *Trans. Amer. Math. Soc.* **103**, 484–494.

Gallagher, P. X. (1962), Group characters and commutators, *Math. Z.* **79**, 122–126.

Gallagher, P. X. (1970), The number of conjugacy classes in a finite group, *Math. Z.* **118**, 175–179.

GAP – *Groups, Algorithms, and Programming* (2006), version 4.4.9 [GAP], (http://www. gap-systems.org).

Gérardin, P. (1977), Weil representations associated to finite fields, *J. Algebra* **46**, 54–101.

Gluck, D. (1984), On the $k(GV)$ problem, *J. Algebra* **89**, 46–55.

Gluck, D. and Wolf, T. (1984), Brauer's height conjecture for p-solvable groups, *Trans. Amer. Math. Soc.* **282**, 137–152.

Gluck, D. and Magaard, K. (2002a), The extraspecial case of the $k(GV)$ problem, *Trans. Amer. Math. Soc.* **354**, 287–333.

Gluck, D. and Magaard, K. (2002b), The $k(GV)$ conjecture for modules in characteristic 31, *J. Algebra* **250**, 252–270.

Gluck, D., Magaard, K., Riese, U. and Schmid, P. (2004), The solution of the $k(GV)$-problem, *J. Algebra* **279**, 694–719.

Goodwin, D. (2000), Regular orbits of linear groups with an application to the $k(GV)$-problem, 1, 2, *J. Algebra* **227**, 395-432 and 433–473.

Gow, R. (1980), On the number of characters in a p-block of a p-solvable group, *J. Algebra* **65**, 421–426.

Gow, R. (1993), On the number of characters in a block and the $k(GV)$ problem for self-dual V, *J. London Math. Soc.* **48**, 441–451.

Green, J. A. (1955), The characters of the finite general linear groups, *Trans. Amer. Math. Soc.* **80**, 402-447.

Griess, R. L. (1973), Automorphisms of extraspecial groups and nonvanishing degree 2 cohomology, *Pacific J. Math.* **48**, 403–422.

Guralnick, R. M. and Saxl, J. (2003), Generation of finite almost simple groups by conjugates, *J. Algebra* **268**, 519–571.

Guralnick, R. M. and Tiep, P. H. (2005), The non-coprime $k(GV)$ problem, *J. Algebra* **293**, 185–242.

Hall, J. L., Liebeck M. W. and Seitz, G. M. (1992), Generators for finite simple groups, with applications to linear groups, *Quart. J. Math. Oxford* **43**, 441–458.

Hering, Ch. (1985), Transitive linear groups and linear groups which contain irreducible subgroups of prime order, II, *J. Algebra* **93**, 151–161.

Higman, G. (1960), Enumerating p-groups. I: Inequalities, *Proc. London Math. Soc.* (3) **10**, 24–30.

Hilton, P. J. and Stammbach, U. (1971), *A Course in Homological Algebra*, [HS], Springer, Berlin.

Huppert, B. (1967), *Endliche Gruppen I*, Springer, Berlin.

Huppert, B. and Blackburn, N. (1982), *Finite Groups III*, Springer, Berlin.

Isaacs, I. M. (1973), Characters of solvable and symplectic groups, *Amer. J. Math.* **95**, 594–635.

Isaacs, I. M. (1976), *Character Theory of Finite Groups*, [I], Dover, New York.

James, G. and Kerber, A. (1981), *Representation Theory of the Symmetric Group*, Addison–Wesley, London.

Jansen, Ch. *et al.* (1995), *An Atlas of Brauer Characters* [B-Atlas], Oxford University Press, Oxford.

Jennings, S. (1941), The structure of the group ring of a p-group over a modular field, *Trans. Amer. Math. Soc.* **50**, 175–185.

Keller, T. M. (2006), Fixed conjugacy classes of normal subgroups and the $k(GV)$-problem, *J. Algebra* **305**, 457–486.

Kleidman, P. and Liebeck, M. W. (1990), *The Subgroup Structure of the Finite Classical Groups*, Cambridge University Press, Cambridge.

Knörr, R. (1984), On the number of characters in a p-block of a p-solvable group, *Illinois J. Math.* **28**, 181–210.

Knörr, R. (1990), A remark on Brauer's $k(B)$-conjecture, *J. Algebra* **131**, 444–454.

Köhler, C. (1999), *Über das $k(GV)$-Problem*, Diplomarbeit TH Aachen.

Köhler, C. and Pahlings, H. (2001), Regular orbits and the $k(GV)$- problem, in *Groups and Computation III* (Columbus, Ohio), de Gruyter, Berlin.

Kovács, L. G. and Robinson, G. R. (1993), On the number of conjugacy classes of finite groups, *J. Algebra* **160**, 441–460.

Külshammer, B. (1987), A remark on conjectures in modular representation theory, *Archiv Math.* **49**, 396–399.

Landazuri V. and Seitz, G. M. (1974), On the minimal degrees of the projective representations of the finite Chevalley groups, *J. Algebra* **32**, 418-443.

Liebeck, M. W. (1985), On the orders of maximal subgroups of the finite classical groups, *Proc. London Math. Soc.* (3) **50**, 426–446.

Liebeck, M. W. (1996), Regular orbits for linear groups, *J. Algebra* **184**, 1136–1142.

Liebeck, M, W. and Pyber, L. (1997), Upper bounds for the number of conjugacy classes of a finite group, *J. Algebra* **198**, 538–562.

Maróti, A. (2005), Bounding the number of conjugacy classes of a permutation group, *J. Group Theory* **8**, 237–289.

Nagao, H. (1962), On a conjecture of Brauer for p-solvable groups, *J. Math. Osaka City Univ.* **13**, 35–38.

Okayama, T. and Wajima, M. (1980), Character correspondences and p-blocks of p-solvable groups, *Osaka J. Math.* **17**, 801–806.

Rasala, R. (1977), On the minimal degrees of characters of S_n, *J. Algebra* **45**, 132–181.

Reynolds, W. F. (1963), Blocks and normal subgroups of finite groups, *Nagoya J. Math.* **22**, 15–32.

Riese, U. and Schmid, P. (2000), Self-dual modules and real vectors for solvable groups, *J. Algebra* **227**, 159–171.

Riese, U. (2001), The quasisimple case of the $k(GV)$-conjecture, *J. Algebra* **235**, 45–65.

Riese, U. (2002), On the extraspecial case of the $k(GV)$-conjecture, *Archiv Math.* **78**, 177–183.

Riese, U. and Schmid, P. (2003), Real vectors for linear groups and the $k(GV)$-problem, *J. Algebra* **267**, 725–755.

Robinson, G. R. (1995), Some remarks on the $k(GV)$ problem, *J. Algebra* **172**, 159–166.

Robinson, G. R. and Thompson, J. G. (1996), On Brauer's $k(B)$-problem, *J. Algebra* **184**, 1143–1160.

Robinson, G. R. (1997), Further reductions for the $k(GV)$-problem, *J. Algebra* **195**, 141–150.

Robinson, G. R. (2004), On Brauer's $k(B)$-problem for blocks of p-solvable groups with non-Abelian defect groups, *J. Algebra* **280**, 738–742.

Sah, C. H. (1977), Cohomology of split group extensions, II, *J. Algebra* **45**, 17–68.

Schmid, P. (1981), On the Clifford theory of blocks of characters, *J. Algebra* **73**, 44–55.

Schmid, P. (2000), On the automorphism group of extraspecial 2-groups, *J. Algebra* **234**, 492–506.

Schmid, P. (2005), Some remarks on the $k(GV)$ theorem, *J. Group Theory* **8**, 589–604.

Schur, I. (1907), Untersuchungen über die Darstellungen der endlichen Gruppen durch gebrochene lineare Substitutionen, *J. Reine Angew. Math.* **132**, 85–137.

Schur, I. (1911), Über die Darstellungen der symmetrischen und alternierenden Gruppen durch gebrochene lineare Substitutionen, *J. Reine Angew. Math.* **139**, 155–250.

Seitz, G. M. and Zalesskii, A. E. (1993), On the minimal degrees of projective representations of the finite Chevalley groups, II, *J. Algebra* **158**, 233–234.

Serre, J.-P. (2003), On a theorem of Jordan, *Bull. Amer. Math. Soc.* **40**, 429–440.

Sin, P. (1996), Modular representations of the Hall–Janko group, *Commun. Algebra* **24**, 4513–4547.

Solomon, R. (2001), The classification of the finite simple groups, *Bull. Amer. Math. Soc.* **38**, 315–352.

Springer, T. A. and Steinberg, R. (1970), Conjugacy classes, in *Seminar on Algebraic Groups and Related Finite Groups*, Lecture Notes in Math. **131**, pp. 167-266, Springer, Berlin.

Steinberg, R. (1967), *Lectures on Chevalley Groups*, Yale University, Yale.

Suzuki, M. (1962), On a class of doubly transitive groups, *Annals Math.* **75**, 105–145.

Tiep, P. H. and Zalesskii, A. E. (1996), Minimal characters of finite classical groups, *Commun. Algebra* **24** (**6**), 2093–2167.

Tiep, P.H. and Zalesskii, A. E. (1997), Some characterizations of the Weil representations of the symplectic and unitary groups, *J. Algebra* **192**, 130–165.

Ward, H. N. (1966), On Ree's series of simple groups, *Trans. Amer. Math. Soc.* **121**, 62–69.

Ward, H. N. (1972), Representations of symplectic groups, *J. Algebra* **20**, 182–195.

Weil, A. (1964), Sur certain groupes d'opérateurs unitaire, *Acta Math.* **111**, 143–211.

List of Symbols

$k(X) = |C\ell(X)| = |\mathrm{Irr}(X)|$; (1.1a)

$\mathrm{Sym}^2(V)$, $\mathrm{Alt}^2(V)$; 2

$Y \backslash X$, $\mathrm{Core}_X(Y)$; 3

$Y \,\mathrm{wr}\, S_n$; 3

$\mathrm{Ind}_Y^X(\theta)$; (1.2b)

$\mathrm{Ten}_Y^X(\theta)$; (1.2e)

ω_χ; (1.3a)

$C\ell(G|\Omega) = \mathrm{orb}(G, \Omega)$; (1.4a)

π_Ω; 6

$\mathrm{Gal}(K|\mathbb{Q})$; 7

$(\mathbb{Z}/e\mathbb{Z})^*$ unit group; 7

$V = C_V(G) \oplus [V, G]$; (1.6a)

$k_g(N) = |C_{C\ell(N)}(g)|$; 10

$I_X(\theta)$, $\mathrm{Irr}(X|\theta)$; 12

$\mu_{K_0 G}(\theta)$; 13

$\mathrm{M}(G) \cong \mathrm{H}^2(G, \mathbb{C}^*)$; 14

$X(\theta) = G(\theta)\Delta_G X$; 15

$R = \mathbb{Z}_{(p)}[\varepsilon]$; 19

ω_B; 20

$|\mathrm{IBr}(X)| = |C\ell(X_{p'})|$; (2.1b)

$d_{\chi\varphi}$, $c_{\varphi\psi}$; 21

e_B; 22

$J(FX)$, $J(Z(FX))$; 25

$k(B)$, $\ell(B)$; 25

$B = b^X$; 26

$d_{\chi\varphi}^y$, $m_{\chi\zeta}^{(y,b)}$; 27

$O_p(G)$, $O_{p'}(G)$, $O_{p'p}(G)$; 29

$V^\sharp = V \smallsetminus \{0\}$; 34

$\delta_V = |V|/\pi_V$; (3.3b)

$H^* = \mathrm{Hom}(H, \mathbb{C}^*)$; 42

$V^* = \mathrm{Hom}_F(V, F)$; 45

$E \cong q_{\pm,0}^{1+2m}$, $\Omega_1(E)$; 47, 48

$C(E) = C_{\mathrm{Out}(E)}(Z(E))$; 48

$\varepsilon(V)$, $I(V)$, $J(V)$; 61

$P_1(V)$; 65

$\beta(G)$, $\beta^*(G)$; 75

$f(g, V)$; 75

$g_V = [d_1, \cdots, d_t]$; 76

$R_0(L)$; 116

$c(L)$; 116

$\mu_{\chi,H}^t$; 117

$Y^{\mathfrak{S}}$; 161

$p(n)$; 170

$P(L)$; 171

Group structures:

$Z_m = m$ cyclic group of order m,

S_m, A_m symmetric and alternating group of degree m, respectively,

Q_m, D_m quaternion and dihedral group of order m, respectively,

q_+^{1+2m} extraspecial group of order q^{1+2m} and exponent $q > 2$,

$2_\pm^{1+2m}$ central product of m copies of Q_8, sign $+$ for even m,

$2_0^{1+2m} = 2_\pm^{1+2m} \circ Z_4$,

$X \times Y$, $X \circ Y$ direct and central products, respectively,

$X \Delta_Q Y$ fibre-product of X, Y over their common quotient group Q,

$G \,\mathrm{wr}\, S$ wreath product of G with the permutation group S.

The notation for the classical groups (groups of Lie type) and sporadic simple groups is standard and follows the [Atlas].

Index